D1621192

THE INSTITUTE OF ANIMAL TECHNICIANS

THE I.A.T. MANUAL OF LABORATORY ANIMAL PRACTICE AND TECHNIQUES

Edited by

DOUGLAS J. SHORT, M.B.E.

and

DOROTHY P. WOODNOTT

CROSBY LOCKWOOD & SON LTD
26 OLD BROMPTON ROAD, LONDON, S.W.7

First published 1963
Second edition,
revised and enlarged, 1969

SBN 258 96739 0

Made and printed in Great Britain by Richard Clay (The Chaucer Press), Ltd.,
Bungay, Suffolk

Foreword to the first edition

SIR CHARLES HARINGTON, K.B.E., F.R.S.

Medical Research Council

It can be claimed with confidence that the very great majority of scientists who have engaged in animal experimentation in this country have done so in a humane manner and with full consciousness of the responsibility involved. The high standards observed by professionally qualified experimenters in their laboratories have not, however, always been matched in the past by equally high standards in the animal house. It is perhaps not surprising that this should be so, for a scientific worker engaged in fairly simple acute physiological or pharmacological experiments on animals might well be scrupulously careful in his work in the laboratory but less mindful of the treatment that the animal had received before it reached him.

With the development of other types of biological work involving the use of animals, however, matters changed. In research on such subjects as nutrition, endocrinology, immunology, and genetics the experimenter found himself frequently involved in daily examination of the animals he was using over long periods. Experience of this kind inevitably brought to the workers concerned the realization of the essential importance of animal care in all its aspects to the success of their experiments, and hence put an end to the idea, which had been prevalent for too long, that the work of the animal technician was an unskilled occupation.

It is fortunate that by the time that professional biologists had come to this realization, there were available men and women engaged in the care of animals for experimental work, who by intelligent devotion to their task over long periods, had acquired the enthusiasm and the breadth of knowledge needed for the formation of an effective Animal Technicians Association. In the relatively short period of its existence, this Association has indeed brought about a truly remarkable improvement in the position of animal technicians, by systematic training to fit them for their duties and by organization to ensure that the importance of these duties is properly recognized. Largely as the result of the efforts of the Association, it has now become evident that the work of an animal technician can offer a satisfying career.

As in any other career, success will depend on adequacy of background knowledge as well as on the practical skills. Hence this manual which, edited as it is and for the most part written by men and women engaged in the practical care of animals, is planned to cover the field of knowledge required by the skilled animal technician of the present day. The extent of this knowledge is self-evident from a glance at the list of contents, and the idea that a fully trained animal technician ought to have even a superficial acquaintance with some of the topics discussed would have seemed totally unrealistic twenty-five years ago.

Nevertheless the idea is certainly right and therefore, as one who has been concerned with biological research for more than forty years, I welcome this book for the influence that I am confident it will exert on present and coming generations of animal technicians, whose part in the development of biological work will certainly grow in importance as the years go on.

C. R. HARINGTON

June, 1963

Foreword to the second edition

SIR PETER MEDAWAR, F.R.S.

Medical Research Council

When Sir Charles Harington wrote the Foreword to the first edition of the Manual in 1963 he probably foresaw the need for a fairly speedy reprinting, but he could hardly have predicted that an almost completely revised and reconstructed edition would be called for within five years. Yet this is just what has happened. In the present edition, eight chapters have been extensively revised and three altogether new chapters have been added.

The smaller improvements to the text, such as we should expect to find in any newly prepared edition of a standard manual, include sections or paragraphs on the incubation of hens' eggs and the rearing of chicks, on the sexing of young animals, and on the diseases of ferrets, hamsters and poultry. In addition, the chapter on mammalian physiology has been enlarged by including a section on digestion.

But the factor which above all others has made it imperative to prepare a new edition is the development of pathogen-free animals—the most important innovation in laboratory animal practice since the development of inbred strains. The present edition accordingly contains new chapters on the production and use of pathogen-free animals and of germ-free animals, and the chapter on the heating and ventilation of animal houses has been appropriately added to and revised.

The 'SPF' revolution was not an altogether peaceable one. Before its principles could be adopted, elder statesmen in the world of medical biology had to be reassured that pathogen-free animals were not in some way abnormal or unrepresentative, and therefore liable to give misleading results. Today we are more inclined to think that it was the animals *they* worked on that were abnormal, and (what is worse) abnormal in unknown ways. SPF animals are in fact healthy animals; the SPF revolution is simply raising the general level of hygiene from nineteenth- to twentieth-century standards. Just one hundred years ago (Mr Bryan Magee has told us) the Metropolitan Board of Works began planning to give London its first modern sewer. The enterprise was violently denounced by *The Times*: 'We prefer to take our chance of cholera and the rest than be bullied into health.' But common sense and common humanity prevailed, and we are now in fact doing for laboratory animals what has been done for human beings over the past one hundred years. It is very satisfactory to record that, just as the establishment of new standards of animal housing and husbandry was largely the work of the founders of the old A.T.A., so their successors in the Institute of Animal Technicians are playing a central part in the revolution that is to guarantee us the use of healthy animals.

If this Manual did not serve a very useful purpose people would not buy it, and a new edition would not have been called for. One does not have to be a clairvoyant to foresee that this second edition will in due course be superseded by a third. What will be its contents? By then, I hope, the SPF regimen will have been established in all major animal colonies, and the principles that underlie it will be taken for granted. In the light of experience, the third edition I foresee will bring up to date many of the recommendations relating to the use of SPF animals: some standards will be relaxed and others tightened. There is sure to be a new chapter on the nutrition of germ-free and SPF animals. Looking still further ahead, I foresee a time when sophisticated new techniques of cross-breeding will give us laboratory animals that are genetically diverse in the sense in which natural populations are diverse, and yet uniform in the sense that they will vary within precisely known limits. One day, of course, animals will be supplanted by tissue cultures and other *in vitro* systems for many of the routine purposes of safety control and biological assay. When that revolution comes, I expect to find that the I.A.T. will have played an important part in bringing it about, and that this Manual will have detailed and authoritative chapters on the large-scale production of tissue cultures and the setting up of isolated organ systems. The work of the animal technician will thus be one of increasing importance and responsibility, and will require even higher skills and qualifications than those which we know to be necessary today.

September, 1968 P. B. MEDAWAR

Contents

CHAPTER ONE

The Law and Laboratory Animals

Act. 39 & 40. Vict. Ch. 77, 1876 is popularly known as the Cruelty to Animals Act, and it regulates the use of living vertebrate animals (other than man) for experiments calculated to cause pain. It is administered by the Home Secretary in England, Scotland, and Wales. Licences granted by the Home Secretary are not valid in Northern Ireland.

In addition to the Act referred to above, laboratory animals enjoy the protection of the Protection of Animals Act 1911, the Protection of Animals (Anaesthetics) Acts of 1954 and 1964 and the Veterinary Surgeons Acts of 1948 and 1966. The 1911 Act makes provision that no procedure which is lawful under the Cruelty to Animals Act 1876 shall be illegal under that Act.

The 1966 Act, among other things, defines veterinary surgery and medicine. It is illegal for anyone other than a veterinary surgeon or veterinary practitioner to carry out these procedures (with certain exceptions). One exception is the carrying out of any experiment duly authorized under the Cruelty to Animals Act 1876. The Protection of Animals (Anaesthetics) Acts 1954 and 1964 provide for the use of anaesthetics in animal surgery. This aspect is covered by the main provisions of the Cruelty to Animals Act 1876 in respect of experimental animals.

I DEFINITIONS

The terms listed below are not defined in the Act of 1876; the following is only a guide to their interpretation.

(i) EXPERIMENT—although the Act does not define the term 'experiment' in practice any procedure upon a living animal (other than veterinary diagnosis on that particular animal) designed to find the answer to a problem is regarded as an experiment. It is considered subject to the Act if it presents a risk of pain or discomfort to the animal, or of any departure from its state of normal health or well being. Other types of procedure that come under this heading include those adopted in uncertainty whether they will answer a particular purpose, and those undertaken to discover something new, test a hypothesis or establish and illustrate a known truth.

(ii) LIVING—an animal is regarded as living so long as it is breathing (or can be artificially ventilated), its heart beating and its cerebrum and basal ganglia or any parts thereof are structurally and functionally intact.

Experiments on pithed frogs or on cats in which the cerebral hemispheres and basal ganglia are destroyed are outside the scope of the Act. However, the pithing of warm-blooded animals (but not frogs) is regarded as an experiment under the Act; this is by recommendation of the Second Royal Commission of 1906.

(iii) VERTEBRATES—this covers all FREE-LIVING vertebrate forms, larval and adult. It does not include NON-FREE-LIVING embryonal forms, such as avian and reptilian embryos before hatching. A mammalian foetus *in utero* is not free living, but in this case the mother is counted as an experiment.

In cases of doubt the advice of the Home Office Cruelty to Animals Inspectors should be sought.

(iv) CALCULATED TO CAUSE PAIN—a procedure is deemed to be 'calculated to cause pain' if it presents a risk of discomfort, fear, distress, or any interference with the animal's normal state of health, comfort, and well being. These are additional to the usual idea of pain. E.g. pregnancy tests in which the injection of urine into an animal may induce ovulation or spermatogenesis (normal, physiological processes which do not cause pain) are outside the scope of the Act. The inoculation of calves with material which may be tuberculous—and which may therefore interfere with the animal's health—comes within the scope of the Act.

II REGISTRATION OF PREMISES

Registration of premises where experiments are to be performed is a statutory requirement. There is no application form for this purpose. Written application, requesting premises to be registered under the Cruelty to Animals Act, 1876, should be made to the Under Secretary of State, Home Office, Romney House, Marsham St., London, S.W.1, by the person or body in authority in the establishment, e.g. vice-chancellor of a university, senior officer of a government department, chairman or secretary of a management committee. A memorandum entitled *Experiments on Living Animals—Registration of Premises* is obtainable from the Home Office and gives information about the conditions which must be fulfilled before premises may be registered.

III LICENCES

Licences permitting the holder to perform experiments are granted by the Home Secretary. Every experiment under a licence alone is subject to restrictions, among which are the following:

(i). Throughout the experiment the animal must be under the influence of an anaesthetic of sufficient power to prevent the animal feeling pain. A local anaesthetic *may* satisfy this requirement.

(ii). The animal must be killed while still under the influence of the anaesthetic if any serious injury has been inflicted upon it or if pain is likely to continue after the effect of the anaesthetic has ceased.

In practice, recovery is rarely permissible in experiments under licence alone.

(iii). The experiment must be for the advancement by new discovery of physiological knowledge or knowledge which will be useful for saving or prolonging life or alleviating suffering in man or animals.

IV CERTIFICATES

Certificates are given by the statutory signatories, and the Home Secretary has the power to allow, disallow, or suspend (in whole or in part) certificates, but the Home secretary has no power to extend the scope of certificates. Possession of Certificate A releases a licensee from restriction (i) above attached to the licence, Certificate B from restriction (ii), and Certificate C permits an experiment to be performed in illustration of a lecture.

CERTIFICATE A releases a licensee from the requirement to anaesthetise an animal if the production of insensibility by means of anaesthetics would frustrate the object of the experiment. On account of a limitation condition imposed, this concession is limited to operative procedures no more severe than simple injections and superficial venesections.

CERTIFICATE B allows the recovery of an animal from an anaesthetic where this is necessary to achieve the object of the experiment, providing the animal is killed at the end of the experiment.

CERTIFICATE C allows animals under general anaesthetic, without recovery, to be used in demonstrations before students and learned societies. However, there is no objection to *bona fide* interested colleagues witnessing experiments performed in accordance with the provisions of the Act, whether under licence alone or under licence and any certificate.

CERTIFICATE E in conjunction with Certificate A and CERTIFICATE EE in conjunction with Certificate B are required for experiments with cats and dogs.

CERTIFICATE F is required for all experiments with horses, asses, or mules.

V CONDITIONS ATTACHED TO LICENCES

Ten conditions are attached to and printed upon all licences. Additional conditions may be attached at the discretion of the Home Secretary.

CONDITION 1 lists the registered place(s) where the licensee may perform experiments. Animals may not be moved from one premises to another without the permission of the Home Office.

CONDITION 2 states that the licensee may not perform any experiment until he has been notified that the certificate covering the experiment has not been disallowed.

CONDITION 3, known as the 'pain condition', epitomizes the purpose of the Act, and the administration of the Act depends upon its strict observance. It applies to all experiments under certificates (except C) and states that:

(*a*) If, at any time, an animal is suffering pain which is *either* severe *or* likely to endure, and if the main object of the experiment has been attained the animal shall be painlessly killed forthwith.

(*b*) If, at any time, an animal is suffering *severe pain which is likely to endure* the animal shall be painlessly killed forthwith (whether or not the main object of the experiment has been attained).

(*c*) If an animal appears to an Inspector to be suffering considerable pain and the Inspector directs such an animal to be destroyed it shall be painlessly killed forthwith.

CONDITION 4, the 'limitation condition', states that under Certificate A (and E or F) no procedures more severe than simple inoculation and superficial venesection may be adopted.

CONDITION 5 applies to all experiments under Certificate B (and EE or F) and requires that all operative procedures shall be conducted under strict antiseptic or aseptic conditions and with adequate anaesthesia. If these precautions fail and pain results the animal shall be painlessly killed forthwith.

CONDITION 6 applies to all experiments under Certificate C, and requires that on the completion of the experiment the animal shall be killed forthwith by, or in the presence of, the licensee.

CONDITION 7 states that no experiments using curare (or substances exerting a similar action*) shall be performed without the special permission of the Secretary of State, and that forty-eight hours notice of the performance of any such permitted experiment, or series of experiments, shall be given to the District Inspector.

This condition arises from the fact that curare is not for the purposes of the Act deemed an anaesthetic.

CONDITION 8 states that the licensee must keep a written record of all experiments which shall always be available for examination by an Inspector, and he (the licensee) shall, annually, report the number and nature of his experiments to the Secretary of State.

CONDITION 9 states that copies of descriptions of experiments performed under the Act and printed for publication or private circulation shall be submitted to the Secretary of State together with a letter stating when and where the experiments were performed.

CONDITION 9a states that cinematograph films of experiments on animals may not be made and/or exhibited without the consent of the Secretary of State. Films of apparatus used in such experiments are exempt from this condition. Consent under this condition may be sought in general terms and not with reference to a particular film.

The object of this condition is to prevent the exhibition of such films to non-scientific audiences.

* 'Curare-form' substances are those which, in the dose used, produce motor paralysis without anaesthesia.

VI OTHER PROVISIONS OF THE ACT

i. SECTION 6

Any exhibition to the general public of experiments on living animals calculated to give pain is illegal. This prohibition does not apply to a licensee's colleagues or assistants, or to a scientist engaged on similar work elsewhere who had been specifically invited by the licensee to witness the experiment, but otherwise only the Home Office Inspector as a legal right to see animals under experiment.

Penalties for infringement—£50 for first offence.
—£100 or imprisonment for a period not exceeding 3 months for second and subsequent offences.

ii. SECTION 10

The Secretary of State appoints Inspectors who must visit, periodically, all registered premises to ensure compliance with the Act. These inspections are unannounced.

Inspectors must hold either medical or veterinary qualifications.

iii. SECTION 21

Protects licensed persons from prosecution under the Act except after the written assent of the Secretary of State. It is doubtful if the Home Secretary has ever given this permission, since his powers of revocation and cancellation are such as to render the need to prosecute unlikely. The effect of this section is to protect licensees from irresponsible or malicious prosecutions.

iv. SECTION 8

The Secretary may license any person whom he considers qualified to perform experiments upon living animals. Graduate scientists are normally licensed to do any experiment which their duties demand and their skill permits. Senior students may be licensed, usually with a supervision condition attached. Technicians may be licensed for procedures of a repetitive nature, or to cover emergencies, and may have a supervision condition attached.

VII ON FILLING IN FORMS

The following notes are complimentary to, and should be used in conjunction with, those printed on the various forms of application.

Application forms for licence (price 4*d.* each) and certificates (3*d.* each) are obtainable through any bookseller or direct from H.M. Stationery Office at:

49 High Holborn, London, W.C.1.
423 Oxford Street, London, W.1.
13a Castle Street, Edinburgh 2.
109 St Mary Street, Cardiff.
Brazenose Street, Manchester.
50 Fairfax Street, Bristol.
33 Smallbrook, Ringway, Birmingham.

i. APPLICATION FOR LICENCE

(*a*) PREMISES—these must be registered.

Where it is necessary to conduct experiments in the field or an unregistered place the licence may be made available provided the Inspector is given sufficient notice of the performance of such experiments to enable him to be present if he so desires. As and when additional or alternative places are needed the licence must be endorsed by the Home Office before it is valid and the new place(s).

(*b*) NATURE OF PROPOSED EXPERIMENT. A broad description only is needed here, since the licence covers all vertebrates, except equidae, for experiments conducted wholely under anaesthesia and from which there is no recovery.

ii. CERTIFICATE A

For experiments in which an anaesthetic would necessarily frustrate the object of the experiment. The procedure involved should be described in terms such as: injection; inoculation, withdrawal of body fluids; administration of substances by enteral or parenteral routes; inhalation; external applications; feeding experiments, the animal being allowed to satisfy its hunger and thirst; exposure to infections, to rays,* to variations of temperature* and/or atmospheric pressure.*

ANIMALS TO BE USED—vertebrates other than cats, dogs, and equidae.

iii. CERTIFICATE B

For experiments under anaesthetic from which the animal is allowed to recover. The certificate is appropriate to minor procedures, such as biopsy, or if an anaesthetic is administered to immobilize an animal for a procedure specified under Certificate A. However, its main purpose is to cover surgical operations which must be accurately described under 'description of experiments to be performed', and the words used should not be open to a broad interpretation—if biopsy only is intended, state 'biopsy'. The species or class of animals to be used must be specified.

iv. CERTIFICATE C

Covers experiments to illustrate lectures in medical schools, colleges, etc. No experiment or demonstration may be performed on a conscious animal. A description in broad terms suffices, e.g. 'experiment to demonstrate the fundamental facts of physiology and pharmacology'. As Certificate C applies also to demonstrations before learned societies, it is convenient to apply for permission to conduct experiments at: (i) the normal place where teaching is done, and (ii) 'Meetings of learned societies held in premises registered under the above Act'.

Further (under description of proposed experiments) to specify 'Demonstrations (i) to students of science/medicine at the place first named above, of the fundamental facts of physiology/pharmacology, and

* The circumstances necessitating these procedures should be explained.

(ii) to members of learned societies, of newly discovered physiological facts or facts which will be useful to them in alleviating suffering or saving or prolonging life'. The Home Office will always require notice of the licensee's intention to perform experiments before learned societies, and this will be written into Condition I of the licence.

v. CERTIFICATE D

Has been obsolete for many years.

vi. CERTIFICATES E, EE, AND F

The purpose of these certificates is well explained in the notes printed upon them.

It is *essential* that the description of proposed experiments be the *same* as that on the accompanying A or B certificate.

Certificate E accompanies Certificate A when it is proposed to use dogs and cats; Certificate EE similarly accompanies Certificate B. Certificate F is required for any type of experiment on Equidae and must accompany the licence, either alone or with Certificates A, B, or C as appropriate. Though one Certificate F can accompany both Certificates A and B, in practice it is better to submit two F's, one to cover A and one B.

vii. AN UNDERTAKING

Is generally required on behalf of all applicants from overseas. The undertaking is in set form (obtainable from the Home Office) and should be signed by some senior member of the department, who thus makes himself responsible for the proper observance by the applicant of the provisions of the Act.

TABLE SUMMARIZING THE REQUIREMENTS FOR LICENCE AND CERTIFICATES IN DIFFERENT CIRCUMSTANCES

Procedure	Equidae	Cats and dogs	Vertebrates other than cats, dogs, and equidae
Under anaesthesia without recovery	Licence + Cert. F	Licence	Licence
Under anaesthesia with recovery	Licence + Certs. B & F	Licence + Certs. B & EE	Licence + Cert. B
No anaesthesia employed	Licence + Certs. A & F	Licence + Certs. A & E	Licence + Cert. A
Lectures and demonstrations, under anaesthesia without recovery	Licence + Certs. C & F	Licence + Cert. C	Licence + Cert. C

VIII ANNUAL RETURN OF EXPERIMENTS

In mid-December all licensees receive a printed form for the return of experiments. This form must be completed and returned to the Home Office *by* January 14th.

(*a*) Normally, *one animal counts as one experiment*. Some trivial procedures under Certificate A ONLY may leave the animal entirely normal; if such an animal is returned to stock and subsequently used for another experiment, then it is counted as another experiment.

(*b*) An experiment involving procedures under more than one certificate is returned as *one* experiment under the more severe certificate.

(*c*) An experiment begins with the first interference with an animal's health, comfort, or integrity, and ends at its death or complete recovery and (under Certificate A *only*) return to stock.

(*d*) An experiment conducted by more than one licensee is returned as a conjoint experiment by all the licensees concerned.

Returns from licensees reporting conjoint experiments should tally.

Note. A useful memorandum entitled *Notes on Plurality of Experiments* may be obtained from the Home Office, and also see following section on Delegation.

IX DELEGATION

Licences and Certificates are legal documents, personal to the holder, and delegation of authority under them is expressly forbidden. It is stressed that there is no relaxation of this prohibition under Certificate C.

The Home Office gives the following guide to the interpretation of the term 'delegation':

(i) There is no delegation where two or more persons, each holding authority under the Act to perform a particular experiment, carry out conjointly the operative or other procedures involved.

(ii) Where necessary a licensee may permit anyone to administer anaesthetics to an animal subject to his experiment.

(iii) He may allow another person to carry out mechanical duties. Thus, a licensee may, for instance, employ an assistant to hold an animal while he gives an injection or to administer a diet he has prescribed, or, while he carries out operative procedures, to control haemorrhage, hold retractors, or to undertake equivalent subaltern duties.

(iv) Subject to the above, the prohibition on delegation is absolute, and a licensee may not allow another person, licensed or unlicensed, to take part in his experiment, even under his supervision or when he himself is present.

X OBSERVANCE TO THE ACT

The Home Office looks to licensees to give strict observance to the Act, the conditions attached to licences, and to the wording of certificates. Infringement can lead to revocation of a licence or even the cancellation of registration of premises. From neither of these decisions is there any appeal.

XI FARM ANIMALS

Farm animals which are used for experimental purposes are protected by the Cruelty to Animals Act, 1876.

The Diseases of Animals Act, 1950, requires that the existence or suspected existence of certain diseases of animals and poultry must be notified to the police.

These diseases are:

Anthrax in four-footed mammals
Cattle Plague in ruminating animals and swine
Epizootic Lyphangitis in equine animals
Foot and Mouth Disease in ruminating animals and swine
Fowl Pest in poultry
Glanders or Farcy in equine animals
Parasitic Mange in equine animals
Pleuro-pneumonia in cattle
Rabies in ruminating and equine animals, swine, canines, and felines
Sheep Pox in sheep
Sheep Scab in sheep
Swine Fever in swine
Tuberculosis (certain forms only) in cattle, i.e. Tuberculosis of the udder; Indurated udder or other chronic diseases of the udder; Tuberculosis emaciation and chronic cough accompanied by definite clinical signs of tuberculosis.

If any of the above diseases is suspected the Ministry of Agriculture, Fisheries, and Food's veterinary staff will determine (without charge to the owner) whether or not the disease actually exists. It is only necessary to notify the police; they will contact the Divisional Veterinary Officer, who will arrange for the animals to be examined.

Other diseases—Veterinary Investigation Service

Veterinary Investigation Officers are stationed at various centres in England and Wales. The main task of the service is to assist veterinary surgeons in the diagnosis of disease among farm animals by laboratory examination of material and the field investigation of local problems in collaboration with them.

Licensing of Bulls, boars, and stallions

These animals must be licensed on reaching certain ages, unless, exceptionally, permits are granted to keep them for special purposes. The ages are:

Bull—10 months
Boar—6 months
Stallion—2 years

Applications for licences should be made to the Divisional Offices of the Ministry of Agriculture, Fisheries, and Food.

XII ON PURCHASING CATS AND DOGS FROM DEALERS

Dogs seized by the police under the authority of the Dogs Act of 1906 may not 'be given or sold for the purpose of vivisection'. They could conceivably be handed over for laboratory procedures outside the Act of 1876 (e.g. the preparation of distemper serum), but in practice this has never been done. This ban does not apply legally to cats, but in effect stray cats are equally inaccessible. There is consequently an ever-present danger that cats and dogs offered by dealers may be stolen animals, and laboratory workers are advised to take every precaution against being incriminated in this way. The practice in many laboratories is to require the dealer to sign a statement to the effect that the animal which he is selling is his own property; the following is a suggested form of undertaking for such a guarantee:

'I certify that these..................are my own property and have been obtained by legal means.

Signed.......................... '

If a further safeguard is thought necessary the dealer may be asked to state the source of each animal.

PUBLICATIONS RELEVANT TO THE FOREGOING SECTION

Pamphlets entitled *Experiments on Living Animals—Registration of Premises* and *Notes on the Plurality of Experiments* are both obtainable from the Home Office.

The final Report of the Royal Commission on Vivisection (1912), the Report of the Howitt Committee (1951), and the Report of the Departmental Committee on Experiments on Animals (1965) are obtainable from H.M.S.O.

Notes on the Law Relating to Experiments on Animals in Great Britain is issued by the Research Defence Society, 11 Chandos St., London, W.1.

Diseases of Animals Law, 1965, Police Review Publishing Co. Ltd., 67 Clerkenwell Road, London, E.C.1.

CHAPTER TWO

Animal Houses

PART ONE

ANIMAL HOUSE DESIGN

Let us begin by stating the obvious, lest it be forgotten. The animal house is where laboratory animals live and animal technicians work: it must therefore be built to accommodate both these functions as efficiently and conveniently as possible.

It is a matter of general policy, outside the scope of this chapter, whether the animal house be attached to the laboratory or in a separate building; on the top storey, in the basement, or somewhere in between; a single floor or several; serving all departments of a university or research institute collectively, or only one. Other considerations will also determine whether the animals used will be bred on the premises or imported from elsewhere. However, all these factors affect the layout, and they must therefore be known before work begins on the design of the animal house.

In Chapter 2 of the *UFAW Handbook*, 2nd edition, it is explained that there are four basic subdivisions of all animal houses. These are:

(1) Accommodation for normal animals, whether bred or held between importation and use.

(2) Accommodation for animals under experiment.

(3) Stores for clean materials—food, bedding, clean cages, and utensils.

(4) Dirty area—for cage cleaning, disposal of soiled bedding, etc.

(3) and (4) can be accommodated in 40 per cent of the floor area of an animal house, but may take up to 60 per cent.

These four subdivisions must be physically separate, and will in nearly all cases be themselves subdivided. Their siting, in relation to each other, must take into account the pattern of traffic between them, so that, for example, routes for the conveyance of clean material are not crossed, at least at the same time of day, by dirty material—an elementary precaution of hygiene all too easy to overlook; or store rooms are not far from the goods-delivery entrance; or the largest and heaviest cages are not to be used in rooms farthest from the cage-cleaning area.

This concept of clean and dirty areas within the animal house is today beginning to look a little old fashioned. It was essential for the proper control of animals colonies that always carried some pathogenic infection and were always liable to acquire (often with catastrophic results) others from unsterilized food, bedding, or careless technicians or visitors, not to mention newly introduced animals. But the standard of health of more and more animal colonies today has been raised to the point that epidemic disease, while still feared as much as ever, is a rare rather than a regular event. Animal colonies are healthier: sometimes their standard of health is so high that to preserve it intact necessitates a peripheral barrier against infection, such as is discussed in detail in the chapter on SPF conditions. The truth is that *all* animal colonies are tending towards that standard of health, hygiene, and the associated degrees of isolation that are implied by the term SPF. In consequence, the animal house is no longer to be regarded as comprising clean and dirty areas: rather, the whole animal house must be regarded as a clean area, including that part of it that handles such things as used cages with their contained soiled bedding, while the world outside the animal house is regarded as dirty.

If this revised concept be correct, then all animal houses must be surrounded by some sort of barrier. A barrier is some discouragement to the entry of pathogens; since pathogens travel on vectors such as food, bedding, equipment, and, above all, people, this means that every person and everything that enters the animal house must be subjected to some sort of decontamination.

When the decontamination includes sterilizing things, and showers and change of clothing for people, the conditions are usually thought of as SPF. But decontamination may entail much less than this. For example, intelligent care in the purchase and storage of food and bedding, a high standard of ordinary cleanliness with sensible use of soap, water, and disinfectants, and the changing of footwear and other garments, together with careful hand washing, constitute a barrier that for many purposes is remarkably effective.

The use of more recently developed ideas, such as filter cages, may also contribute to reducing or eliminating the spread of infection within an animal house. This emphasizes the point that barriers are normally only hindrances to the entry of infections, and are not absolutely impenetrable. Defence against invasion, therefore, should be in depth also, so that penetration at any point by a pathogen is met with a further and a further obstacle. Put yourself in the position of a pathogen bent on gaining entrance to a susceptible colony of animals, and establishing an empire of infection and disease within, and you will see that at any stage in your career of depredation an obstacle can defeat you. You get into the food, only to be sterilized before the food leaves the bag. You lie in the bedding, and undergo the same fate. Attach yourself to boot or clothing, and find you are left behind in the changing room. Get on to hands that are about to handle your destined prey, and be washed down the sink in a lather of sopy water. Get into a cage with the idea of riding on dust particles round the room, and find yourself frustrated by the filters on all the cages, which neither let you out of your own cage nor into the others. If you do get out you may quickly be blown away through the ventilation exhaust, never to return.

Such precautions as indicated are, therefore, matters of degree. The more and better the precautions, the less likely will the pathogen's success be in overcoming them. Thus, 'clean' and 'dirty', in the contexts so far used in animal houses, become only relative terms. Some parts will demand more precautions, other need only nominal defence. For the clean and dirty concepts, therefore, we should substitute degrees of security against infection, from maximum to minimum.

Thus, in the area of greatest security (maximum precautions that are practicable) will be kept breeding colonies and all stock animals awaiting use. The rule here is that animals only go out, to the experimental side; no animal will enter, except imported animals, and then only after adequate quarantine or other precautions. No experimental animals, not even long-term ones, must ever be given house room in the clean side.

The experimental-animal area normally draws all its animals from the clean area, although exceptionally, imported animals may go directly into the experimental area after suitable quarantine. Here, precautions may have to be just as strict, especially for long-term observations; but generally speaking, the need for investigators to pay frequent visits to their animals, and carry out experimental procedures and manipulations on them, necessitates some relaxation.

The store area has to accommodate supplies of food and bedding, and to store clean cages, water bottles, and other utensils. Here may also be found storage for special clothing, stationery, and, most important, a small office for the head technician, where records are kept and the general running of the unit is controlled.

The washing and sterilizing area is devoted to cleaning cages and utensils, sterilizing, and disposal of soiled bedding and carcasses, by incineration or otherwise. Cages after treatment are then returned to the store.

Thus we have a key to the general layout of any animal house.

Because the movement of all material will be on wheeled vehicles, steps are ruled out. If difficulties of level are inevitable gentle ramps must be provided, bearing in mind that a heavily loaded trolley can run away with a technician on a steep or awkward slope. If the animal house is on more than one storey a lift is essential. Corridors should be wide enough for the traffic expected. In most cases it should be easy for two trolleys to pass without obstruction, even though part of the corridor width may be taken up by bins or piles of cages. The animal technician has the duty of working out the volume of this traffic, so that he can advise the architect on the requisite width of corridors.

ANIMAL ROOMS

Wherever possible, rooms should be designed for use by any species, rather than specifically for mice, guinea pigs, rabbits, and so on. However, certain animals, such as dogs, cats, farm animals, and monkeys, are likely to require special accommodation, such as soundproof rooms, fixed stalls and water troughs, exercise runs, and internal subdivisions. This must be accepted, but rather as a penalty: it limits the flexibility of the animal house. For the common species, animal rooms should be adaptable and interchangeable.

In considering the size and shape of animal rooms, it is useful to visualize something that has for most purposes many advantages. Departures from this notional animal room can then be worked out for specific reasons. The dimensions of the module must be such that a whole number of the largest-sized cages may be housed on racking along one wall, and there is sufficient space between the racks to work with such cages. Thus, racking and cage dimensions must be decided at an early stage in planning. The notional module may be a room 8–10 ft wide and 12–20 ft long. It can be economically racked along the two long sides, with a door at one or both ends, giving free working area down the middle. This is an economical use of available space, and suspended shelving has the great advantage of leaving a clear floor. But it is more difficult to keep the walls clean than when mobile or fixed island racks occupy a central space in the room. Breeding rooms may well be much larger, and rooms to accommodate island racking, rather than wall- or ceiling-mounted racking, may be square and probably two or three times the size of the notional module. Cats, rabbits, and larger animals may require much larger rooms. On the other hand, isolation animal rooms have usually to be small and numerous.

A useful rule of thumb for relating the length of racks to gangway widths is the 1:2:3 rule. If the distance between adjacent peninsular or island racks is a single unit (about 33 in.) the width of the gangway between rows of racks is two units and the greatest distance from any part of a rack to the gangway three units.

All animal rooms have to be cleaned down from time to time. This governs the structure and finish. Floors must be inpervious to water, to disinfectants, urine, and ideally, to acids, alkalis, and organic solvents, as well as to trolley wheels and racking feet. There is no flooring that meets all these requirements, and the choice lies between asphalt, granolithic, terrazzo, quarry tiles, smooth cement finished with a sealing compound containing sharp sand or carborundum, and possibly rubber or some of the new plastics. If they are level they have to be mopped and squeegeed; if they slope the gradient must be enough to ensure a good run away—not less than $\frac{1}{8}$ in./ft (1 in 100). Gulleys or drains are convenient for hosing down, with either a glazed half channel covered with a removable metal grill or a corner or centre drain; in every case well trapped. But they introduce a real danger of infection coming up the drain from outside. The tendency today is to have level floors, without drains, that are easily mopped down.

Corners should be coved, with a skirting integral with the floor and, at its upper edge, flush with the wall, rather than standing proud of it.

Walls may be of fair-faced brick or concrete block, hard plaster or cement, covered with a suitable paint; or of granolithic, terrazzo, or tiles, either all the way up or as a dado up to about 6 ft, with plaster, brick, or other surface above. The corners of all walls must be rounded.

Ceilings do not have to be capable of being hosed, although this is an advantage, but it must be possible to mop them down. The choice of paints is wide. Preference should be given to something that is washable, very durable, resistant to chipping, and pleasant in colour. Some paints are markedly antistatic: that is, they do not readily collect a coating of dust; and this is a great advantage in the animal room.

Plumbing, conduits, and other services may be either chased into the walls or carried in an impervious service duct. Where this is not possible, or is not desired, pipes should be held an inch or two out from the wall, so that it is easy to clean behind them. Pipes coming through the walls must be closely sleeved as a precaution against insect pests. Service pipes run along corridor ceilings are readily accessible for servicing or alteration with minimum disturbance to the animals.

Doors should be close fitting and proof against even a young mouse. This means that not more than ¼ in. gap can be tolerated. Steel doors have certain advantages, but they are heavy and noisy. Wood is lighter, but requires to be covered over the lower 4 ft or so with metal. The door frames can be of metal or metal clad like the doors. Steel door-frames are particularly suitable because they do not shrink or warp, and are resistant to knocks from trolleys. All doors should have observation panels at eye level. Lever-operated door handles are a great advantage; also rising butt hinges. Sliding doors can often save much space. Sliding doors must be suspended from above rather than running in grooves in the floor, because such grooves may become obstructed with dirt or grit.

The provision of windows is open to argument. Bearing in mind that artificial lighting is no disadvantage, at least to rodents, and that ventilation must in all modern houses be ducted in, or out, or preferably both in and out, windows merely complicate the problem of adequate heating and ventilation, as well as letting in direct sunlight, which is usually undesirable. But technicians may prefer to work in daylight, and many people, including the Home Office Inspector, regard natural daylight as highly desirable if not obligatory for dogs, cats, and large animals. If animal-room windows face north-east direct sunlight will not be a serious problem, but if they face south, sun-blinds will be necessary. Venetian blinds are the most efficient, but are serious dust traps; they should therefore be fitted in the space between double windows. It is a great asset to fit airtight speaking windows into the clear glass observation panels of a closed unit. Experimental animals may not be exhibited to the public, so it is advisable to fit obscure glass, or glass bricks, in all windows through which the public may be able to view the animal house.

A window-less room can look like a prison cell, but this effect may be mitigated by painting the surfaces different colours. A bold use of some stronger colours gives a more satisfying effect than the timid use of all-pastel shades. Blue is often recommended for ceilings, to give an illusion of the sky.

The provision of windows, then, is a debatable point. Artificial lighting is no disadvantage to most species of laboratory animal, and in breeding rooms a controlled light cycle, rather than the seasonal variation in length of daylight, may be necessary to regulate breeding programmes, and therefore windows, if present, will need to be screened. But whatever decision is come to, it should be based on functional considerations, and the architect instructed accordingly.

SERVICES

Animal rooms, corridors, and most other parts of the animal house need to be frequently washed down. Drains should be trapped to collect sludge, and it must be impossible for wild rats or mice to gain access to the animal rooms through the drains. (This would seem an elementary precaution, but it has not seldom been overlooked, with unfortunate consequences.) A macerator may be needed for the disposal of carcases which are radio-active, or in smokeless zones.

Plumbing is costly, and careful consideration should be given to the number and positioning of taps and sinks. Ideally, each conventional animal room should have a small sink for washing hands and filling water bottles (*not* for washing cages and utensils, which should be done elsewhere), hot and cold water, a hose point and, if necessary, a separate cold drinking-water supply from the mains. High-pressure water for hosing down may sometimes be required, especially for rooms containing monkeys, dogs, and large animals, where cages or stalls may be inconveniently large or fixed. In a new building consideration should be given to piping high-pressure water to each room. A closed unit may have different requirements, e.g. hose points are not needed where there are no floor drains; acidified water may be fed to animals; ready-diluted disinfectant may be piped to each room. In the latter case plastic piping should be installed so that corrosive solutions, such as hypochlorite solution, may be used in the system.

The lighting intensity, both natural and artificial, should be a good working light—25 lumens at the darkest part of the room—but it is often useful to have a movable source of more intense light, such as can be provided by an Anglepoise lamp. It is wise to fix a time switch to the lighting circuit. It may never be used; on the other hand, it is an inexpensive device, and may prove to be invaluable. One time switch can control several rooms, though this economy can reduce the flexibility of the rooms. Switches should be in the corridors to leave the room walls clear.

There should be at least one electric power outlet in each room, for electrical balances, and other pieces of equipment. At the time of building twin power outlets can be installed for a few shillings more than the cost of a single outlet, and at a fraction of the cost of adding a second outlet at a later date.

HEATING AND VENTILATING

Modern animal accommodation is expensive, all the more so if extravagant use is made of the available space. To rely for ventilation solely on opening the window will reduce the usefulness of an animal room to the point of extravagance. For this reason, mechanical ventilation of some kind is obligatory in the modern animal house.

Full air-conditioning, which means making provision for heating and cooling, humidifying and dehumidifying, and controlling the quantity of incoming air, is the best of all methods, and also the most expensive. In the equable climate of the British Isles there will be a great economy at some sacrifice in working and living conditions if the cooling and the humidity control are omitted. However, air cooling (and, thus, some measure of dehumidification) is possible at reasonable cost if the establishment has an emergency water supply which can be used to run through the necessary cold-water batteries. The cost of air cooling should be considered against the increased number of animals which may then be housed in a room; the reduced breeding performance of animals exposed to high temperature and high humidity; the occurrence of heatwaves at times of maximal breeding combined with minimal usage—because workers are on holiday; and the moral obligation not to subject persons in closed units to well-nigh intolerable conditions. Furthermore, certain experimental requirements may make full air conditioning obligatory, even to the point of duplicating some of the equipment and providing a stand-by generator in the event of power failure.

It is a sound principle to provide background heating for the whole animal house up to the normal comfort zone, which is usually taken (in England) to be about 65–70°F (18·3–21·1°C). Animal rooms can then be equipped with supplementary heating devices, controlled by individual thermostats, which can raise the temperature to 75° or 80°F (21·1° or 26·7°C). The background heating may be by incoming air, radiators, heated panels, heated floors, walls or ceilings, or by convectors, and the supplementary boosting by extra heating batteries—steam, hot water, or electric—in the air input duct, by electric, water or steam tubular heaters, or by convectors. There is a wide choice of method, and this is not the place to discuss it in detail, but it should not be forgotten that the heat produced by the animals and the technicians working in the animal rooms may raise the temperature as much as 8°F (4·5°C), or that the method chosen must be capable of maintaining the desired temperature range whether the animal population density be high or low.

Heating and ventilation is a highly technical subject, and the following points, therefore, are merely a guide for discussion with the heating and ventilating engineer or consultant.

Every part of the animal room must be equally ventilated. Draughts are to be avoided, and also stratification of the air. The conventional system is for warm air to enter at a low level (by the window, the coldest part of the room) and to be removed at a high level at the opposite side of the room (often through the door and into the corridor), but, for the sake of added human comfort, some air may be brought in at head level. Recently, air diffusers have been used and are usually placed in the ceiling to save wall space. The diffuser is both the inlet and outlet for air, and incoming air is channelled outwards from the apparatus. Air circulation from diffusers is good, as can be demonstrated by passing smoke through the system. This check will also reveal whether or not there is a cone-shaped dead-space where air does not circulate immediately beneath the diffuser, and the dimensions of that dead-space. Another recent development is that of 'laminated air', where very thin layers (laminae) of air enter from top to bottom of one wall and are extracted at the opposite wall or from the centre of the room. The air speed of this system requires careful adjustment, but in a well-balanced system even germ-free animals will remain germ-free for considerable periods because the moving, clean air carries all micro-organisms away from the cages. This is a refinement of another method, in which air enters the room from louvres all down one wall.

In breeding and other clean animal rooms the pressure of air in the room should be higher than outside, to reduce the possibility of infection being sucked into the room. Conversely, infective animal rooms should have a negative pressure. As an additional precaution, an air-lock may be built between each infective room and the corridor.

The size of ducts should be such as to reduce the air noise to a negligible level, and the openings should be so fitted that they can be hermetically sealed, in the event of it being necessary to close a room for fumigation. It must be remembered that ducts may also need fumigation and should be designed to allow for this. Incoming air should be filtered to remove dust particles and, in some cases, smaller particles including micro-organisms. The higher the efficiency of the filters, the more they will cost. Expensive, high-performance filters should be protected by using coarser, pre-filters, which may be replaced frequently at relatively small cost. Outgoing air from infective rooms may need to be sterilized. This may be achieved by the use of filters, or by incineration, which is expensive, or by the use of ultra-violet light. The latter is not recommended, as air must be dust-free (and has, therefore, to be filtered) if this process is to be effective, and the wavelength of the ultra-violet light must be checked frequently. Short-circuiting of air currents between input and output, whether in the rooms or in the main duct system, should be virtually impossible. Recirculation of air is not advised. There may be a penalty in the heating bill, but the risk of spread of infection is a more important consideration, and it would be necessary to install dust filters, a dehumidifier, and carbon filters to prevent the circulation of dust, and moisture and smell from animals.

Unfortunately, not all heating and ventilating consultants are yet fully aware of the problems encountered in animal houses or of the progress made in recent years through attempts to overcome these problems. Whatever the system proposed by the consultant, it should be examined critically, and a second opinion should be sought if the estimates for the cost of heating and ventilating are much in excess of 10–12 per cent of the total cost of the building, for heating and ventilating but no refrigeration; or 25 per cent for full air conditioning. A good system should result in the adequate circulation of heat and air to all parts of the animal rooms, comfortable living and working conditions, and the absence of unpleasant smell in animal rooms—provided, of course, that a high standard of hygiene obtains.

SANITATION

Sanitation includes the washing and sterilizing of cages and utensils, and the measures that have to be taken to keep the animal house generally clean.

The provision of mechanical means of cage and bottle washing should be the normal practice. Only very small animal houses can afford to do without them, for they represent a great saving of labour, and of an unpleasant kind. Most such machines require a supply of steam, water, and electricity, which must be foreseen in the planning of the animal house.

Autoclaves, which are, for practical purposes, the only sure means of bacteriological sterilization, are expensive. Autoclaving is the most acceptable method of sterilizing food and bedding for a closed unit, though treatment with ethylene oxide or by irradiation are possible alternatives. Autoclaves are necessary in an animal house where dangerous pathogens are being used, but in many animal houses the only pathogens are likely to be those associated with accidental intercurrent infections, and free steaming or treatment in an efficient mechanical washer will destroy all of these.

However, if a mechanical cage washer is to be relied upon for cage sanitation constant attention must be paid to the functioning of the pump and jets. The jets in particular easily become partially blocked, and the efficiency of washing is thereby seriously impaired.

HYGIENE

The animal house is a dense population of susceptible creatures, an ideal subject for epidemic disease. The standard of hygiene must therefore be high as high as in a children's hospital.

Ample provision needs to be made throughout the animal house for washing hands, when going from room to room, or from animal room to clean store, for example. Foot trays containing disinfectant outside the door of each room are an added hygienic precaution if used correctly.

Indeed, the maintenance of a high standard of health in an animal house depends as never before on the faithful observance of certain rules of hygiene, more than on the provision of showers, sterilizers, and so on. Such facilities are always useful, and for SPF conditions are essential, but they do not guarantee to keep the pathogen out. The intelligent application of a discipline of hygiene is, above all, the greatest deterrent to the pathogen. One careless or ignorant technician, investigator, or visitor can cause a major epidemic. Working in a disease-free animal house is rather like being on top of a volcano; you never know when disaster is going to overwhelm you. But whereas man cannot control the eruption of volcanoes, he can do a great deal to prevent the outbreak of disease. It is the responsibility of the staff working in such an animal house to see that a proper discipline of hygiene is established and never, never relaxed.

AMENITIES

Work in the animal house is sometimes arduous, and attendance at weekends is an inescapable duty. A technician's room is seldom an extravagance, and may be a great asset. This is especially so in SPF units, where the internal environment can be oppressive, and a rest room with tea-making facilities will be more than ordinarily welcome.

Toilet accommodation is also a necessity. Showers, which are not absolutely necessary except in strictly isolated units, are nevertheless a much appreciated amenity in almost all animal houses.

CONCLUSION

There is no such thing as the ideal animal house. Each one must be designed to serve the particular purposes for which it is going to be used. In all cases, however, certain general principles must be observed, which have been referred to in this chapter.

No reasonable man would plan a kitchen without seeking the advice of his wife, who will spend much of her time working there. In planning an animal house it would be equally unwise to fail to seek the advice of the animal technician who has to work there. For his part the animal technician must be prepared to give the sort of advice that only he can give, namely on the practical considerations that will affect the efficiency of his daily work.

Certain aspects of animal-house design have been omitted or only lightly touched on in this chapter, because they are more the concern of others than of the animal technician. However, the competent animal technician will not be unaware of the existence of these topics, which are dealt with elsewhere more particularly in the publications listed in the bibliography.

BIBLIOGRAPHY

WORDEN, A. N. and LANE-PETTER, W. (Editors), *The UFAW Handbook on the Care and Management of Laboratory Animals*, Universities Federation for Animal Welfare, London, 3rd Edition (1967).

Comfortable Quarters for Laboratory Animals, Animal Welfare Institute, New York (1958).

LANE-PETTER, W., *Provision of Laboratory Animals for Research: A Practical Guide*. Chapter 8, 'Physical Environment'. Elsevier Publishing Co., Amsterdam (1961).

LANE-PETTER, W. (Editor), *Animals for Research: Principles of Breeding and Management*, Academic Press, London, 1963.

Various Articles in the *Journal of the Animal Technicians Association* and *Proceedings of the Animal Care Panel.*

Various references in *Federation Proceedings*, 1960, 19, No. 4, Part III, Supplement No. 6; also in *Federation Proceedings*, 1963, Vol. 22, No. 2, Part III.

Publications of the Institute of Laboratory Animal Resources, 2101, Constitution Avenue, Washington, D.C., 20418.

PART TWO

HEATING AND VENTILATION OF LABORATORY ANIMAL ACCOMMODATION

INTRODUCTION

The comfort and health of a population of animals confined within the restricted space of a room or building makes it essential to provide adequate means to control the supply, quality, and temperature of the air which they breathe; that is to say, their environment must be controlled. To some extent the same problems of environmental control occur both with buildings for human occupation and with greenhouses for the indoor cultivation of plants. While it is only to be expected, therefore, that there will be similarity in the methods used in these different fields, it is important to recognize that both technical and economic factors pose special problems in animal-house control, and accordingly demand the application of techniques which are correctly tailored to these particular circumstances.

Environmental control is a term which embraces a number of quite distinct factors which may be summarized as follows:

(*a*) Temperature control

This requires a combination of heating and cooling, both to ensure that the environment is sufficiently warm even under the most rigorous winter conditions and to make certain that the natural body heat produced by the animals is fully dissipated. Care is needed to ensure that the temperature control is effective throughout the animal house so that each animal enjoys the ideal temperature range appropriate to its own species.

(*b*) Humidity control

As with temperature, provision should ideally be made either to raise or lower the humidity of the air so as to balance the varying quality of incoming air from outside the building, on the one hand, and the humidity evolved by the animals, on the other. Humidity control is important not only for the comfort of animals but also with certain species so as to prevent disease.

(*c*) Odour removal

While probably the main purpose for minimizing the inevitable smell in animal houses is for the comfort of the technicians who look after the animals, it is interesting to note that evidence has been presented to show that odour control is apparently of importance to pregnancy block in mice.[1]

(*d*) Air cleanliness

While it is obviously desirable to minimize the amount of dust and soot contained in the incoming air, in some animal houses it is of much greater importance to ensure that the air is free from harmful organisms. Similarly, it is essential to make certain that any such organisms originating from one or a few of the animals are rapidly removed from the house before they are able to infect the remaining population. Furthermore, in some circumstances it is equally important to avoid such dangerous organisms being discharged with the exhaust air into the surrounding atmosphere in view of the hazard this might constitute to both animals and human beings.

The basic method whereby these environmental factors are controlled is to provide an adequate flow of incoming air of the desired quality, taking care to ensure its even distribution from the point of view of heating or cooling without causing draughts which may adversely affect the health or reproduction of the animals. The rate of air flow into a room is normally expressed in terms of the number of air changes per hour. In practice, this does not necessarily mean that this same volume of fresh air is drawn in from outside, since there are strong economic reasons for recirculating as much of the air as possible, thereby conserving heat. With either the once-through or recirculation system the air is passed through a series of heaters, coolers, and filters to ensure that it is of the desired quality before it is discharged into the animal house; similarly, in some instances the air being exhausted to atmosphere is first filtered to eliminate from it any dangerous organisms.

For maximum efficiency and economy it is important for the equipment for environmental control to be planned and engineered as a composite whole. In order that each stage may be more readily understood, however, the several different factors summarized above are separately discussed in more detail below. Before doing this, however, it is worth commenting that whereas ideally an animal house should be provided with full air conditioning, the high costs involved have so far resulted in this being a rare practice in the United Kingdom; fortunately the normal climatic conditions in this country generally permit the adoption of a cheaper compromise solution, although it must be realized that such a course inevitably involves facing the possibly disastrous effects of the occasional extreme conditions which may occur in both summer and winter. Thus it is that humidity control is in practice generally limited to variations in the numbers of air changes, rather than depending upon the type of humidifier used in the air conditioning of buildings for human occupation. A final point to note is that for Specified-Pathogen-Free (SPF) animals care is taken to maintain the animal house at a slightly elevated pressure so as to ensure that only filtered air can enter, and infected animal rooms are maintained at a slight negative pressure to prevent airborne organisms spreading to other parts of the accommodation.

AIR FLOW AND CONTROL

The following separate aspects of the flow and control of ventilating air may be distinguished:

(*a*) The rate of air change for a given house or room;
(*b*) the division of the air supply between two or more rooms;
(*c*) the distribution of air within each room;

(*d*) the location of the fresh-air intake duct;
(*e*) the use of air recirculation;
(*f*) the use of air locks.

While to a considerable extent local conditions and individual preferences will determine what course is adopted in any particular instance, each of these aspects is briefly commented upon below.

(*a*) Air-change rate

Although the air-change rate can be a critical factor in determining the overall capital and running costs, there is regrettably no firm basis yet available for making this important decision. Alschuler[2] has recommended no less than eighteen air changes per hour using 100 per cent fresh outside air; while no doubt this is a safe figure from the point of view of the animals, it would be a very extravagant one to adopt as a general standard in view of the very high costs which it would involve. Obviously, in fact, any such arbitrary figure has little general validity, since in practice the need must vary widely with such factors as the number and species of the animals to be housed in a given space.

The correct economic approach would appear to be to design the accommodation so as to house satisfactorily the desired number of animals, and then to calculate the ventilation rate on the basis of the minimum air volume required per animal to maintain it in a healthy condition. Unfortunately the very limited amount of data available regarding the air volume required for laboratory animals seems to specify relatively high figures; if these are applied literally it is probable that an unnecessarily large and expensive system will be selected.

In practice, therefore, probably the best generalization which can be given is to select the lowest air-flow rate which will restrict the animal smell to a reasonable level; this, of course, assumes that the cages are regularly cleaned. If the smell level is low it is probable that the environment will be satisfactory for the animals' welfare, providing some form of humidity control is exercised and that a reasonably constant controlled temperature is maintained throughout the animal room.

(*b*) Division of the air supply between rooms

Where the one air supply is being used to feed two or more rooms, the flow is divided as required by means of suitable dampers in the distribution ducts. These dampers also serve to permit one room to be isolated from the system without interfering with the others. The setting of the dampers is of critical importance under normal operating conditions, since their adjustment controls the proportions of air flowing through each room; to facilitate this adjustment, it is common practice to fit each room with an inclined manometer (0–0·5 in. w.g.), to indicate the extent to which the pressure within the room differs from that outside it.

(*c*) Air distribution within each room

The even distribution of the total incoming air within the room is of great importance. The way in which this is achieved will vary considerably according to the local conditions of each application, but it is worth keeping in mind the following key points:

(1) The even distribution should extend throughout the whole room and should avoid the leaving of any dead pockets or stagnant corners.
(2) Care must be taken to avoid placing any animal in a draught.
(3) The inlet and outlet grilles should be positioned so as to promote the rapid removal of airborne organisms or smells.

The extent of variation which may occur in practice in pursuit of these objectives may be seen by comparing the normally preferred designs for two animal houses for different duties. Thus, where the house contains small animals, such as rats, mice, guinea pigs, etc., which are accommodated in relatively high tiers of cages, experience has indicated a preference for admitting the air at low velocity through inlets sited at low level with the extract grilles placed in the ceiling; by contrast, in SPF chicken houses, in which the droppings may accumulate beneath a grid floor, it appears to be preferable to admit air through the ceiling and remove it through the grid floor, whence it passes into ducts which discharge to atmosphere at high level.

(*d*) Location of fresh-air intake duct

It is worth taking some care in selecting the location for the fresh-air inlet duct with a view to minimizing the dirt load reaching the filters and thus maximizing their life. In practice, this means to site the duct up-wind of smoke stacks and other sources of fume and smell, and also locate it at the highest possible level.

(*e*) Air recirculation

In the air conditioning of buildings for human occupation it is normal practice to recirculate a large proportion of the air so that the volume of fresh air drawn in is minimized and the costs of heating or cooling consequently reduced. While this system has been used with animal houses, it generally tends to be avoided from fear of the possible recirculation of pathogenic organisms and of smells; instance where troubles of this sort have occurred have, in fact, been reported from the U.S.A. None the less, risks of this sort can be avoided by the use of ultra-high-efficiency filters, the success of which has been conclusively proven as a means for removing sub-micronic organisms; furthermore, it has also been clearly demonstrated that animal smells can be removed from air by passing it through filters of activated carbon. Provided this type of precaution is taken, there is no reason why the economies arising from air recirculation should not be enjoyed in animal-house systems; already at least one such system is being installed in the United Kingdom, while undoubtedly others will soon follow.

(*f*) Air locks

For duties such as the housing and breeding of SPF animals, it is particularly important to ensure that only filtered air can enter the house. To achieve this, the pressure within the house is maintained at about 0·1–0·2 in. w.g. above the outside pressure, and airlock doors are used to prevent the loss of this pressure when staff are entering or leaving the building. This elaboration is not generally used for normal animal accommodation, where the standards of air cleanliness do not have to be so high, although even here it is advisable to minimize the opportunity for air to enter the animal house other than through the purification system.

HEATING AND COOLING

The cost of heating an animal house represents a significant proportion of the total running costs of an animal establishment, and therefore merits very careful study at an early stage in planning. Heat can be fed into an animal house by two main methods, either by heating the incoming ventilating air or by installing space heaters. The first method can also be used to cool the house during the summer, since the temperature of the inlet air can then be reduced by passing it through a cooler instead of through a heater. While heating the inlet air is undoubtedly the better method from the point of view of closeness of temperature control and evenness of distribution of the heat, the use of local heaters may well be preferred on economic grounds.

With either system, several different types of fuel and heating devices are available. The optimum system will vary widely with local conditions, so that it is impossible to lay down a set of hard and fast rules to follow; the following points will none the less provide some general guidance.

(*a*) Radiators and convector heaters

These are not recommended, both because they occupy valuable space within the animal house and because they tend to give rise to convection currents and to uneven distribution of heat.

(*b*) Underfloor hot-water heating

While fairly costly to install, this is an effective system when used in conjunction with a heater battery in the inlet-air duct.

(*c*) Underfloor electric heating

Although cheap to install, this method suffers from potential disadvantages in either running cost or flexibility, depending upon whether the power is paid for at the normal or the off-peak rate. The off-peak tariff is by far the cheaper, but since it cannot be used throughout the whole 24 hours of the day, control can be a real problem. While this disadvantage is avoided with the normal tariff, the much higher power cost is a serious drawback. With either tariff, underfloor electric heating has the difficulty of being inflexible in control, so that it is best used in conjunction with a heater battery in the inlet air.

(*d*) Radiant heating

For duties where the ventilating rates are low, such as in chick units, radiant ceiling panels or infra-red electric heaters have considerable advantage. The method is not recommended for more general use, however, in view of the uneven heat distribution which it entails.

(*e*) Heater battery in inlet air

This method is the optimum for animal houses, and may be used either alone or in conjunction with one of the others already considered. Heater batteries may be either of the electric or hot-water types; steam may also be used, but it is not recommended, because close temperature control is more difficult than with hot water, and because maintenance costs on control valves also tend to be relatively high. With any type of heater battery it is important to install an over-riding thermostat in the duct to ensure that, irrespective of room temperature, the temperature of the air fed into the room from the duct does not fall below about 60°F.

From the design aspect, several generalizations may be made, since they apply regardless of which form of heating system is selected. Thus, care should be taken to avoid designing to maintain the temperature of the animal house at an unnecessarily high level, since this can result in large increases in running costs; as an illustration of this, it should be noted that for each degree above 65°F the fuel cost will increase roughly 5 per cent. On the other hand, consciousness of this cost factor should not be allowed to result in the heater battery being undersized to deal with extreme winter conditions; in fact, whereas with a normal building the heating plant is rated for an assumed lowest ambient temperature of 32°F, in the case of an animal house it is essential to base the design on the lowest actual temperature which is reasonably likely to occur, which usually will be distinctly less than 32°F. In practice, an outside temperature of 20°F is frequently used as the basis of the heating calculations.

Attention should also be paid to the thermal insulation and general structure of the building, since these factors may have an important bearing on both running costs and on the suitability of the building for its intended purpose. For example, from the structural point of view, allowance should be made for the fact that an air space in the roof will often result in a significant increase in the temperature inside the main part of the building due to solar heat gain; likewise, animals are likely to be overheated by the direct rays of the sun if windows are so placed that cages are exposed to sunshine. In general, it is recommended that heat loss and condensation should be minimized by the use of double glazing of the windows and thermal insulation of walls and ceilings. The degree of lagging should be such as to give a maximum heat loss rate of 'U value' or 0·2 BTU/sq. ft./deg. F/minute, although it is often economically justified to design to an even lower figure than this.

FILTRATION

Before air is admitted into an animal house, whether from a once-through or a recirculation system, it should be filtered to remove both ordinary atmospheric dirt and also pathogenic organisms. The degree of filtration may be varied according to the needs of each particular application in view of the wide variety of filters which are commercially available.[3,4] These filters range from the one extreme which can remove only very coarse particles of dirt down to the other extreme, represented by the so-called 'ultra-high-efficiency' filters, which can arrest virus particles of the smallest size known; this latter type of filter is sometimes referred to as an 'absolute' filter, a term which the writer prefers to avoid, since it disguises the fact that it is not really 100 per cent effective against all and every size of particle, even though in practical application it may approximate to this ideal.

While there is regrettably no universally recognized classification used to characterize the very wide range of air filters, Table I presents a system which the writer has devised and found to be of considerable practical value. It will be seen that the filters are divided into four groups according to their performance in two British Standard tests (BS 2381: 1957 and 3928: 1965), one test being either against sodium chloride or methylene blue particles ranging from about 0·02 to 2 microns in diameter, while the other is an aloxite powder with a mean particle size of about 5 microns. Performance is quoted as the percentage of the particles which are trapped by the filter, the figures being expressed as percentage trapped or 'filtration efficiency' for values up to about 90 per cent, or as 'percentage penetration' (i.e. 100—filtration efficiency per cent) for higher values. Aloxite is really intended for the measurement of the dust-holding capacity of the lower grades of filter but is also often used to indicate the efficiency of these same grades, whereas sodium chloride and methylene blue are more appropriately restricted to the higher grades.

TABLE 1. CLASSIFICATION OF VENTILATION AIR FILTERS IN RELATION TO THEIR SODIUM FLAME OR METHYLENE-BLUE EFFICIENCY, WITH AN APPROXIMATE COMPARISON OF THEIR EFFICIENCY AGAINST BS TEST DUST NO. 2

GRADE OF FILTER	SODIUM FLAME OR METHYLENE-BLUE TEST		ALOXITE POWDER NO. 50 TEST DUST NO. 2	
	Filtration efficiency, %	Penetration, %	Filtration efficiency, %	Penetration
Roughing	10–40	90–60	70–99	30–1
Medium efficiency	40–95	60–5	99–100	1 to NIL
High efficiency	95–99·95	5–0·05	100	NIL
Ultra high efficiency	Above 99·997	Less than 0·003	100	NIL

Where the higher degrees of filtration efficiency are required, it is normal practice to pass the air through a series of filters of progressively higher retentive power. By this means the life of the more expensive high-efficiency filters can be greatly extended, since the main dirt load will be collected by the relatively cheap roughing or medium-grade filters. Both the life and efficiency of the high- and ultra-high-efficiency filters are frequently further augmented by operating them at only 20–30 per cent of their maximum recommended rate of throughput.

For normal animal accommodation it is usual to employ only roughing or medium-efficiency grades of filter. In the case of Specified-Pathogen-Free

(SPF) and infected-animal accommodation, however, it is necessary also to utilize high-efficiency or ultra-high-efficiency filters, in which case the system must be designed with special care to ensure that optimum performance is maintained over long periods. For these more rigorous duties the following principles should be given particular attention:

(*a*) the filters must not require to be changed too frequently;

(*b*) it is desirable that roughing filters can be changed without shutting down the ventilation plant;

(*c*) ventilation air must remain sterile regardless of the amount of organic contamination either in the ventilating ducting or in the outside atmosphere;

(*d*) it must be possible to shut off the ventilation system to one room without affecting adjacent rooms;

(*e*) in the event of plant failure, it must be impossible for airborne organisms to pass from one room into another.

It has been stated earlier, one of the features of SPF accommodation is that the pressure within it is maintained at about 0·1–0·2 in. w.g. above the outside pressure, thereby making it impossible for any unfiltered air to leak into the room. Where, on the other hand, the accommodation houses infected animals, it is then necessary to make certain that no air can escape from the room into the surrounding atmosphere without first passing through suitable high-efficiency and/or ultra-high-efficiency filters. Under these conditions the room is maintained at a negative pressure of about 0·1–0·2 in. w.g. by means of an exhaust fan which draws air from the room and through appropriate filters before discharging it to atmosphere.

INSTRUMENTATION

Since the heating and ventilating system for an animal house must continue in operation for very long periods, it is important for it to include adequate instrumentation to permit regular checks upon performance.

Instruments which are recommended are:

(*a*) Manometers (0–2 in. w.g.) to indicate the pressure drop across the filters, thereby making it easy to see when the filters need to be cleaned or replaced.

(*b*) An inclined manometer (0–0·5 in. w.g.) to show the pressure difference between the animal room and the surrounding areas; this is of particular importance with SPF and infected-animal accommodation.

(*c*) A thermometer or sometimes a temperature recorder showing the dry-bulb temperature within the room.

(*d*) Sometimes it is advisable to have also a relative-humidity indicator or recorder for each room. Where this is done, it may well be convenient to use a combined humidity/temperature recorder in which both functions are recorded on one chart.

(*e*) An air-flow-rate indicator is very useful, both to assist in determining when filters need cleaning or replacing and also to maintain the correct balance between inlet and outlet air flows.

SAFETY MEASURES

With any form of equipment which is in constant use there is inevitably the risk of failure of one or other components in the circuit. It is therefore essential for suitable provision to be made to minimize the effects of failures. Similarly, in some instances it is important to guard against operator error and to ensure that operations are carried out in the correct sequence. The main hazards and the appropriate preventive measures may be summarized as follows:

(*a*) Overheating may occur due to failure of the thermostat control. To combat this danger, an over-riding thermostat should be installed so that it can take control of the heating source and at the same time actuate a visual or audible alarm.

(*b*) Low temperatures may occur due to failure of the thermostat control while the fans continue to discharge into the room what is then unheated air. This also can be avoided by the use of an over-riding thermostat, but in this case it is set at a low level and is wired to shut off the fans while also actuating an alarm.

(*c*) Humidity may fall below or rise above the desired level. In the case of a fully air-conditioned system a suitable humidistat would be provided so as to over-ride the normal controller. For the more normal animal-house system, in which no separate provision is made to control the humidity, it may be considered prudent to install a humidistat to actuate an alarm in the event of the humidity straying outside the desired range.

(*d*) Freezing fog can clog the inlet filters. This is a common source of trouble with temperatures below 32°F, and instances have been reported at temperatures as low as 14°F. The possibility of such a blockage may be avoided by means of a thermostatically controlled heater in the air-inlet chamber, thereby ensuring that the incoming air is always above the danger temperature.

(*e*) Maloperation is most likely to occur with systems in which separate fans are provided for the inlet and outlet air; this applies to both SPF and infected-animal accommodation, but the hazards are not the same in these two cases. With SPF accommodation it is essential for the inlet fan to be switched on first and off last so that the room is at a positive pressure; by contrast with infected-animal accommodation, the exhaust fan must be started first and switched off last so as to maintain a negative pressure. In both cases the desired sequence can be ensured by the use of electrical interlocks.

CHOICE OF FANS

It will be appreciated that with all ventilation systems where filters are used to clean the air, the dirt arrested tends to clog the filters, which causes their resistance to air flow to increase and the air volume flowing in the ventilation system to be reduced. In time the increase in filter resistance will reduce the air volume to below the design level, and in respect of animal accommodation the rooms may become over-warm and smelly due to the lack of fresh air. It is then necessary to change the air filters.

The rate at which the air volume falls as the filter resistance rises depends upon the characteristics of the fan. Propeller fans and some curved-forward centrifugal fans are not designed to work against any significant variable resistance, and if used in a ventilation system incorporating air filters the air volume will fall rapidly as the filter resistance rises. Such fans are not therefore the most suitable choice for this type of system, as the filters would have to be changed relatively frequently. On the other hand, generally speaking,

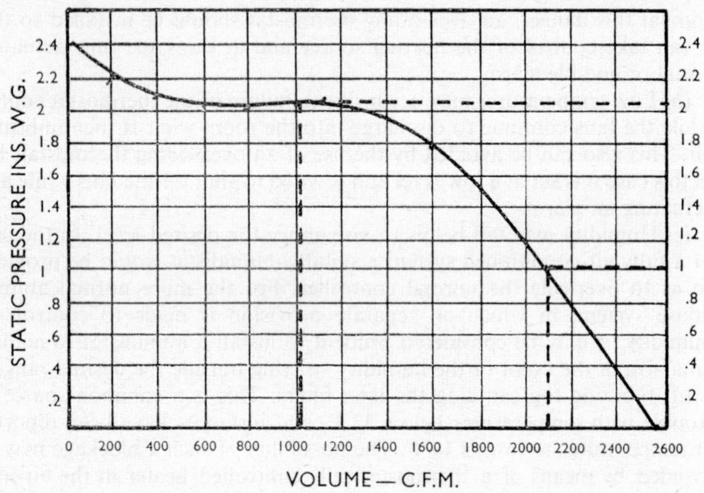

FIG. 1. Typical characteristic curve of curved-forward centrifugal fan.

axial fans with aerofoil impellers and curved-back centrifugal fans have a steep characteristic curve, and therefore as the filter resistance rises the fall-off in air volume is relatively small, consequently a longer filter life can be obtained.

Fig. 1 is a typical characteristic curve of a curved-forward centrifugal fan and Fig. 2 shows a typical characteristic of an aerofoil fan. On both graphs are marked an assumed increase in filter resistance from 1 in. water gauge (w.g.) to 2 in. w.g. Comparison of the curves shows that, when the filter resistance has increased by this amount the reduction in the air volume with the centrifugal fan would be from 2,100 c.f.m. to about 1,000 c.f.m. (over 50 per cent), whereas with the aerofoil fan the reduction would be from 2,100 c.f.m. to only 1,800 c.f.m. (i.e. 14·2 per cent).

As another exercise, let us assume that the ventilation system designer had allowed for filter-resistance increase to reduce the air volume by 15 per cent (i.e. to 1,750 c.f.m.) before changing filters. Referring to the graphs, it will be seen that with the centrifugal fan the filters would have to be changed when the system resistance was about 1·5 w.g. (an increase of 0·5 in. w.g.), whereas with the aerofoil fan the system resistance could rise to 2·1 in. w.g. (an increase of 1·1 in. w.g.), which means that the aerofoil fan may afford about twice the filter life obtainable with the curved forward centrifugal fan.

GENERAL POINTS

An important factor to bear in mind with any type of forced-ventilation system is the amount of noise generated by the fans. While there is little information available as to the effect of such background noise upon the health of the animals, it would seem desirable to keep the noise down to the minimum reasonable level, even if only for the comfort of the animal technicians. For

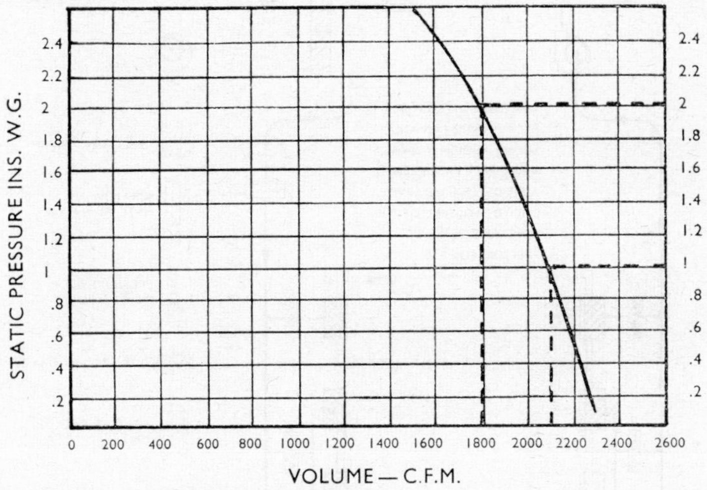

FIG. 2. Typical characteristic curve – axial fan.

this reason it is recommended that the maximum noise level originating from the ventilating plant should be not more than 45 decibels as measured within the animal room.

Comment should also be made on the use of ultra-violet bactericidal tubes, since they find use for other purposes, and might therefore well be thought to be applicable in this context. These tubes are not, in fact, recommended as a method for sterilizing either the inlet or the outlet air on grounds of cost, efficacy, and safety. None the less, they can be useful for irradiating the atmosphere in animal rooms in those circumstances where the animals may disperse organisms likely to be a hazard to the animal-room staff. Even then, however, it should be remembered that their effect is only to reduce the background count of organisms (and incidentally the smell), but not to achieve complete sterility.

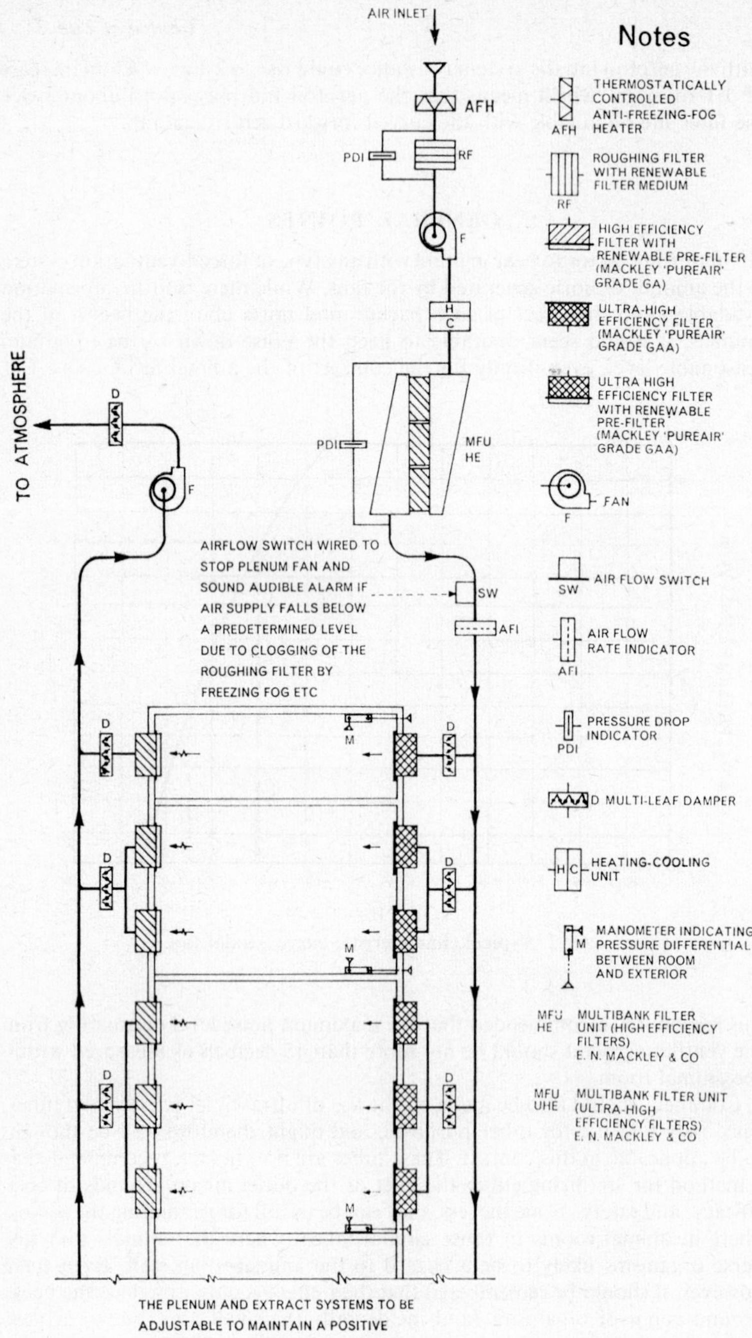
AIR INLET
AFH
PDI
RF
F
H
C
PDI
MFU
HE
TO ATMOSPHERE
D
F
AIRFLOW SWITCH WIRED TO STOP PLENUM FAN AND SOUND AUDIBLE ALARM IF AIR SUPPLY FALLS BELOW A PREDETERMINED LEVEL DUE TO CLOGGING OF THE ROUGHING FILTER BY FREEZING FOG ETC
SW
AFI
M
THE PLENUM AND EXTRACT SYSTEMS TO BE ADJUSTABLE TO MAINTAIN A POSITIVE PRESSURE IN EACH ROOM OF 0.1" TO 0.2" WG
Notes
AFH THERMOSTATICALLY CONTROLLED ANTI-FREEZING-FOG HEATER
RF ROUGHING FILTER WITH RENEWABLE FILTER MEDIUM
HIGH EFFICIENCY FILTER WITH RENEWABLE PRE-FILTER (MACKLEY 'PUREAIR' GRADE GA)
ULTRA-HIGH EFFICIENCY FILTER (MACKLEY 'PUREAIR' GRADE GAA)
ULTRA HIGH EFFICIENCY FILTER WITH RENEWABLE PRE-FILTER (MACKLEY 'PUREAIR' GRADE GAA)
F FAN
SW AIR FLOW SWITCH
AFI AIR FLOW RATE INDICATOR
PDI PRESSURE DROP INDICATOR
D MULTI-LEAF DAMPER
H C HEATING-COOLING UNIT
M MANOMETER INDICATING PRESSURE DIFFERENTIAL BETWEEN ROOM AND EXTERIOR
MFU HE MULTIBANK FILTER UNIT (HIGH EFFICIENCY FILTERS) E. N. MACKLEY & CO
MFU UHE MULTIBANK FILTER UNIT (ULTRA-HIGH EFFICIENCY FILTERS) E. N. MACKLEY & CO

FIG. 3

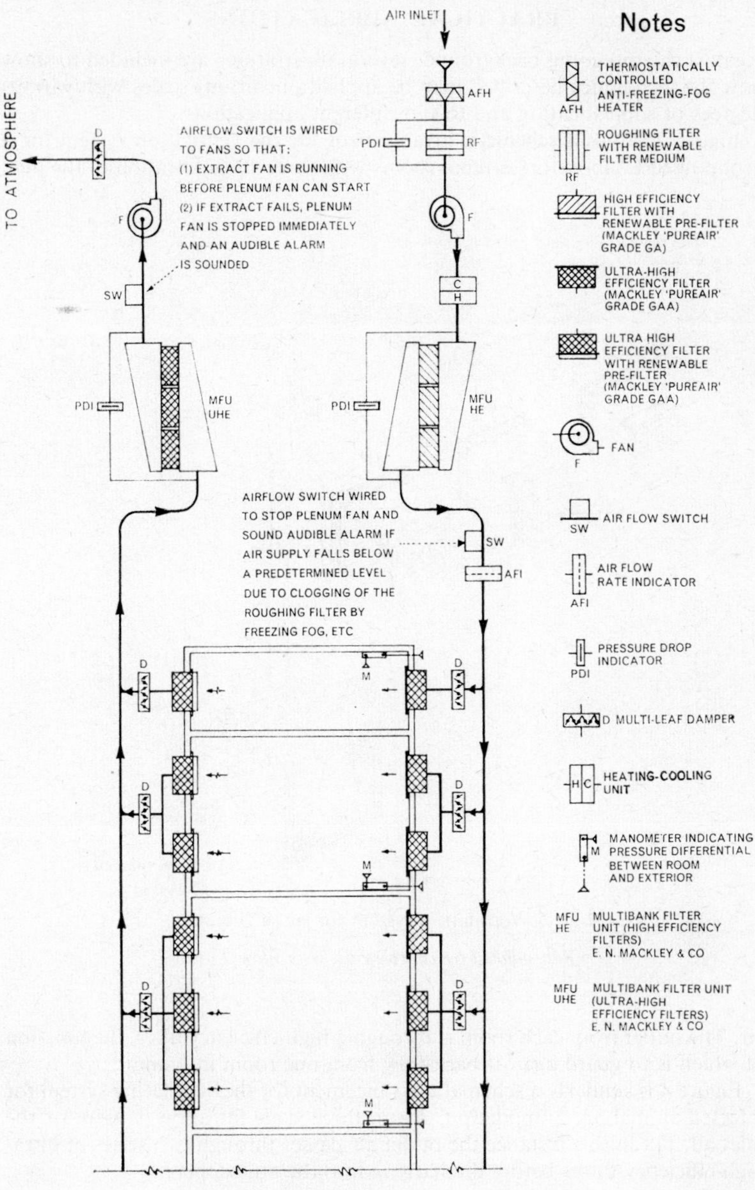

AIR INLET
Notes
AFH
PDI
RF
F
C
H
SW
PDI
MFU
HE
SW
AFI
D
TO ATMOSPHERE
AIRFLOW SWITCH IS WIRED
TO FANS SO THAT:
(1) EXTRACT FAN IS RUNNING
BEFORE PLENUM FAN CAN START
(2) IF EXTRACT FAILS, PLENUM
FAN IS STOPPED IMMEDIATELY
AND AN AUDIBLE ALARM
IS SOUNDED
MFU
UHE
AIRFLOW SWITCH WIRED
TO STOP PLENUM FAN AND
SOUND AUDIBLE ALARM IF
AIR SUPPLY FALLS BELOW
A PREDETERMINED LEVEL
DUE TO CLOGGING OF THE
ROUGHING FILTER BY
FREEZING FOG, ETC
M
THERMOSTATICALLY
CONTROLLED
ANTI-FREEZING-FOG
HEATER
ROUGHING FILTER
WITH RENEWABLE
FILTER MEDIUM
HIGH EFFICIENCY
FILTER WITH
RENEWABLE PRE-FILTER
(MACKLEY 'PUREAIR'
GRADE GA)
ULTRA-HIGH
EFFICIENCY FILTER
(MACKLEY 'PUREAIR'
GRADE GAA)
ULTRA HIGH
EFFICIENCY FILTER
WITH RENEWABLE
PRE-FILTER
(MACKLEY 'PUREAIR'
GRADE GAA)
FAN
AIR FLOW SWITCH
AIR FLOW
RATE INDICATOR
PRESSURE DROP
INDICATOR
MULTI-LEAF DAMPER
HEATING-COOLING
UNIT
MANOMETER INDICATING
PRESSURE DIFFERENTIAL
BETWEEN ROOM
AND EXTERIOR
MFU HE
MULTIBANK FILTER
UNIT (HIGH EFFICIENCY
FILTERS)
E. N. MACKLEY & CO
MFU UHE
MULTIBANK FILTER UNIT
(ULTRA-HIGH
EFFICIENCY FILTERS)
E. N. MACKLEY & CO
The Plenum and Extract systems to be
adjustable to maintain a negative
pressure in each room of below 0.1 wg

FIG. 4

PRACTICAL APPLICATIONS

Against the foregoing background, several illustrations are included to show how the techniques described may be applied on various scales with various degrees of sophistication and to suit different applications.

Figure 3 shows a schemetic arrangement for the ventilation system for a group of SPF laboratory animal rooms with three-stage filtration of the inlet

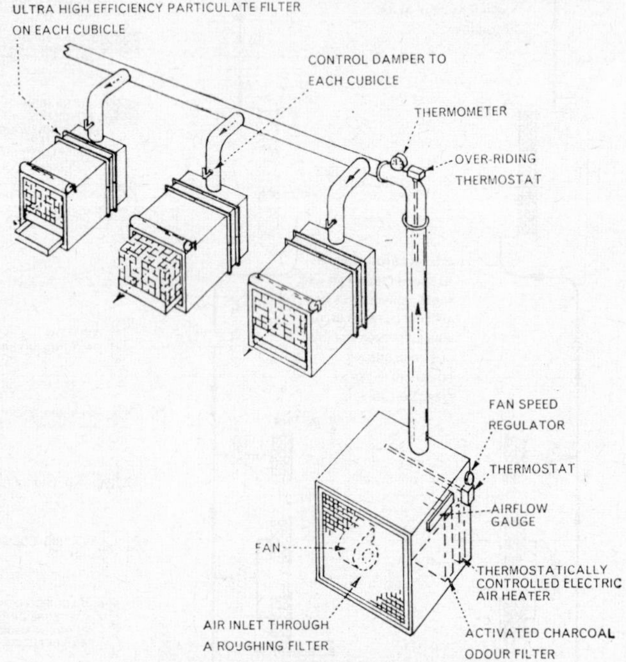

FIG. 5. Ventilation system for small cubicles.
Reproduced by courtesy of Microflow Ltd.

air. The outlet from each room is through a high-efficiency filter, the function of which is to guard against back-flow from one room into another.

Figure 4 is similarly a schematic arrangement for the ventilating system for infected-animal accommodation. Again three-stage filtration is used for the inlet air, but in this instance the outlet air passes through two series of ultra-high-efficiency filters before discharging into the atmosphere.

In Figs. 3 and 4 air locks have been omitted for clarity.

Figures 5 and 6 show how the same principles can be applied to the building of small cubicles and trolleys so that the controlled environment is restricted to the minimum necessary volume.

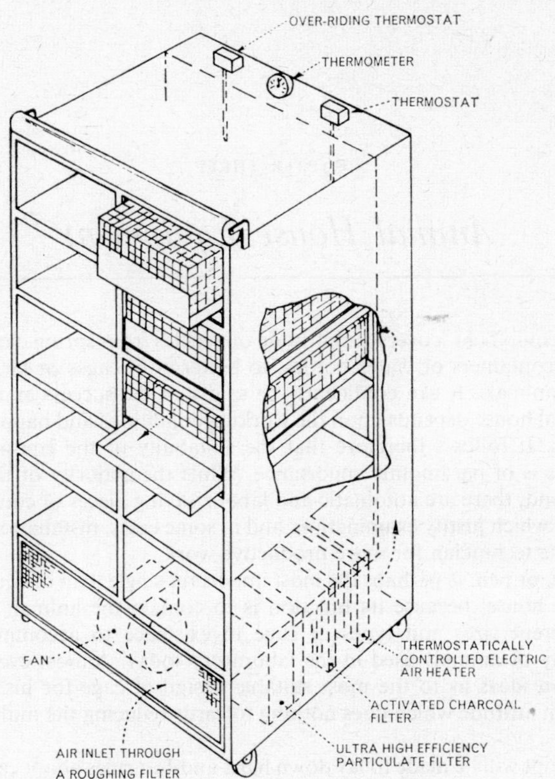

FIG. 6. Ventilation system for trolleys. Positive pressure used for SPF animals; negative pressure for infected animals.

Reproduced by courtesy of Microflow Ltd.

REFERENCES

1. BRUCE, H. M., *J. Reprod. Fert.* **3**, 310–11 (1960).
2. ALSCHULER, J. H., *Lab. Animal Care*, Pt. 2, **13**, 321–31 (1963).
3. FIRMAN, J. E., *Filtration and Separation*, **1**, 31–5 (1964).
4. FIRMAN, J. E., *Filtration and Separation*, **2**, 102–7 (1965).

CHAPTER THREE

Animal House Equipment

The term 'equipment' covers a multitude of articles from spring clips, used to hold food containers or water bottles, to batteries of cages or large pens to house the animals. It can confidently be said that the success and efficiency of an animal house depends upon the working conditions and happiness of its technicians. It follows therefore that the suitability of the equipment they have to use is of paramount importance. While the majority of the work is done by hand, there are automatic and labour-saving pieces of equipment on the market which justify examination, and in some cases, installation; thereby relieving the technician for more productive work.

The cage, or pen, is perhaps the most important single item of equipment in the animal house, because its function is to contain the animal. There are many different sizes and types of cage in existence to accommodate the wide variety of animals used in the laboratory today. Almost every worker has his own ideas as to the most suitable design of cage for his particular purpose; an attitude which does nothing towards reducing the multiplicity of cages.

No attempt will be made to lay down hard-and-fast rules about cage design; rather an attempt will be made to offer constructive suggestions of a general nature and to draw attention to some of the pitfalls which may be encountered. Before discussing cage design it would be convenient to consider some of the materials available for their manufacture.

MATERIALS USED IN THE MANUFACTURE OF CAGES

(i) Galvanized iron

This is iron covered with a protective layer of zinc, and is probably the most popular material in use for cage making in Britain. Galvanized iron (or steel) is resistant to attack by alkalis but not by acids, including those found in urine.

Iron is quickly destroyed by oxidation (the formation of rust or iron oxide) when it is exposed to air and water. The protective zinc is also destroyed by oxidation, but the decay of zinc to zinc oxide on exposure to air and water is a much slower process than the rusting of iron. Furthermore, the formation of zinc oxide protects the underlying zinc from further oxidation. As the zinc oxide layer is worn away, fresh oxide is formed. This continues until all the zinc has been oxidized, after which oxidation of the underlying iron begins and proceeds rapidly until holes appear in trays and the mesh of cages breaks. Items made from galvanized iron can be easily and cheaply regalvanized, providing this is done before the iron work has been allowed to rust. Regalvanized items can be expected to wear as well as new, galvanized items. All equipment made from galvanized iron should be inspected regularly for signs of wear and should be set aside for regalvanizing immediately rusting appears.

Iron may be galvanized either by dipping in molten zinc or by an electrolytic process. The latter method is not suitable for items intended for the animal house, as the layer of zinc deposited is too thin to offer adequate protection to the metal. Although galvanized-iron sheet and wire are readily available, it is customary to construct animal cages and trays from iron and to galvanize the articles after manufacture. Such a procedure ensures that all joins, seams, and bends are properly protected by the zinc coating.

Mild steel (which is iron mixed with a small amount of carbon) is the metal commonly used for cages and trays. Twenty-two-gauge mild-steel sheet is suitable for trays, and 16-gauge mild-steel wire on 10-gauge wire frames is suitable for grids of moderate area. Orders for cages and trays should specify 'galvanized after manufacture'.

(ii) Sheet steel

This is plated with tin (usually by a dipping and rolling process) and is known as *tinplate*. It is an easily worked material, very useful for 'mocking up' some new design of cage, but unsuitable for permanent apparatus, as the surface is soft and easily damaged, so that rusting of the exposed steel soon occurs, and spreads rapidly.

(iii) Sheet zinc

This is used for some equipment. It has all the desirable corrosion-resistant characteristics of galvanized iron, but it is soft and can be chewed through by rodents.

(iv) Aluminium

This is used widely for making trays, mouse boxes, and racking. It is important that 99 per cent pure aluminium is used, as some of its alloys corrode quickly in the presence of urine. Aluminium is a soft metal which is easily formed, but it cannot be joined by the common methods of welding. The special equipment needed for working this metal may not be available in cage-making plants which do not specialize in aluminium work. Aluminium has excellent resistance to corrosion by water and the acids of urine, but it is attacked by alkalis, e.g. common washing soda. An inhibited cleanser should therefore be used for washing aluminium equipment. The resistance of aluminium to corrosion is derived from the protective layer of aluminium oxide (about 0·0000005 in. thick) which forms on aluminium immediately it is exposed to air. This protective film may be increased to more than 600 times this thickness by an electrolytic process called anodizing.

The softness of this metal, which permits easy forming, also means that aluminium articles can be damaged easily. Thus, aluminium has only a limited application to equipment in an animal house.

Ninety-nine per cent aluminium of 16 gauge is recommended for mouse boxes.

(v) Stainless steel

This is a material which is likely to become more popular for cage manufacture. Its present high price arises from its being a difficult material to work and an expensive one to manufacture. It is highly resistant to atmospheric and chemical corrosion, and is extremely hard-wearing. If a number of cage designs became generally accepted and were available to buy 'off the shelf', then stainless steel would be the material of choice for these cages. Production costs would be much reduced by the methods of manufacture applicable to a steady market, and the durability of stainless steel would make the cost/life of such cages comparable with that of cages made from conventional materials.

Not all types of stainless steel are highly resistant to corrosion; the chromium–nickel types are most suitable for cages (Hoeltge, 1961).

(vi) Wood

This material still finds some favour, particularly for the construction of mouse boxes and rabbit hutches. It has a few advantages over the other materials in common use—it is warm, and the animals prefer it to metal. Its disadvantages are many. Wood must be treated with preservative if it is not to be rotted by urine. Wet sterilizing warps the wood, loosens glued joints, and rusts nails and screws. Animals chew at projecting or irregular parts of wooden cages, so much time has to be spent on the repair of such cages.

Mouse boxes made from resin-bonded plywood have proved fairly successful in use. They are the only type of wooden box which can reasonably be recommended.

(vii) Plastic materials

There are five basic plastic materials, three of which, fibreglass, polycarbonate, and polypropylene, can be autoclaved. Styrene–acrylonitrate copolymer and linear (high-density) polyethylene should not be exposed to high temperatures.

The low thermal conductivity of all plastic materials is suitable for most laboratory animals. They are light in weight, and boxes made from fibre glass, linear polyethylene, and polycarbonate will stand repeated impacts from rough handling, and they will not dent.

Most plastic cages can be stacked, and they are not attacked by cleaning agents, most disinfectants (except the cresol group), or animal waste.

Fibreglass reinforced polyester

This material has exceptional thermal stability and will not distort or deteriorate after repeated exposures to heat in the 250–290°F range (120–140°C). Impact resistance is good. The material can be machined.

Polycarbonate

This is a clear material with a very high impact strength and the only transparent plastic which can withstand temperatures of 280–290°F (138–145°C). It contains the optical and thermal properties of glass with the strength properties of the new thermoplastics. Unlike most plastics, it does not shatter under high-energy impact by sharp objects. Even thin wall sections will not break when subjected to repeated hammer blows. (See Fig. 1 for specimen boxes.) After many periods of autoclaving in high temperatures, this material tends to cloud, thus spoiling the visibility and the appearance of the box. Chemical sterilization with the ampholytic surface-acting agents, e.g. Tego, has no ill effect; phenol and its derivatives produce a cloudy appearance of the polycarbonate.

Polypropylene

This translucent material has good impact strength and a glass-smooth surface. The heat distortion point is in the range 215–230°F (101–109°C). It has excellent chemical properties and is resistant to most chemicals and solvents.

FIG. 1

Linear (high-density) polyethylene

This material is similar to polypropylene in looks and has excellent impact strength, but the heat distortion point is only 180°F (82°C), and if repeatedly exposed to high temperatures tends to warp. It cannot be autoclaved.

Styrene (acrylonitrile copolymer)

A clear material with a medium–low impact strength. The heat distortion point is about 180°F (82·2°C). It cannot be autoclaved.

Polystyrene

This is a clear material with low impact strength. The heat distortion point is about 150–170°F (65·5–76·7°C). Useful for rigid disposable cages.

If agreement could be reached on the standardization of basic cage designs better use could be made of modern materials such as stainless steel and plastics, for the cost of making but a few cages of any one design in these materials is prohibitive.

(viii) Glass

Tanks or jars of glass are sometimes used to house small animals such as mice, reptiles, and insects, but glass has a limited use as an animal container. It is reasonably cheap, can be cleaned easily, and it allows direct observation of the creature it surrounds. The breakage hazard is considerable, and should always be borne in mind.

The foregoing remarks about the suitability of various materials for cage manufacture apply equally to other items of equipment, such as food baskets and hoppers.

CAGES

All cages have some features in common, whether they are used to house animals for experiments, or for breeding, or for stock. They must be strong enough to stand the day-to-day wear and tear and repeated temperatures of 250°F (123°C) in steam sterilizing and/or the rigours of chemical sterilization. They must also be practical in construction and easy to assemble and clean. A cage must be escape-proof; the animal should not be able to gnaw or break through the fabric of the cage, or be able to undo the door fastening. Doors and other openings should be well fitting, yet easy to open and close by using only one hand. Some animals—dogs, rabbits, and especially monkeys—have the ability and cunning to open cage doors unless these are firmly secured by fastenings placed out of reach of the animals or by locks.

The dimensions of ventilation holes and the mesh of floor grids and the distances between the upright bars forming the top and sides of a cage should be carefully chosen. The sizes selected may be suitable for adults and the weaned young of a particular species, but this could be large enough to permit the escape of sucklings. A day-old rat is a very active creature which can, and will, crawl through a $\frac{1}{2}$-in. square; and, having escaped from the nest, such a creature would die from starvation and exposure.

Wire-mesh grid cage bottoms are best made from welded wire and then galvanized. The size of mesh and the gauge of the wire are important, and should be appropriate to the animal which is to be supported by the grid. 12 s.w.g. wire and $\frac{3}{4}$-in. mesh is suitable for rabbits, and 20 s.w.g. wire and a $\frac{3}{8}$-in. mesh for rats and other small rodents. It should be noted that mesh size is measured from centre to centre of the holes in the mesh; the mesh size therefore *includes* the width of one strand of wire. Floor grids made from square mesh of a size larger than this is unsuitable for guinea pigs. The hind-legs of these animals easily slip through square mesh, and are often broken as the frightened animals struggle to free themselves. Weldmesh grids of 3 in. × $\frac{1}{2}$ in. mesh are recommended for guinea pigs or a $\frac{1}{4}$-in. mesh with the corners of the floor grid cut away.

It is impossible to give definite advice about the sizes of cages. Lane-Petter (1957) published a formula as a guide to the minimum dimensions for a cage in which laboratory animals from a rabbit to a mouse may be confined for a matter of weeks, but this would not be applicable for a cat, dog, or monkey. However, all the suggestions are empirical, because little or no serious investigation has been carried out on the subject. The following table gives the sizes of cages in common use.

CAGE SIZES

SPECIES	NO. OF ANIMALS	FLOOR AREA	CAGE SIZES
Cat	Doe and litter	6 sq ft	3 ft × 2 ft × 18 in.
Ferret	Doe and litter	4–6 sq ft	3 ft × 2 ft × 12 in.
Guinea pig	Doe and litter	150 sq in.	Various
Hamster	Doe and litter	48–60 sq in.	12 in. × 5 in. × 6 in.
Mouse	Breeding	48–60 sq in.	12 in. × 5 in. × 6 in.
Rabbit	Doe and litter	$4\frac{1}{2}$–6 sq ft	4 ft × 18 in. × 18 in.
Rat (Norvegicus)	Doe and litter	140 sq in.	14 in. × 10 in. × 10 in.
Rat (Norvegicus)	Breeding pair	168 sq in.	14 in. × 12 in. × 10 in.
Rat (cotton)	Doe and litter	140 sq in.	14 in. × 10 in. × 10 in.
Other small rodents	Doe and litter	140 sq in.	14 in. × 10 in. × 10 in.

For breeding purposes it is customary to use cages having solid floors and one or more solid, or partly solid, sides. Such a cage offers maximum comfort and protection to the young. Nesting material is trampled through grid floors, leaving the young exposed on the bare grid. Young animals can be badly injured if their limbs are caught in a grid. It is not uncommon to use a box, only the lid of which is of perforated material, for breeding small animals such as rats and mice. Fibreglass boxes for guinea-pigs are shown in Fig. 2.

FIG. 2. Fibreglass boxes for guinea-pigs, as used at Roche Products Ltd., Welwyn Garden City, Herts.

These boxes are often designed to slide in and out of a rack, much as the drawers of a chest slide open and close. Such an arrangement, known as a 'battery' of cages (Fig. 3) uses the available space in an animal room to best

FIG. 2*a*. Outdoor cat run at Roche Products Ltd., Welwyn Garden City, Herts.

FIG. 3

advantage. The outstanding disadvantage of a battery is that the whole unit may have to be replaced in order to house a different species. Batteries are now available for housing dogs, cats, rabbits, guinea pigs, and small rodents. Some batteries are designed for suspended cages with grid floors (Fig. 4); this arrangement permits the trays to be removed from beneath the cages without handling the cages. Several cages can be hung over one tray, thus further reducing the amount of work and time needed to clean the tray.

There are some special-purposes cages in use in laboratories, such as the metabolism cage. This cage is used when it is necessary to make quantitative

FIG. 4

collections of urine and faeces from an animal. The cage is usually suspended over some device which separates urine and faeces and permits the separate collection of both. Most cage manufacturers have models of this type of cage in their catalogues, but few completely fulfil their proper purpose of collecting urine and faeces without mixing with food and water. Fig. 5 shows a cheap but effective home-made metabolism cage for small rodents which is made up from a retort stand, a glass funnel, a test-tube basket, and the necessary clamp with a glass device for separating the urine and faeces.

Cages for housing radioactive animals can be standard cage (Fig. 6) tops which fit on to special funnel bottoms. This allows excreta to be collected into a glass vessel.

Fig. 7 shows a three-tier cat breeding cage designed for the cat colony at Mill Hill. This cage, which is of light construction, has a separate diet tray which can easily be removed for cleaning; this avoids the necessity of opening the cage door and saves time when a litter of kittens are romping around. This battery of cages, being made from light material, is easily moved from place to place. See also Fig. 2*a*, which shows a cat run.

FIG. 5

Cages for a particular use, such as the exercise wheel shown in Fig. 8, with electric recording, is an example of 'do-it-yourself'. This piece of useful apparatus was made out of bits and pieces at an eighth of the cost of bought

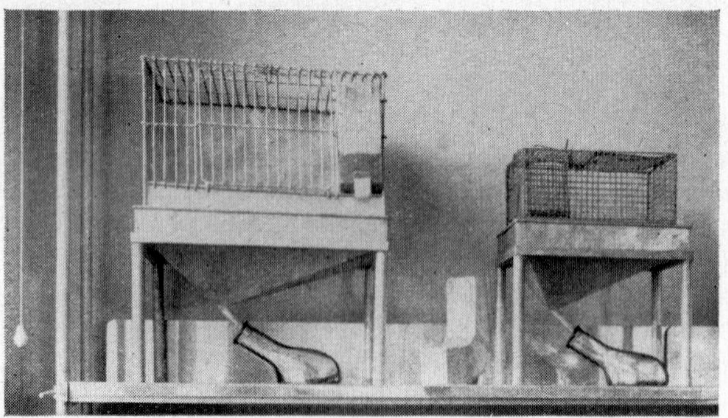

FIG. 6

equipment, but will do the same job. Another special cage is used to retain an animal in one position for a period of time, as, for example, during the continuous recording of temperature. Animals, particularly rabbits, can be trained to stand still for quite long periods without any restriction being applied.

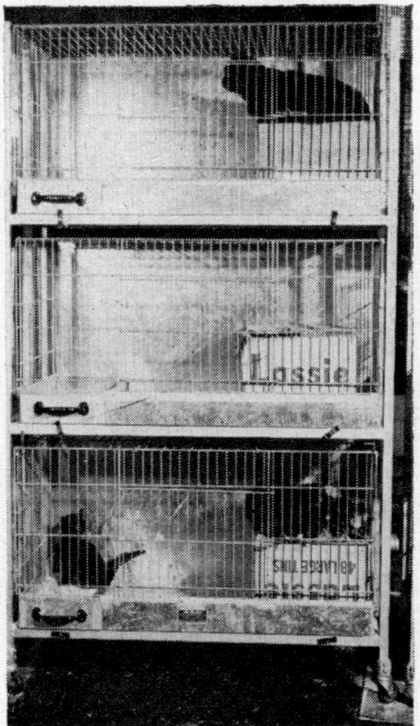

FIG. 7

FIG. 8

Short (1960) reported an automatic battery for housing rabbits. Cleaning, feeding, and watering were done mechanically with the minimum of human attention. The initial cost for housing 672 rabbits was soon recovered by the saving in labour costs. It is possible that large numbers of guinea pigs could be housed using a similar system (Fig. 9).

Braby's (November 1961) described the primate quarters attached to the Medical School, University of Birmingham, Department of Anatomy. The primate quarters were designed for a capacity of 250 monkeys. They consist

FIG. 9

of a very large fully air-conditioned room (Fig. 10) housing some 125 cages, and two additional smaller rooms for the quarantine and treatment of suspected or known sick animals (Fig. 11). It will be seen from the illustrations of the interior of the large monkey room that full advantage of space has been taken by building a tubular structure supporting a grill-floor to provide a staging for many of the cages (Figs. 12 and 13).

FIG. 10

FIG. 11

Fig. 12

Both single and double cages (provided with a removable partition) are used and shown in the photographs (Figs. 11 and 14). All are provided with a sliding door moving on a guarded runner, a metal perch, water dispenser, and a pan. They are bolted to a solid backrail to prevent them being 'walked about' by the monkey; doors are secured by spring padlocks.

Fig. 13

Some type of rack or shelf is essential for the economic accommodation of the cages, and, like the cages, these racks may be seen in a variety of forms. They can be built on to walls, suspended from the ceiling, or be free-standing and fitted with castors so that they may be moved about the room.

The materials that are suitable for the manufacture of cages are also generally suitable for the construction of racks; galvanized iron and aluminium are popular. Wood is an undesirable material for several reasons, particularly that of hygiene. Racks should be fully adjustable and easy to dismantle. The ability to adjust the distance between the shelves enables both large and small cages to be accommodated on the same rack. To be able to dismantle a rack completely allows thorough cleaning, and the rack can be stored in the minimum of space.

FIG. 14

Some establishments have used racking made from ready-drilled and slotted steel or aluminium members (e.g. 'Dexion' and 'Handy Angle'). These materials are supplied in standard lengths with holes and slots stamped out at regular intervals. A special cutter is available for use with these materials and the angle members are marked at appropriate distances to ensure that the holes and slots match up when cut lengths are bolted together.

The shelves used on the racks can be either solid (sheet metal) or open (rods or bars). While the open type of shelf has the advantage of presenting a small surface area on which dust and dirt can collect, the solid shelf has the advantage that faeces, food, and bedding cannot drop down to contaminate the cages below.

Whatever type of rack is used, good clearance should be allowed between the lowest shelf and the floor so that it is easy to clean beneath the rack. Suspended racks are ideal from this point of view, as there is no floor obstruction whatsoever (Figs. 15, 16). The mobile type, running on castors, is probably the most useful general rack. While ground clearance is important, such racks can be moved to clean the floor beneath them. Castors for these racks should be of a reasonable diameter and be fairly wide. Castors make the rack easier to move (a loaded rack can be very heavy) and also minimize pitting in soft floors if the loaded rack is kept in one position for any length of time. The

FIG. 15

castors must be of good quality and be protected from water and dirt; they should be serviced regularly.

Wall racks are best fixed by means of slotted brackets, so that the projecting members can be lifted down and the wall cleaned or painted. Animal rooms should have the minimum of permanent fixtures so that cleaning and redecoration may be easily accomplished.

WATERING DEVICES

Bruce (1950), comparing the relative water requirements of human beings and laboratory animals, says, 'Even allowing for a very large margin or error in the estimate, it is evident that laboratory animals fed on a dry diet only, drink much larger amounts of water than human beings.'

The watering of small animals by means of inverted bottles and glass drinking-spouts is attended by two small problems. The first is that of attaching the drinking-spout to the bottle so that there is no leak and dismantling it for filling and cleaning is easy. The second problem is that although glass drinking-tubes and bulbs are easy to clean and airlocks are visible, the glass is fragile and the breakage rate is high, with the possibility of serious

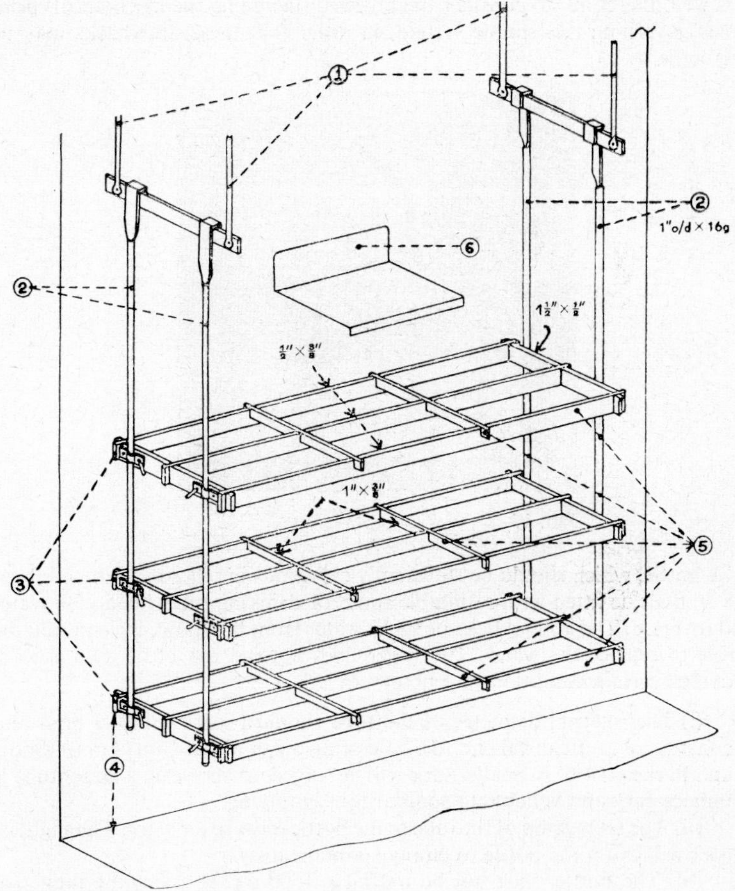

FIG. 16

injury to the technician. Short and Parkes (1949) described a metal drinking-spout which could be screwed on to the standard blood transfusion bottle, and a similar spout to fit the 'medicine flat'. Lane-Petter (1951) described a soft plastic cap which fitted over the neck of the bottle, with a Pyrex glass tube for the spout. Short (1952) described a cheap, hard plastic combined screw cap and drinking-spout designed to fit a 4-oz screw-top glass bottle.

The plastic caps and drinking-spouts are much cheaper than their metal counterparts, but the metal outlasts the plastic by many years. Fig. 17 shows a selection of water-bottles and spouts.

Lane-Petter (1952), writing about the mechanics of the animal water bottle, said, 'The increasing popularity of the inverted bottle and drinking spout for watering small laboratory animals has brought to light potential drawbacks: it is well, therefore, to consider the physical (including the mechanical) principles governing this simple system, in order that these drawbacks may be overcome. . . .

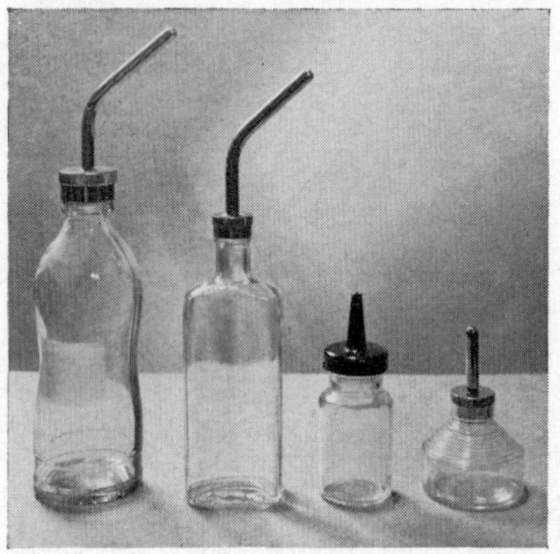

FIG. 17

'A bottle, which should be sufficiently capacious in relation to the needs of the animals, is fitted with a suitable spout or drinking tube, filled with water and inverted. The animal licks or sucks water from the spout, air entering the bottle to replace the water. The water does not run out of its own accord, provided certain conditions are observed:

'(i) The internal diameter of the tube should be $\frac{1}{4}$–$\frac{3}{8}$ in. (6–9 mm) and constricted at the distal end to a hole of diameter about $\frac{1}{8}$ in. (3 mm) (Short and Parkes 1949). A smaller tube will be subject to air locks: a larger tube is unnecessary and is liable to spontaneous emptying.

'(ii) The connexion of the tube to the bottle must be air-tight. The slightest leak will cause the bottle to empty spontaneously.

'(iii) The bottle must not be too large: 500 c.c. is about the limit. An increase in the volume of air inside the bottle will lead to expulsion of water, and there are several ways (apart from leakages around the bung) in which this can happen when the animal is not drinking: (*a*) an increase in temperature of 1°C will cause an increase in volume of approximately 0·3 per cent, that is 0·3 c.c. for every 100 c.c. of air in the bottle. (*b*) A fall in atmospheric pressure of 0·1 in. (2·5 mm) mercury will cause a similar increase of volume; neither of these results in serious leakage normally. (*c*) Shaking the bottle; movement of the bottle may cause the volume of water expelled to exceed the volume of air admitted: a further volume of air will then be drawn into the bottle, and the process may be repeated until the bottle is empty.

'Leakage from any of these causes depends on the increase in the volume of the air in the bottle being greater than the drop of water which can hang on the end of the tube; if the drop does not fall off, a subsequent decrease in volume (such as will immediately follow an increase resulting from shaking) will draw the drop back into the tube, provided the drop covers the aperture in the end of the tube.

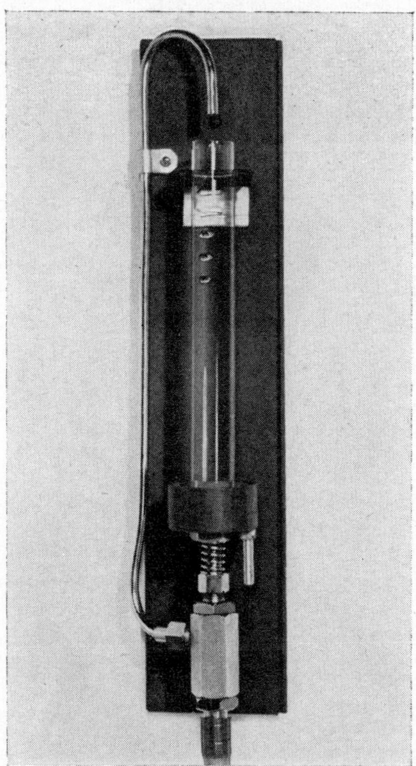

FIG. 18

'Drop-size depends on a number of factors, the most important of which are the material comprising the tube and the shape of its extremity. Glass has been found to carry a larger drop than other materials, metal being the least satisfactory: while a small terminal expansion can increase the drop-size considerably.

'If the tube, instead of being vertical, is at an oblique angle (as frequently happens when the bottle is held outside the cage) the drop of water will not cover a centrally placed aperture. It is necessary, therefore, to make the terminal aperture at the most dependent part of the tube, so that contraction of the air inside the bottle will suck back the drop of water, rather than air, (which may lead to spontaneous emptying).'

Watson (1961) described an automatic drinking system for rabbits which, with the aid of Schrader tank valves, was non-leaking and could be used for animals on solid floors. Gray and Carter (1960) described a drinking-fountain for dogs that would: (i) fill automatically to provide fresh water at all times;

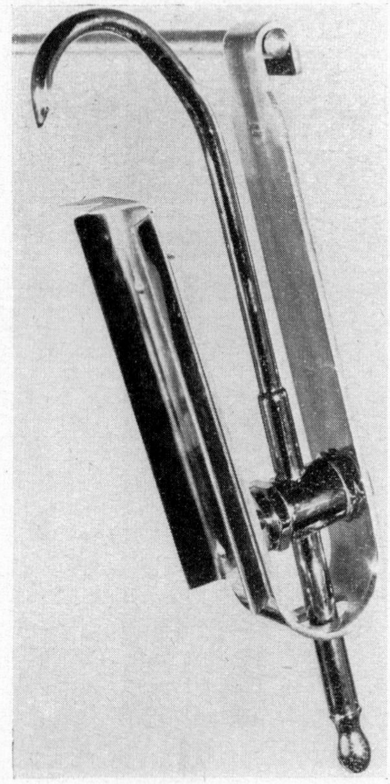

FIG. 19

(ii) the water-bowl was so designed that it would easily be removed for cleaning; (iii) there would be no back-siphoning into water-supply system; (iv) the valve would be protected against urination; (v) the whole fountain could be fitted into a space bounded by sides of 4–6 in.

There are many automatic bottle-filling devices; two are shown in Figs. 18 and 19. They have the advantage that they can: (i) be taken to the cages so that bottles can always be replaced in the cage from which they came, and (ii) the bottles need not come into contact with the filler pipe. The apparatus shown in Fig. 18 was described by Gaunt (1961), and that shown in Fig. 19 has been in use at Mill Hill since 1953.

FIG. 19*a*. Method of fitting valve to box.

FIG. 19*b*. Pipe-line assembly.

Automatic watering devices for laboratory animals have exercised the minds of those connected with animal husbandry for some years. The conventional water bottle and spout which is widely used has many disadvantages, e.g. the constant labour of filling, washing, and sterilizing, especially in large establishments, where large numbers to be serviced daily. Elimination of water bottles would represent a considerable saving in labour. The reason why an automatic system has not been generally adopted was the lack of a cheap, non-leaking, simple valve.

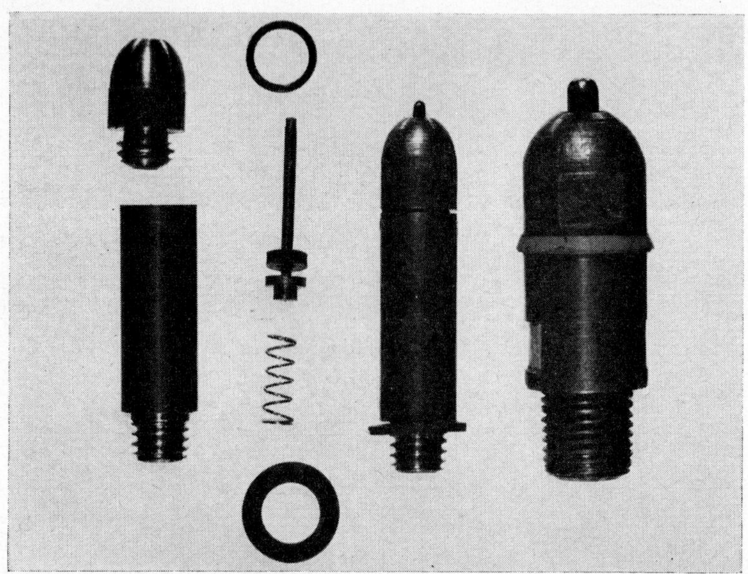

FIG. 19*c*. The large and small valves, showing the component parts.

To allow the smallest laboratory animal to obtain water, automatic valves must operate at low pressure, but leakages easily occur with low-pressure systems, and this could not be tolerated with solid-bottom boxes or cages.

Millican and Short (1965) described a valve which is suitable for most laboratory animals, but stressed the fact that these valves are a piece of fine machine work, but, made in suitable quantities, could be produced at a price comparable to the cost of glass or plastic bottles and spouts. Williams, Coley, and Humphreys (1967) describe the development of an automatic watering system for laboratory animals using the 'Chinchilla' watering unit. Fig. 19*a*, *b* and *c* shows the device of Millican and Short.

There are a number of automatic watering devices available which can be connected to the mains water supply through a storage tank. They are fitted with an internal constant-level device which allows the animal a constant supply of water.

These automatic units require some plumbing and are rather expensive to instal, but they soon repay their cost by the labour saved.

FEEDING HOPPERS, BASKETS, AND POTS

These are the common receptacles for food for animals. Diets in the form of dry powders or wet mashes or gruels are presented in open dishes. Hoppers and baskets are used for the cubed (or pelleted) diets, which are now the standard form of diet for stock colonies.

FIG. 20

Hoppers and baskets exist in several forms; some are built into the top of the cages; others hang on the front or sides of the cage. If possible, it should be so arranged that the food containers can be replenished without opening

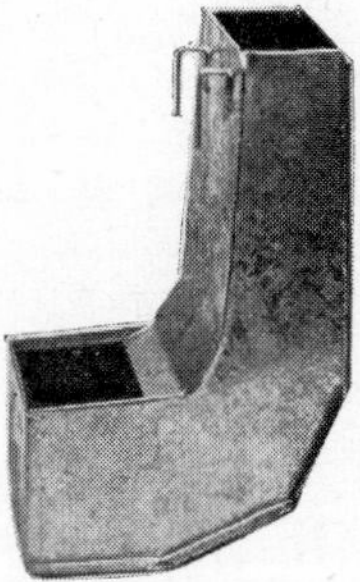

FIG. 21

the cage; this saves much time where many animals have to be fed. Hoppers and food baskets should be designed so that the animal can neither get inside them nor climb up the outside and foul the food with their excreta (Fig. 20). They should hang with a clearance of 1 in. above the bedding. The base of a rabbit food hopper (Fig. 21) should be 2 in. above the floor grid; this will prevent the animal from scratching the food out with its front feet.

The materials for the construction of hoppers or baskets are similar to those suitable for cages, and galvanized iron is probably the most popular material.

Rats and mice can be fed with cubes contained in baskets, but the wires of the basket should be arranged to give a long (about 2 in.) vertical opening about $\frac{5}{16}$ in. wide. This makes it much easier for the animal to chew the cubes. The use of hoppers and baskets effects considerable economies in the cost of food and labour.

Small ointment jars or glass or china pots are suitable for feeding powdered diets to small rodents; they can be secured in the corner of the cage by means of a suitably sized 'Terry' spring clip. Spillage can be collected and weighed on a piece of filter-paper placed beneath the pots under the wire grid.

WEIGHING MACHINES

A balance suitable for weighing such things as small animals and the constituents of diets is a piece of apparatus which should be available in any animal house.

The most suitable type of balance for weighing small animals (such as rats, guinea pigs, and rabbits) is a double-pan, automatic-lever balance. These can be obtained with a scale of 250 gm graduated in 1-gm divisions and weighing up to 5 kg with extra weights. They should have an hydraulic damper to prevent undue hunting of the pointer with every slight movement of the animal. Scales with luminous dials showing the exact weight are available, and when large numbers of animals have to be weighed these are great time savers.

Similar models are obtainable for weighing food, etc., but scoop or helmet shaped, rather than a flat, weighing pan is desirable for a food balance. A scale of 20 or 25 gm graduated in 0·1 or 0·2-gm divisions is useful if its range can be extended with the addition of extra weights. A balance of this type usually has a maximum capacity of about 500 gm.

Balances with larger capacities are readily obtainable to weigh up to 25 or 50 kg, but the scale is usually of 500 or 100 gm, graduated in 5-gm divisions.

The domestic-types of spring balance should be avoided, as they do not stand up to the continual work of a laboratory. On the other hand, it has been found that a relatively cheap pair of 'sweet' scales with a capacity of up to 2 kg will, if used carefully, give very good results with an accuracy of 1–2 gm in 500 gm.

ELECTRIC HAIR CLIPPERS

These are essential in the animal division, and care is necessary in selecting them. They should be suitable for clipping any type of hair from horse to mouse without causing discomfort to the animals. The clippers should be suitably insulated as proof against electric shock. Most clippers run hot after constant use (e.g. clipping a large batch of guinea pigs), but Fig. 22 shows hair clippers operated by compressed air, either from direct supply or bottle, which may be used for several hours without overheating, and which does not carry the risk of electric shock should the operator be compelled to work in wet conditions.

FIG. 22

ROUTINE CLEANING OF DIRTY BEDDING FROM ANIMAL CAGES

This is one of the most time-consuming jobs in the animal house, as well as being the most irksome duty of the animal technician. The traditional dirt bin, cage scrapers, brooms, and brushes can be possible sources of disease unless constantly sterilized. Experience has shown that a central vacuum system could be used for the general cleaning of dirty animal bedding from cages, sweeping floors, brushing cage racks, and walls, thus dispensing with all the items of equipment listed above. Suitable points can be arranged in the animal rooms where flexible hose can be attached, and all dirty cage bedding would be instantly conveyed through sealed pipes to a container situated outside the animal house. The container or hopper could be erected directly over the incinerator and, by the aid of controlled doors, fed directly on to the fire without being touched by human hand. Such installations are already in use, and reports on their usefulness are eagerly awaited. Charles, Poppleton, and Stevenson (1962) reported a cheap flexible vacuum system for cleaning out mouse cages which could be adapted for other animals. They considered that such a system would improve the standard of hygiene and the efficient use of labour in the animal house.

CAGE-WASHING MACHINES

These machines are an accepted piece of equipment in the animal house. Machines vary from small upright models which one technician can operate to large tunnel-type machines which require two persons. The tunnel type vary in size from 16 ft 6 in. in length, 4 ft 6 in. in width, and 7 ft 6 in. in height to 14 ft in length, 2 ft 8 in. in width, and 5 ft 6 in. in height. The weight of these machines varies from 18 cwt. to 2 tons. They are constructed from mild steel plate and stainless steel. The steam pressure required to operate the machines is from 40 to 80 lb per sq in., and the method of treatment is:

(i) hot detergent at 140–180°F (60–82°C);
(ii) hot water wash at 160°F (71°C);
(iii) hot rinse at 180°F (82°C).

Water consumption varies. In large machines it is from 100 to 150 gallons per hour, and the washing speed is 5 to 10 ft/min and is variable. The aperture for cages varies. Usually it is 3 ft × 2 ft, but this can be altered to suit requirements. The price of these machines are from £700 for an upright one to £2,500 for a tunnel type; but again, prices will vary according to individual requirements. The delivery time is from three to six months.

BIBLIOGRAPHY

HOELTGE, E. J., *Proceedings of Animal Care Panel.* **11**, No. 1 (1961).

SPIEGEL, VONA, and GONNERT, R., *Ischr. Versuchstierk.* Bd. 1. 5. 38.46 (1961).

LANE-PETTER, W., *UFAW Handbook* (1957).

SHORT, D. J., *Journal of Animal Technicians Association*, **11**, No. 1 (1960).

BRUCE, H. M. *Journal of Animal Technicians Association*, **1**, No. 3 (1950).

SHORT, D. J. and PARKES, A. S., 'Drinking Spouts for Laboratory Animals', *Nature*, **163**, 292 (1949).

LANE-PETTER, W., 'Soft Plastic caps for Water Bottles', *Journal of Animal Technicians Association*, **2**, No. 3, 13 (1951).

SHORT, D. J., 'Some Items of Animal Equipment', *Journal of Animal Technicians Association*, **2**, No. 4, 13 (1952).

LANE-PETTER, W., *Mechanics of Water Bottles* (1952).

WATSON, S. C., *Journal of Animal Technicians Association*, **1**, No. 4 (1961).

GAY, W. L. and CARTER, J. L., *Journal of American Veterinary Association*, **137**, No. 9 (1961).

GAUNT, W. T., *Journal of Animal Technicians Association*, **1**, No. 4 (1961).

CHARLES, R. T., POPPLETON, W. A. R., and STEVENSON, D. E., *Journal of Animal Technicians Association*, **13**, No. 1 (1962).

BRABY, F. & CO., *Braby News* (November 1961).

LANE-PETTER, W., 'Animal House Equipment', *UFAW Handbook on Care and Management of Animals*, 2nd Edition (1957).

MILLICAN, K. G. and SHORT, D. J., *J. Inst. anim. Tech.* **16**, No. 4. (1965).

WILLIAMS, J. H. O., COLEY, R., and HUMPHRIES, L., *J. Inst. anim. Tech.* **18**, No. 3 (1967).

CHAPTER FOUR

Measurement of Temperature and Humidity

The vital processes of the animal body are accompanied by considerable energy changes. The energy produced is used for mechanical work, but a large proportion of the energy is evolved as heat.

The interior of the body must be maintained at a constant temperature, and this is achieved by physiological mechanisms which govern the rates of heat production and heat loss by the body.

Heat is produced by oxidation of food. Heat is lost by *radiation* from the skin to the boundary surfaces of the room, by *convection* from the skin to the air surrounding the body, and by *evaporation* of moisture or sweat from the skin. Increasing the rate of air movement, e.g. by fans, will increase the convective and evaporative but not radiative heat losses.

Although the internal body temperature is relatively constant, the skin temperature varies with the temperature of the environment. When this is low, blood flow to the skin is reduced, the skin temperature falls, and heat loss by radiation and convection is reduced; at high environmental temperatures the reverse occurs.

The object of maintaining suitable external environmental conditions is to minimize physiological stress required to maintain the constancy of the internal environment of the body.

In order to assess the environmental conditions one should measure the temperature and humidity of the air in the room, the air speed, and the temperature of the boundary surfaces of the room.

Animals should not be placed in draughts, and if the ventilation is adjusted so that no draughts are apparent to the human the rate of air movement will be low, i.e. not more than 50 ft per minute.

In many rooms there is little difference between the temperatures of the air and the boundary surfaces of a room under steady-state conditions, so that in practice it is usually necessary only to measure the temperature and humidity of the air.

But the effects of radiation must be remembered. For example, in one room when the air temperature was 74°F (23·3°C) the mean radiant temperature (MRT) was 102°F (38·9°C), owing to the sun shining through the windows. Closing the blinds reduced the MRT to 80°F (26·7°C), but away from the window the MRT was only 75°F (23·9°C).

Animals should not be placed close to sources of radiant heat.

Humidity

The atmosphere contains water in the form of water vapour. The amount of water vapour which can be contained in a given volume of air depends only on the temperature of the air. The hotter the air is, the greater the amount of water vapour it can contain. For any given temperature there is a maximum amount of water vapour that is able to be contained by a given volume of air (unless supersaturation occurs). Air carrying its maximum amount of water vapour is said to be at saturation point. If such air is cooled (e.g. by cold window glass) condensation of the water vapour occurs and liquid water is deposited.

All animals lose water vapour from the skin as well as from sweat. The amount of water vapour lost in this way depends, initially, on the temperature of the skin; the hotter the skin temperature, the higher the rate of vaporization. Factors which affect the skin temperature are the environmental temperature (see above) and the physical state of the animal, e.g. whether it is exercising or resting; is normal or is in ill-health. The ability of the animal to lose water vapour from the skin to the atmosphere depends on the amount of water vapour which the atmosphere already contains, which is, in turn, dependent on the temperature of the atmosphere. The nearer the atmosphere is to saturation-point, the slower will be the rate of evaporation of water from the body surface. This process of evaporation requires heat-energy, which is provided by the heat of the animal body. Thus, in losing water vapour from the skin an animal also loses body heat. This is an important factor in the process of maintaining a constant body temperature. It should never be forgotten that the animals' immediate, cage environment may differ considerably from the animal-room environment. The temperature and humidity recorded at the centre of a mouse nest lying in the darkest corner of a box having a half-solid lid will be higher than the temperature and humidity recorded in the gangway between two racks of mouse boxes. The temperature of such a nest will also be affected by the material from which the mouse box is constructed, since metal has a higher heat conductivity than either plastic or wood. On the other hand, the temperature and humidity in an all-mesh rat cage will be comparable to that observed in the gangway between racks of rat cages.

THE MEASUREMENT OF TEMPERATURE IN THE ANIMAL HOUSE

A THERMOGRAPH gives a continuous record of room temperature and shows not only the maximum and minimum temperatures attained but also the *duration* of any variation from the desired temperature. A thermograph consists of a bimetallic spiral, having the more expansible metal on the inside, which is fixed at one end and has a lever (for magnifying the movement) connected to the free end. When the spiral is heated it uncoils gently and the temperature is registered, by a pen attached to the lever, on a paper-covered, rotating drum. A good instrument is accurate to about $\pm\frac{1}{2}$°F (0·3°C) over most of the range used.

MAXIMUM AND MINIMUM THERMOMETERS are used to ascertain the extremes of temperature reached over a period of time—usually one day. Various designs are available, but the general principle underlying them is as follows. A large bulb filled with alcohol is connected by a capillary tube containing a short column of mercury to a second, smaller bulb which is partially filled with alcohol. Two small, metal markers are placed in the capillary tube, one on

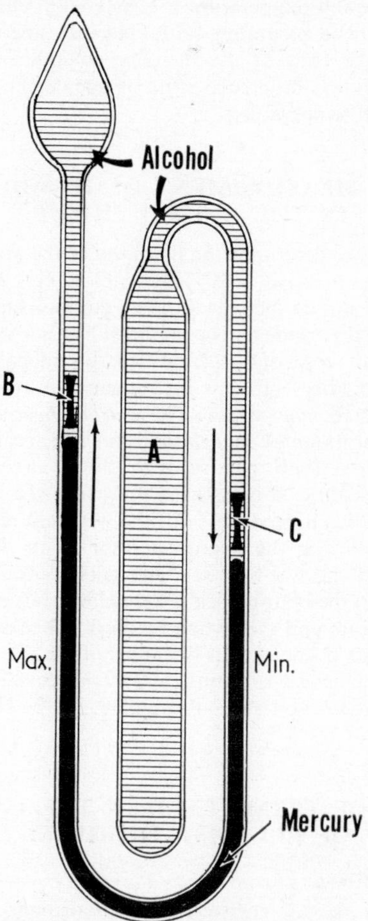

FIG. 1. Diagram to show a maximum–minimum thermometer.

either side of the mercury column. The thermometer is usually bent, often into a U-shape, for compactness, and mounted over two temperature scales.

A rise in temperature causes the alcohol in the large, filled bulb *A* to expand and push the mercury column forward. The mercury column pushes the marker *B* forward. A fall in temperature causes the alcohol in bulb *A* to contract, and the mercury column falls back, pushing marker *C* back along the capillary tube. Marker *B* remains unmoved, as alcohol wets the markers and passes them, which mercury cannot do.

The maximum and minimum temperatures attained over any chosen period can be read from the position on the scales of the ends *nearest* the mercury column of the two markers. The mercury column itself expands with increased temperature, and allowance is made for this when the thermometer is calibrated. The present room temperature is registered at either end of the mercury column itself. On an accurately calibrated instrument the two temperatures registered at the ends of the mercury column are identical. Cheap thermometers may show a difference of up to one degree between these two, theoretically identical temperatures.

THE MEASUREMENT OF HUMIDITY IN THE ANIMAL HOUSE

The instruments used to determine the humidity of the atmosphere are called HYGROMETERS OF PSYCHROMETERS. The simplest *wet- and dry-bulb hygrometer* consists of two similar mercury-in-glass thermometers mounted side by side. The bulb of one thermometer is kept moist by a surrounding cotton wick which dips into a small vessel of distilled water. It is difficult to obtain accurate measurements of humidity with this instrument; to avoid uncertainty it is best to use a ventilated hygrometer. The *whirling hygrometer* is a wet- and dry-bulb thermometer mounted in a rattle-frame which can be rotated rapidly so that the thermometer bulbs pass through the air at considerable velocity. To obtain accurate readings it is essential that: (i) a ventilated hygrometer is used; (ii) the wet-bulb is indeed wet; (iii) the wet-bulb reading is taken *first*, immediately after whirling the hygrometer for 30 or 40 seconds; (iv) the instrument is whirled and wet-bulb readings taken until two successive readings agree closely; (v) the cotton wick is kept clean; (vi) clean, distilled water is used in the reservoir; and (vii) when readings are taken the instrument is held so that the hands do not warm the bulbs of the thermometer.

The *Assmann psychrometer* is a much more expensive instrument in which air is drawn over the thermometer bulbs by means of a clockwork or electrically driven fan.

DEFINITION OF TERMS USED IN THE CALCULATION OF RELATIVE HUMIDITY

(*a*) Dry-bulb temperature

The temperature of the air registered by an ordinary thermometer. An ordinary thermometer with a bulb about 1 in. $\times$ $\frac{1}{4}$ in. in calm air assumes equilibrium between the mean radiant temperature and the true air temperature. The sling thermometer quickly assumes air temperature. If both thermometers are used one may ascertain if thermal radiation measurements are warranted. In other words, the bulb of an ordinary stationary thermometer is affected by thermal radiation.

(*b*) Wet-bulb temperature

The temperature registered by a thermometer the bulb of which is covered by a wetted wick and exposed to a current of rapidly moving air.

(*c*) Dew-point temperature

The temperature to which air has to be cooled (at constant pressure) before it becomes saturated with respect to liquid water; in saturated conditions the dew point and air temperature are equal.

(*d*) Wet-bulb depression

Difference between dry- and wet-bulb temperatures.

(*e*) Dew-point depression

Difference between dry-bulb and dew-point temperatures.

(*f*) Vapour pressure of water

The vapour pressure is a measure of the pressure exerted by water which is in the state of a vapour in air rather than in the liquid state. The saturation vapour pressure is a measure of the pressure exerted by the water vapour when the air holding it is completely saturated with water vapour and condensation of water vapour back to the liquid state is occurring. The pressure exerted by any saturated vapour depends only upon the temperature and the particular liquid used.

(*g*) Relative humidity and percentage humidity

The relative humidity is defined as the ratio of the actual vapour pressure to the saturation vapour pressure at the air temperature. This value multiplied by 100 is the percentage humidity.

(*h*) Absolute humidity

The absolute humidity is defined as the ratio of the mass of water vapour to the volume occupied by the moist air with which it is associated. It is thus equivalent to the density of the water vapour, and can be calculated from the vapour pressure and the temperature.

RELATION BETWEEN DRY-BULB, WET-BULB, AND DEW-POINT TEMPERATURES AND RELATIVE HUMIDITY

Whenever the air is not completely saturated the wet-bulb temperature is lower than that of the dry-bulb, due to cooling by evaporation. Since this wet-bulb temperature is a function of total heat content of the air including moisture, it follows that if both wet- and dry-bulb temperatures are known, relative humidity and dew point can be determined. Stating it more broadly, if *any two* of the properties of air for a given condition are known—dry-bulb temperature, wet-bulb temperature, dew-point temperature, or relative humidity—the other two may be found.

Relative humidity can be determined when dry-bulb and dew-point temperatures are known. The dew-point temperature determines the actual vapour pressure in the air at a given temperature, and the saturated vapour pressure at this dry bulb temperature can be found, both quantities being obtained from psychrometric tables. If the air is saturated no evaporation can take place from the wick of a wet-bulb thermometer, so the dry- and wet-bulb temperatures are the same. Under such conditions the dew-point temperature coincides with the dry- and wet-bulb temperatures.

Conversion of these quantities from one to another can be effected by reference to psychrometric tables. For practical work, however, a psychrometric chart is used, because its degree of accuracy is usually acceptable for ordinary calculations. The chart is concise and convenient, and with its aid conditions may be visualized. A large-scale psychrometric chart is supplied in the Supplement to the MRC War Memorandum No. 17, published by HMSO.

EXAMPLE 1:

Given: dry-bulb temperature, 70°F (21·1°C).
wet-bulb temperature, 60°F (15·6°C).
Find: percentage of relative humidity and dew point.

Locate point of intersection of vertical line representing 70°F (21·1°C) dry-bulb temperature with the oblique line representing 60°F (15·6°C) wet-bulb temperature. By interpolation this point indicates the percentage of relative humidity as 56 per cent and by following the intersecting horizontal line to the left to its intersection with the saturation line the dew point is indicated as 53·6°F (12·0°C).

EXAMPLE 2:

Given: dry-bulb temperature, 80°F (26·7°C),
relative humidity, 59 per cent.
Find: dew-point and wet-bulb temperature.

Locate the point of intersection of the vertical line representing 80°F (26·7°C) dry-bulb temperature with the interpolated position of the curved line which would represent 59 per cent relative humidity.

Reading horizontally to the left from this point, the dew point is indicated as 64°F (17·8°C), and reading obliquely upward to the left, between the wet-bulb lines, to the saturation line, the wet-bulb temperature is indicated as 69·3°F (20·6°C).

EXAMPLE 3:

Given: dry-bulb temperature, 75°F (23·9°C),
dew-point temperature, 55°F (12·8°C).
Find: percentage of relative humidity and wet-bulb temperature.

Locate the point of intersection of the vertical line representing 75°F (23·9°C), dry-bulb temperature with the horizontal dew-point line intersecting the saturation curve at 55°F (12·8°C). By interpolation, this point indicates the relative humidity as 50 per cent and the wet-bulb temperature as 62·6°F (17°C).

HAIR HYGROMETERS which are commercially available are unsuitable as standard instruments for measuring humidity, but as recording instruments they are simple and efficient, provided they are regularly checked and kept in good condition.

PAPER HYGROMETERS may be used as indicators of the humidity.

In modern, lavishly equipped animal houses a continuous, graphed record of environmental temperature and humidity is made by automatic recording devices.

SPECIES	RECOMMENDED		RANGE OF	
	TEMPERATURE, °F	RELATIVE HUMIDITY, PER CENT	TEMPERATURE, °F	RELATIVE HUMIDITY, PER CENT
Mouse	72	50	68–74	45–55
Rat	72	50	65–75	45–55
Cotton rat	72	50	65–75	45–55
Mastomys	72	50	65–75	45–55
Gerbils	72	50	65–75	45–55
Hamster*	72	50	65–75	45–55
Guinea pig	70	50	65–75	45–55
Ferret	63	50	60–65	45–55
Rabbit	65	50	62–68	45–55
Cat	63	50	60–65	45–55
Dog	60	50	55–65	45–55
Monkeys:				
From Africa—				
Cercopithecidae*	80	55	78–85	50–60
Baboons*	75	55	65–78	50–60
Chimpanzees*	80	58	78–82	55–60
From India and Far East—				
Rhesus	70	50	68–72	45–55
Cynamolgus*	80	58	78–85	55–60
From the New World—				
Spider monkey*	80	55	78–82	50–60
Capuchin*	80	55	78–82	50–60

* The temperature and humidity must be closely controlled if these species are to be kept in good health.

THE CLINICAL THERMOMETER is a mercury-in-glass thermometer with a short working range of 95–110°F (35–43·3°C). Each degree division is subdivided into five equal parts. There is a small constriction in the bore of the thermometer between the bulb and the lowest graduation. When the bulb of the thermometer is heated the mercury expands and rises up the bore, past the constriction. When the thermometer is cooled the mercury contracts, but the mercury which has risen above the constriction cannot fall back into the bulb, thus the highest temperature recorded by the thermometer may be read at leisure. The thermometer is reset by shaking, so that the mercury trapped above the constriction is jerked back into the bulb.

Clinical thermometers are made with small bulbs and narrow bores, so that they may be quick to respond to changes in temperature. Such a thin column of mercury is difficult to see, so the glass casing of these thermometers is made to give a convex surface through which a magnified image of the mercury thread may be seen. The glass casing of the bulb is very thin and fragile.

'Minute' and 'half-minute' clinical thermometers are obtainable; the designation refers to the minimum period of time for which the thermometer must be applied to the human body to obtain an accurate record of body temperature. The thermometers are also made with either 'long' or 'short' bulbs. The short bulb is safer in use, as it has the more blunted shape and is somewhat less fragile than the long bulb. However, the short bulb is of wider diameter than the long bulb, and usually requires a little longer time to reach a steady temperature. The dimensions of the bulbs are approximately $\frac{1}{4}$ in. long by $\frac{1}{8}$ in. diameter for the short one, and $\frac{5}{8}$ in. long by $\frac{1}{16}$ in. diameter for the long one. Clinical thermometers having an extended temperature range are available and should be used if an abnormally low body temperature is anticipated, as is the case in some forms of injury and stress.

Sterilization of clinical thermometers

Clinical thermometers may not be sterilized by boiling, because the maximum temperature they are designed to measure is some 100°F (55·5°C) below that of boiling water; nor may they be washed in water from a hot tap. These thermometers may be sterilized by immersion in 70 per cent alcohol, and they should then be rinsed thoroughly in cold, running water before use. A clinical thermometer should not be used without first checking that it has been reset and that the top of the mercury column is $\frac{1}{2}$ in. below the lowest graduation on the thermometer. Resetting is done *after* sterilization so that infections may not be spread by shaking a contaminated thermometer.

Taking the body temperature of an animal

The only safe way to take the body temperature of an animal is to hold the thermometer in the animal's rectum. A thermometer placed in an animal's mouth is certain to be bitten and broken. It is possible, but not usually desirable, to place the thermometer in the axilla (arm-pit), but it must be kept in position for double the time specified on the thermometer and, even then, will record a lower temperature than that found in the mouth or rectum. A long-bulb thermometer must be used for small animals, such as mice; the short bulbs are of too great a diameter for comfortable insertion. Short-bulb thermometers should be used for the larger animals.

For rectal temperatures the thermometer is prepared by smearing the bulb and lower part of the stem liberally with Vaseline, or some other acceptable lubricant. The animal is held so that it can rest quietly and comfortably for a period of up to 2 minutes, but also so that it cannot make any movement which could dislodge the thermometer, e.g. kicking with the hind-legs. The thermometer is then inserted into the rectum slowly and gently until the *whole of the bulb*, and perhaps part of the stem, is within the lumen of the rectum.

Any resistance to insertion should not be overcome by force, as forcing may result in perforation of the gut, or the thermometer may be broken. Very gentle and slight sideways and/or forward and backwards movement of the thermometer will enable it to pass faecal matter.

If the thermometer is observed it will be seen that the mercury rises fairly rapidly over about the first 15 seconds, then, over about the next 15 seconds, a further small and slower rise is registered, and finally—if the thermometer is kept in position for about 2 minutes—another very small (up to 1°F) and even slower rise may occur. It is customary to record the temperature registered after the second, slower rise and not to wait for the final rise.

Factors affecting the observed body-temperature

It is impossible to obtain an accurate measurement of body temperature if the thermometer is used incorrectly. It must be correctly set immediately before use, with the mercury column $\frac{1}{2}$ in. below the lowest graduation mark on the scale. The mercury column must not be fractured. The thermometer bulb must not be warmed by the human hand. The thermometer must be inserted for a sufficient distance into the rectum and be maintained in position for a sufficient period of time (at least 30 seconds) to allow it to attain a steady reading.

The thermometer must be inserted for the same distance in animals of the same species and size if comparable results are to be obtained. The actual length of thermometer which may be inserted varies from species to species and with size of animal in any one specie. It is good practice to site the thermometer bulb in the same, relative anatomical position in all animals.

It is well known that the body temperature has a daily fluctuation about a mean and that the temperature will be highest during periods of activity and lowest during periods of rest. Body temperatures should therefore always be recorded at a set time of day.

Similarly, every attempt should be made to have the animals in the same physical state, preferably just roused from sleeping, when body temperature records are made. An animal which has to be chased round a cage or pen before it can be caught and handled will exhibit a higher body temperature than it would do in the resting state.

The body temperature of an animal under stress may be either above normal (hyperthermia), as in the case of fevers, or sub-normal (hypothermia). A normal, healthy animal which is exposed to an environmental temperature of 5°C for several hours will exhibit hypothermia, because it is losing body heat to the atmosphere. Environmental temperature affects body temperature, either because the animal is losing body heat to a cold environment or because the animal is unable to lose body heat to a hot environment. Death ensues if an animal is exposed for a sufficient period of time to either extreme of thermal environment. A resting animal does not lose body heat to the atmosphere and does not suffer any ill-effect if it is exposed to an environmental temperature which lies within the animal's *zone of thermal neutrality*. In its zone of thermal neutrality a resting and fasting mammal exists at its basal metabolic rate. The *critical temperature* for an animal is the temperature below which it has to increase its heat production to maintain heat balance. Above the critical temperature there is only a narrow range of temperature over which the body can successfully combat (by panting and sweating) the ultimately fatal effects of overheating. It will be noted that, for their comfort, laboratory animals are customarily housed at environmental temperatures below those of their zones of thermal neutrality. In these circumstances the animals always eat some 'additional' food to supply the heat energy needed to maintain a constant body temperature against losses to the atmosphere. It will also be noted that to house animals within their zones of thermal neutrality would create almost intolerable conditions for their associated human beings, as the critical temperature for European man is between 27° and 29°C.

CRITICAL TEMPERATURES AND ZONES OF THERMAL NEUTRALITY OF SOME SPECIES

	BODY WEIGHT* kg	CRITICAL TEMPERATURE* °C	RANGE OF THERMAL NEUTRALITY† °C
White mouse	0·029	29	28–30
Lemming	0·051	14	
Wild rat	0·227	23	
Hamster	0·305	22	
White rat	0·343	28	27–29
Guinea pig	0·640	30	29–31
Opossum	1·12	25	
Rabbit	2·0	15	15–20
Dog	9·7	24	20–26
Canary			34–36
Pigeon			31–36
Chicken			16–28
Sheep			21–25
Goat			13–21
Chimpanzee			20–29

* Hart, J. S. (1963). † Brody, S. (1945).

BIBLIOGRAPHY

AMERICAN PHYSIOLOGICAL SOCIETY, *Handbook of Physiology, Section 4—Adaptation to the Environment.* American Physiological Society, Washington, D.C. (1964).

BRODY, S. *Bioenergetics and Growth.* Reinhold Publishing Corporation, New York (1945).

HART, J. S., in *Temperature—Its measurement and control in science and industry, Vol.* 3, *part* 3—*Biology and Medicine.* Reinhold Publishing Corporation, New York (1963).

GREGORY, H. SPENCER and ROURKE, E. *Hygrometry.* Crosby Lockwood, London (1957).

BODY TEMPERATURES OF COMMON SPECIES

Figures obtained from daily measurements taken over a fortnight. No appreciable difference was found between the body temperatures of males and females

SPECIES AND NUMBERS USED	AVERAGE TEMPERATURE °F	°C	MAXIMUM AND MINIMUM TEMPERATURES RECORDED °F	°C
Cockerels (6)	104·9	40·0	103·1–106·4	39·5–40·5
Dogs (2 m.)	101·2	38·3	100·5–102·2	38·1–38·9
Ferrets (3 m. 8 f.)	101·9	38·8	100·0–104·1	37·8–40·0
Guinea pigs (14 m. 14 f.)	101·0	38·3	99·3–103·0	37·4–39·4
Hamsters, Golden (6 m. 6 f.)	99·3	37·4	97·0–102·3	36·1–38·9
Mastomys (25 m. 25 f.)	99·1	37·2	95·9–102·6	35·6–39·1
Meriones (7 m. 7 f.)	99·4	37·4	96·3–102·8	35·8–39·0
Mice, Albino (60 m. 60 f.)	98·8	37·1	95·0–102·6	35·0–39·0
Monkey, Rhesus (entire stock) over several months	103·4	39·6		
Rabbits (18 f.) (strain variation insignificant)	101·1	38·3	99·1–102·9	37·2–39·3
Rats				
Cotton (12 m. 12 f.)	100·8	38·2	98·4–103·6	36·7–39·6
Hooded (12 m. 12 f.)	99·2	37·3	96·8–102·1	35·6–38·9

CHAPTER FIVE

Handling and Sexing Animals

'The importance of handling animals in the correct manner cannot be overrated. Improper handling may result in injury to the animal, to the technician himself, or, most important of all, to the animal–man relationship' (Short, D.J., *UFAW Handbook*).

The proper method of handling animals cannot be taught by lecturing or learned by reading; and the most satisfactory method of teaching is by practical demonstration, and of learning is by constant practice.

Forceps or leather gloves should never be used for handling animals unless circumstances warrant it, e.g. with infected animals, where there is a danger of the technician becoming infected from a bite.

The smaller laboratory animals quickly become docile if they are handled properly and frequently. Animals should always be held firmly, but not tightly. If an animal feels insecure or uncomfortable it will struggle to get free, and in doing so may injure itself.

Pregnant animals should be handled with great care and only when absolutely necessary.

It is necessary to employ a different technique with each animal, but the handler must always be relaxed, and must approach the animal with queit confidence.

SPECIES

(*a*) **Rabbit**

To take a rabbit out of a cage, hold the ears steady with one hand facing in the direction of the operator, and place the other hand underneath the belly of the animal. Lift both hands together gently but firmly.

To carry a rabbit, place the index finger between the ears, the thumb and other fingers will then control the head. The other hand, which takes all the weight, is placed round and under the tail (Fig. 1).

For artificial insemination, place the animal facing the handler on a non-slippery table. Its head should be tucked into the handler's body, and a hand run down each side of the rabbit's body to grasp the hind-legs, which are held by the first and second fingers of the hand, the other two fingers being placed round and underneath the hind-legs. Hold the legs apart and lift them slightly.

When sexing a rabbit always hold the legs facing *away from operator* (Fig. 2).

FIG. 1 FIG. 2

(*b*) Rat

Do *not* pick a rat up by the tail. Place the hand, palm downwards, high up over the animal's back with the thumb round the neck and under the mouth. Grasp the rat firmly, but not tightly (Figs. 3, 4, and 5). When approaching rats never hesitate or fumble or frighten the animals by waving the hands about.

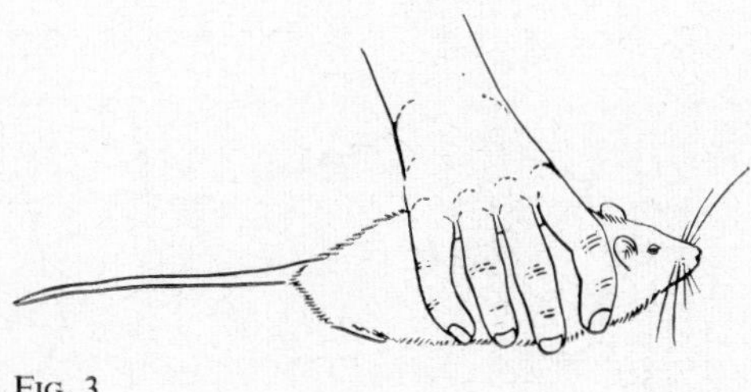

FIG. 3

(*c*) Mouse

The mouse, alone among the laboratory animals, may be lifted by its tail. The base of the tail is held between a thumb and forefinger, and the weight of the body is supported by letting the mouse stand on the top of the cage top

(Fig. 6). Mice may be sexed easily and quickly by holding them in this way. If a mouse is held by the tip of the tail it may climb up its tail and bite the handler.

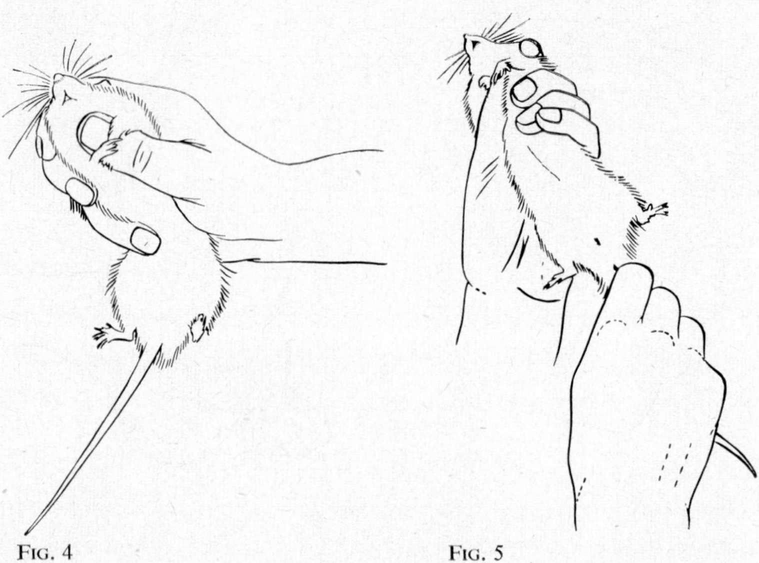

FIG. 4 FIG. 5

To immobilize a mouse for a manipulation, take the loose scruff of the neck between the thumb and forefinger, turn the hand so that the mouse lies, belly uppermost, in the palm and grip the tail between the third and fourth (or fourth and fifth) fingers (Fig. 7).

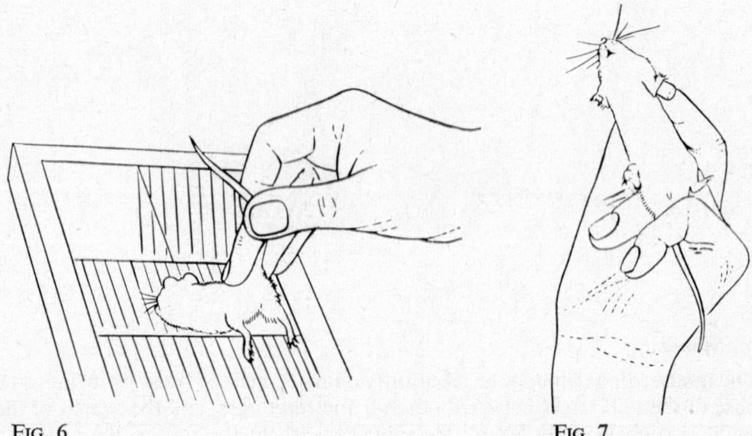

FIG. 6 FIG. 7

(*d*) Guinea pig

These inoffensive animals are much more shy of being handled than are rats, although, when under control, they are completely docile. Handle as for rats, but support the weight of the animal with the free hand. Great care is needed with pregnant animals (Figs. 8 and 9).

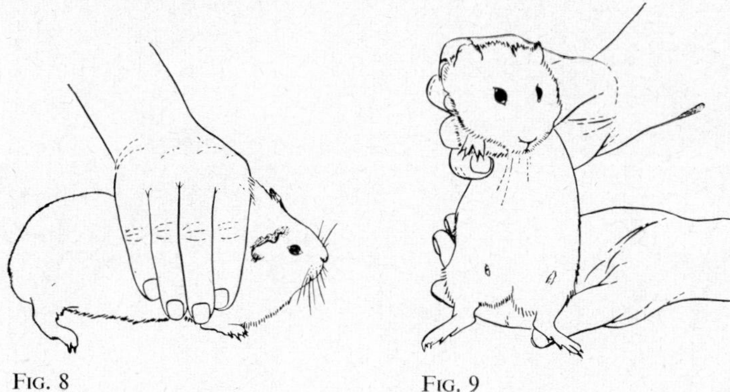

FIG. 8 FIG. 9

(*e*) Hamster

If hamsters are handled frequently they present no problem. Hamsters may be lifted by cupping both hands under the animal. Do not grip them too tightly (Fig. 10). If the animal must be held firmly for manipulation or inoculations grasp the loose skin at the back of the neck with the forefinger and

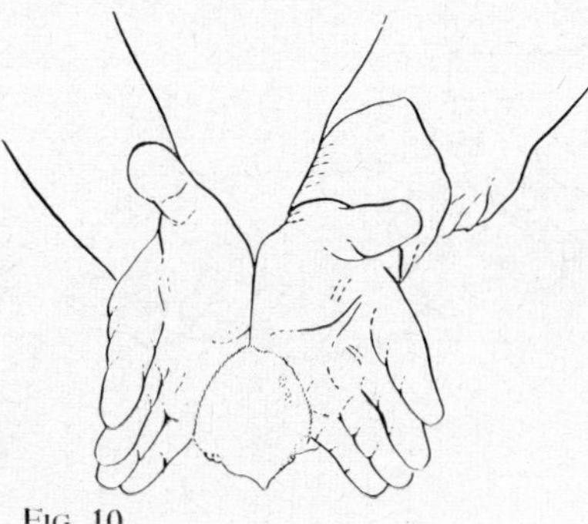

FIG. 10

thumb of the right hand and lift the animal, turn the wrist over, and the animal's back will then be resting in the palm of the hand; the fifth finger of the hand is then placed between the animal's right hind-leg and its body, or this leg is held firmly between the fourth and fifth finger. The position allows complete freedom of one hand (Fig. 11).

(*f*) Ferret

When handling ferrets do not hesitate. Pick the animal up firmly but gently by placing the thumb underneath the mouth and the fingers round the neck (Fig. 12). If the animal does attempt to bite the thumb can be used to close its

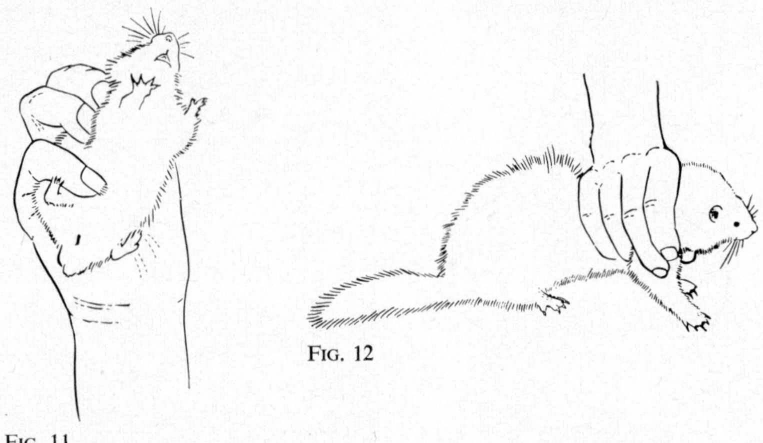

FIG. 12

FIG. 11

mouth. If necessary, support the weight of the animal with the free hand. On no account use stiff leather gloves or tongs to handle ferrets. These animals should be handled frequently if they are to remain docile (Fig. 13).

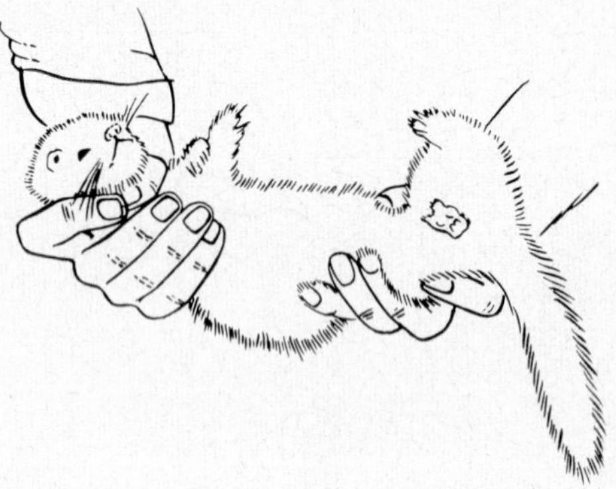

FIG. 13

(*g*) Cotton rat

These animals are not vicious, but they do resent being handled. They are very agile, and can jump 2 feet into the air from a standing start. Never catch a cotton rat by the tail; put its cage into a sanitary bin, or other deep receptacle, and let the animal come out; place the palm of the hand over the rat's back and grasp the skin at the back of the head with the thumb and forefinger. Much practice is necessary to perfect this technique (Figs. 14 and 15).

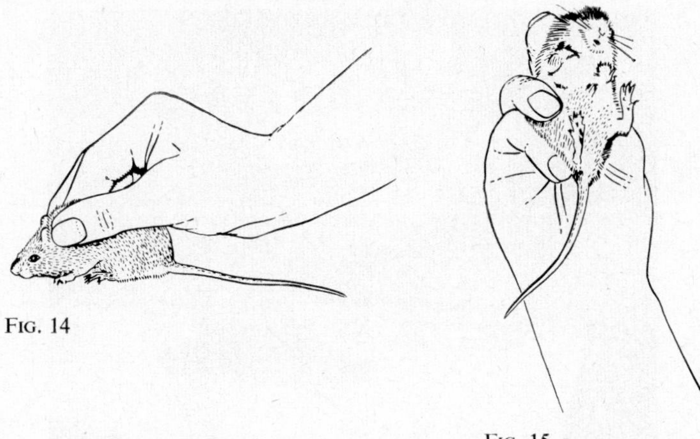

FIG. 14

FIG. 15

(*h*) Fowl

Always keep the wings close to the bird's body to prevent fluttering. Face the bird and slide both hands, with fingers outspread, down and under the breast, placing the first and second fingers of one hand between the bird's legs. Grip both thighs and wing tips with the thumbs and third and fourth fingers, and then lift bird. Be calm and deliberate, and do not make quick movements (Figs. 16 and 17).

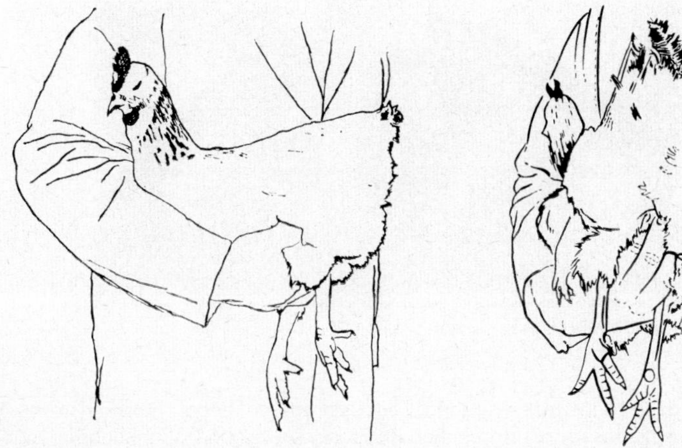

FIG. 16

FIG. 17

(*i*) Frogs and toads

With the fingers pointing towards the rear of animals, place a forefinger between the hind-legs and close the hand round the animal's body (Fig. 18 and 18 [*a*]).

FIG. 18

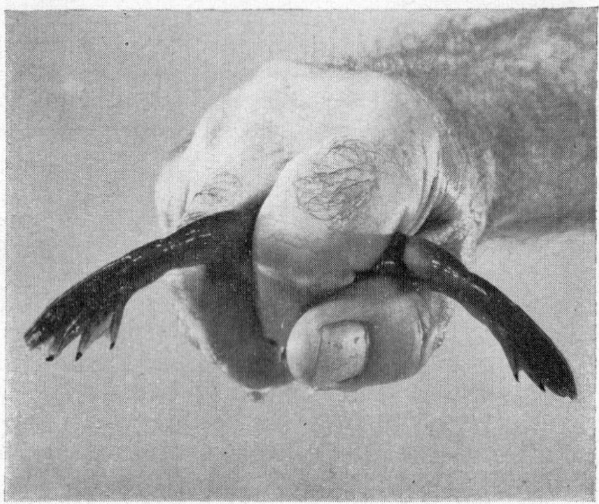

FIG. 18 [*a*]

(*j*) Dog

The approach to a strange dog must be quiet and confident. Nothing arouses a dog's suspicions more rapidly than diffidence on the part of the handler, and from suspicion it is but a short step to active resentment. Make the dog aware of your presence by sight as well as sound before attempting to handle it. Let the dog see your hand; place the back of the hand in front of and below the muzzle and gradually work along the side of the face until the animal has general confidence.

Do not make a sudden grasp for the collar or loose skin of the neck. Whichever is grasped, the movement should be slow and deliberate, and the holder's forearm should be kept in line with the dog's spine to avoid bites (Fig. 19).

FIG. 19

If it is impossible to approach the dog use a dog stock, but with care.

To prevent the dog biting during manipulations it may be necessary to tie a tape round the muzzle. This is done by winding a 2-in. bandage round the jaw once or twice and tying it at the back of the neck. This method is useful for nervous dogs.

When applying a tape muzzle it may also be necessary to restrain the fore-egs, as the dog may try to pull the tape off with its feet.

(*k*) Monkey

Because of the possibility that newly imported monkeys may be infected with some disease which is transmissible to man, great care must be observed if the animals have to be handled, especially during the first few weeks. Face masks and protective gloves must be worn. It may sometimes be necessary to anaesthetize newly imported monkeys before handling them, in which case sodium pentobarbitone or nembutal may be given by the intraperitoneal route.

Method I. Use a box, preferably metal, large enough for the monkey to get into, but not big enough for it to turn round in. The box should be fitted with a sliding door at each end, one door being of transparent Perspex. Induce the animal into the box. To remove the monkey, pull up the sliding door at the tail end of the monkey, grasp the animal around the loins, and pull it gently out of the box until its shoulders emerge; then grasp the two arms together behind the monkey, which is then under perfect control. This method is recommended for the smaller monkeys weighing 3–9 lb. (Fig. 20).

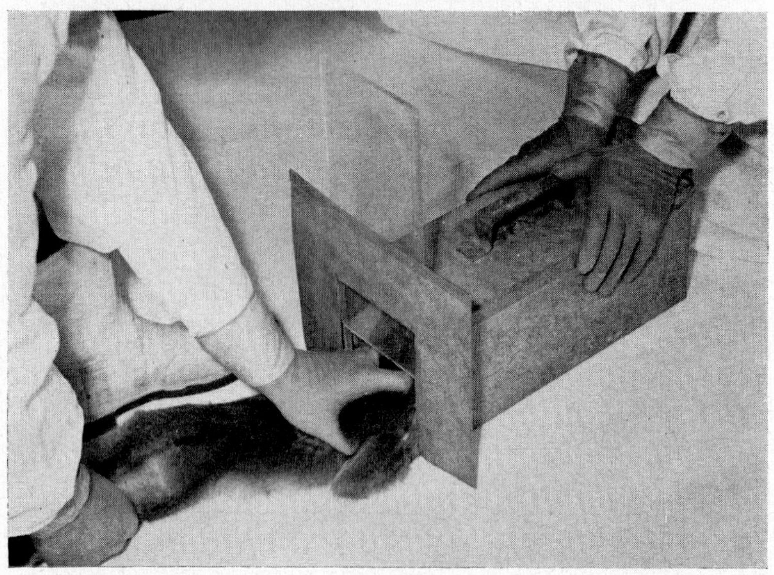

FIG. 20

Method 2. This method requires a larger box of light wood, 16 in. × 13 in. × 20 in. high fitted with a sliding door at the top and bottom, and a net fitted to a stout iron frame 13 in. × 10 in. with a handle attached. The frame and net fit snugly into the box. Hold the box, with one end open, up to the cage and slide open the cage door. When the monkey has entered the box replace the sliding end and put the box, with the monkey in it, on the floor. The monkey may have to be encouraged to enter the box by pulling forward the movable back-panel of the cage. The net, on its frame, is then placed over the top sliding door and that door removed, so that the net may be pushed down into the box and over the top of the monkey. The rim of the net is then resting on the bottom sliding door, which is removed. The remaining four sides of the box are lifted away over the net handle, leaving the monkey trapped in the net. To take the monkey out of the net, face the monkey's head away from the handler, put a hand under the net, and run it up the back of the monkey until the arms can be grasped firmly high up and behind the animal (Fig. 23). The net may then be drawn from over the monkey.

FIG. 21

Monkeys of 50–60 lb in weight can be handled easily by this method (Figs. 21 and 22). The handler has perfect control, without any emotional upset between the animal and the handler.

It is difficult to recapture an escaped monkey by using a net. It is better to keep the animal moving until it tires, it will generally end by catching hold of a

FIG. 22

FIG. 23

cage door or window guard. It is then an easy matter to grasp the animal by the neck and tail and replace it in its cage.

Make sure the monkey's head is facing away from its handler at all times.

(*l*) Cat

The cat can be the most difficult of the small laboratory animals to handle. The typical wild-cat temper can be provoked in many animals as a result of ordinary handling and restraint. The cat has four sets of claws and a set of teeth with which it is always ready to defend itself. For these reasons it is desirable at all times to have complete control over the cat, even when merely lifting it or placing it in a basket. Lift a cat by taking a firm grip of the scruff of its neck high up, being careful to keep the forearm in line with the cat's spine.

To carry a cat for a short distance, tuck its head and the right hand holding the scruff under the left arm, and use the left forearm to hold the cat's body close to the handler's body.

To give the cat treatment it is often necessary to control all danger points, i.e. four sets of claws and the teeth. In this case place the cat on a table with its left flank towards the handler. The left hand is placed on the front of the fore-legs with the fingers gripping them just above the middle joint. The right hand does the same with the hind-legs. This immobilizes the four sets of claws. The left arm, which now is resting against the neck of the animal, is pressed firmly into the handler's leg or body. This prevents the animal moving its head and using its teeth. The animal may be held down on a table in a similar manner. When releasing the animal let go with both hands at the same time (Fig. 24).

(See *UFAW Handbook* for more detailed chapter on this subject.)

FIG. 24

(*m*) **Coypu**

The coypu is a large rodent of South American origin introduced into this country in the early part of this century by fur farmers. The fur was popular at one time and is known as Nutria. The animals are very much at home in water, being splendid swimmers, and have webbed hind-feet. Adults can weigh as much as 12 kg and more. The length of an adult is between 2 and 3 ft. The body is compact, with a short neck, a broad head, humped hind-quarters, and the teeth are orange-yellow in colour. The tail is like that of a rat, rounded and shorter than the length of the body. The fur is agouti coloured and is very thick and waterproof.

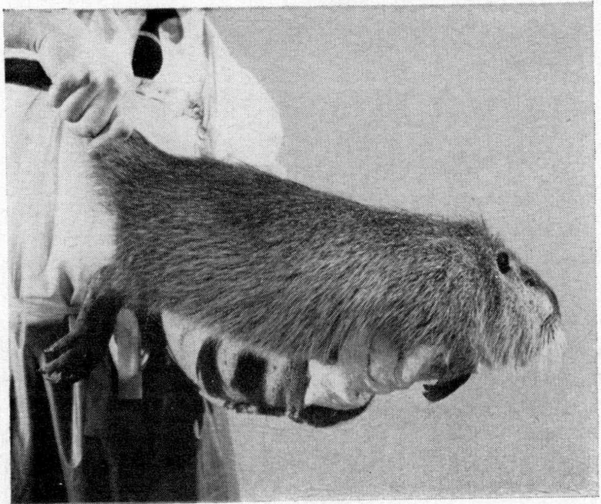

FIG. 24 [*a*]

Coypu bred in captivity are easily tamed and handled. Grasp the tail just behind the body and put the other hand between the front legs and under the chin, so that the animal's front feet rest on the forearm (Fig. 24 [*a*] and [*b*]). An elbow length cuff of padded sacking and a stout but pliable glove protect the arm and hand from scratches while the coypu is being held.

Coypu received direct from the wild are nervous and may bite. Grasp the tail, as before, but a broom handle may have to be used to support the front legs. The broom provides a surface for biting. It is important to keep the animal's head pointing away from the handler at all times.

FIG. 24 [*b*]

SEXING LABORATORY ANIMALS

Sexing animals correctly is a skill and has to be learned through constant practice. All laboratory animals can be sexed at birth; the rabbit is the most difficult, and the task requires good eyesight as well as much practice.

Rabbits

The animal should be placed so that the legs are facing away from the handler. Control the head by holding the ears, and with the forefingers of the free hand exert a little pressure just above the genital region. In the male the rounded tip of the male organ (the penis) will be seen (Fig. 25), and in the female a slanted slit (the vaginal orifice) will be seen (Fig. 26). Rabbits are difficult to sex between the ages of 1 and 42 days, and the ability to sex day-old rabbits is a hard-won skill. Fig. 27 shows day-old rabbits; (male on left; female on right) in the male the penis appears as a rounded rube with the top folded in, and in the female the vaginal orifice appears as a slit-like line running between two small orifices. The female may also be recognized by the negative sign of the absence of a penis. The anogenital distance is somewhat greater in the male. The baby rabbit should be placed in the palm of the hand face upwards and with the back legs extended. The tail should be bent backwards and held by the fingers.

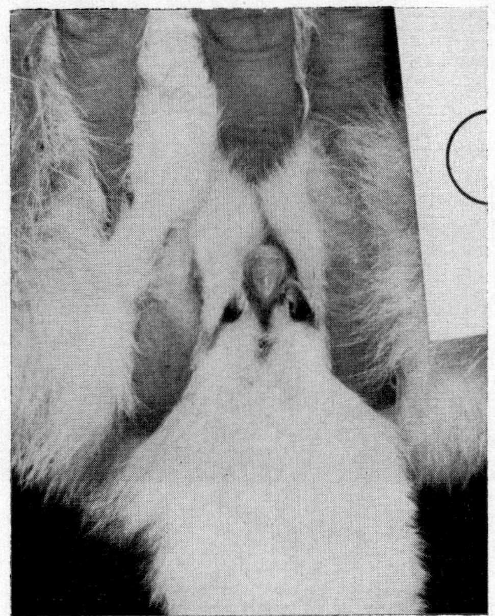

Fig. 25

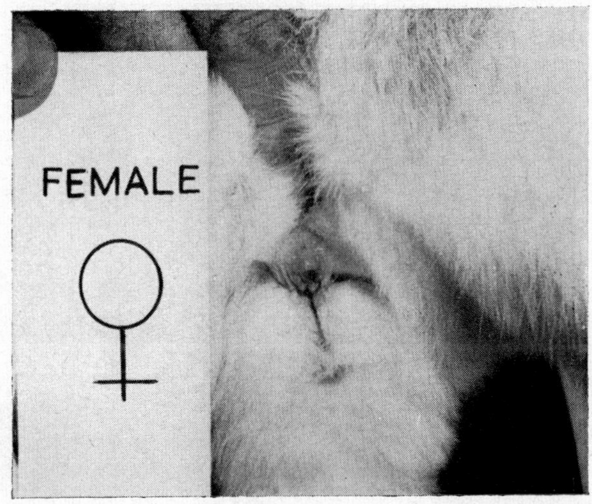

Fig. 26

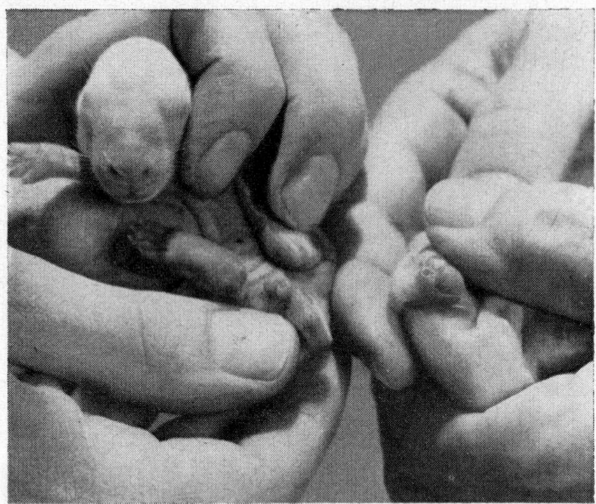

FIG. 27

Cats

The sex of adult cats may be recognized easily by the difference in the ano-genital distances in the two sexes. In the kitten this difference is less marked but is still quite obvious (Fig. 28). To sex a cat or kitten, place it on a table or hold it in the arms and hold the tail bent over the animal's back—the sex may then be seen at a glance.

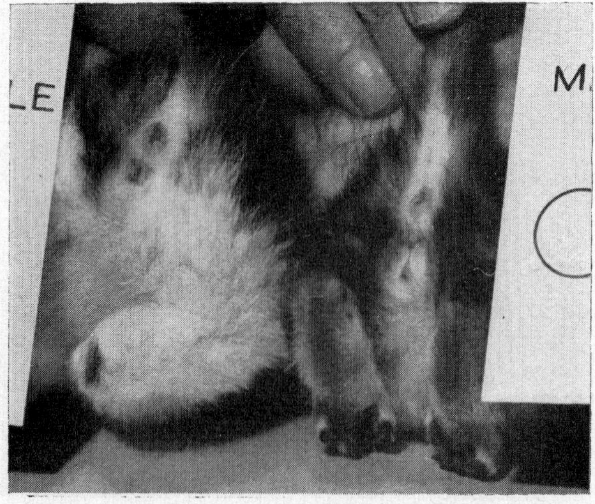

FIG. 28

Monkey

The determination of sex in monkeys of all ages is as easy as it is in human beings, and can be done by observing the animal without removing it from the cage.

Hamsters

These animals resent being held tightly. Put the hamster facing you on a table and put a hand, with the fingers closed together, on either side of the animal,

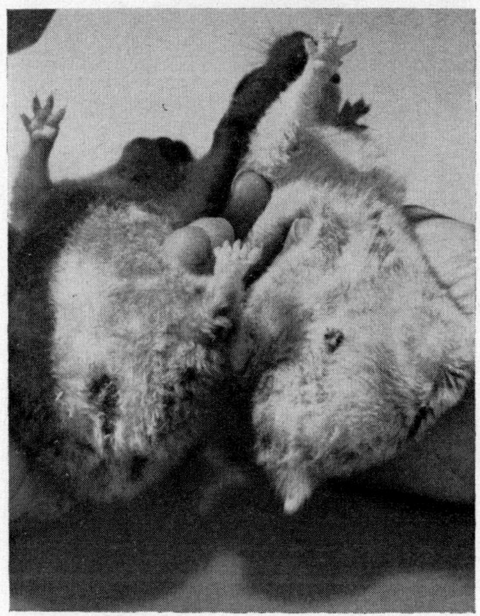

FIG. 29

which is then cupped into the hands by rolling the wrists; the animal may be turned over and sexed by inspection of the anogenital distance (Fig. 29, male on the right).

Ferrets

Hold the animal so that its feet are turned away from you. There is a great difference in the anogenital distance in the male and female of this species, and ferrets are even easier to distinguish by this method than are rats and mice (Fig. 30, male).

Guinea pig

The anus and genitalia are close together in both the male and female guinea pig, and young animals can be difficult to sex.

Support the hind-quarters in the palm of one hand and apply light pressure with a finger or thumb in front of the genital region. This causes the male penis to be extruded slightly or the female vaginal orifice to be revealed. It should be remembered that the vaginal orifice is closed by a thin membrane (the hymen), which breaks down spontaneously only when the mature animal approaches oestrus or parturition and regrows afterwards.

To sex young animals, use both hands and lay the animal in the cupped palms. Place both thumbs and forefingers beside and above the genital region and press, downwards and outwards, a little harder than in adult animals (Fig. 31, female; Fig. 32, male).

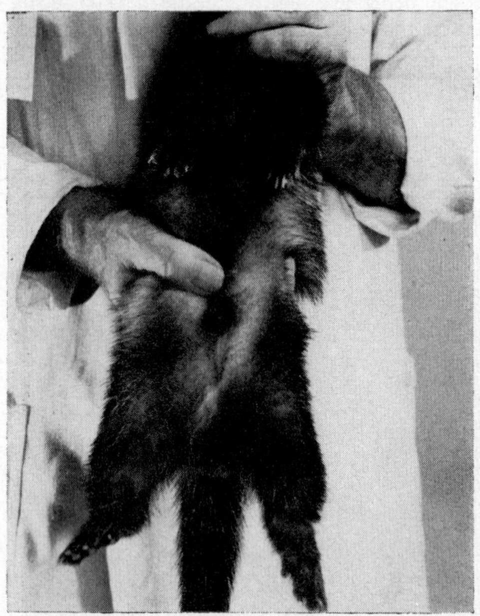

FIG. 30

Rat

Rats are among the easiest of laboratory animals to sex owing to the difference in the anogenital distance in the two sexes; this distance is considerably greater in the male, and the difference is obvious even in day-old animals if they are held with the tails bent slightly backwards (Fig. 33, female on the right).

Common frog (Rana temporaria)

The male can be identified by the colour of its belly which is white or pale yellow; the belly of the female is yellow to orange or even green. During the breeding season only, the male may be distinguished by the horny growths (nuptual pads) on the fingers.

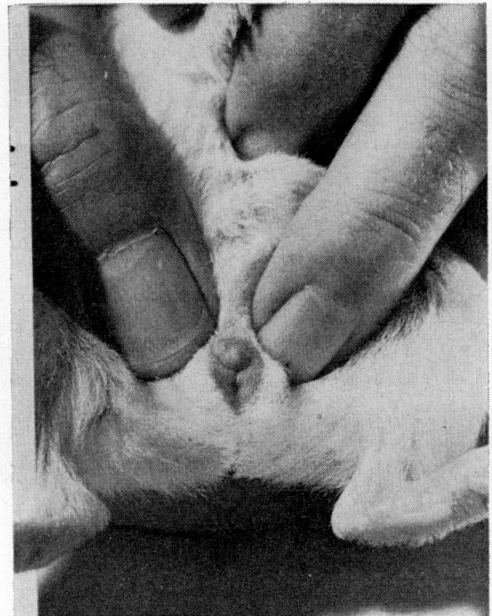

FIG. 31

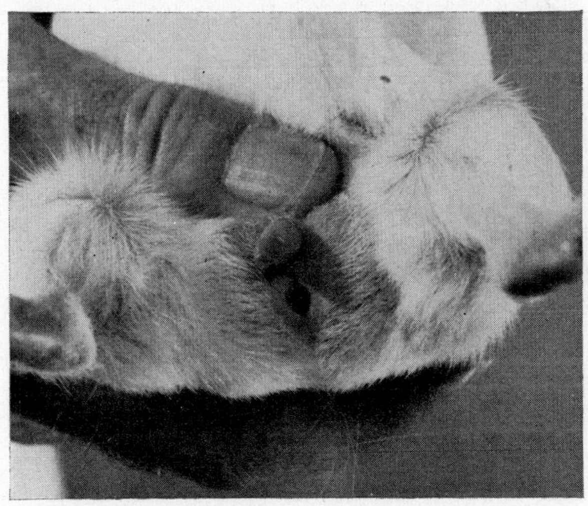

FIG. 32

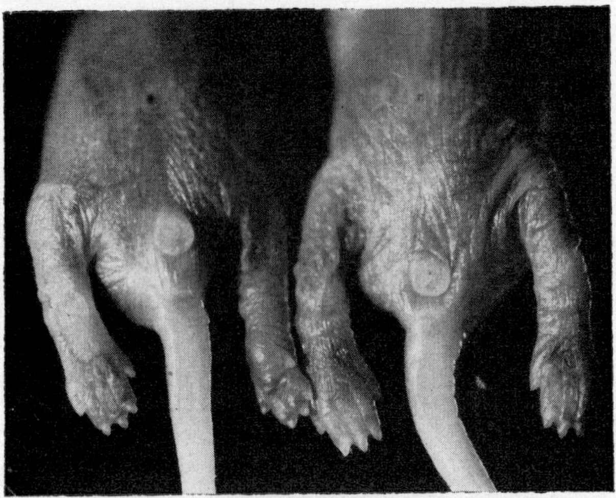

Fig. 33

Clawed toad (Xenopus laevis Daudin)

The females may be distinguished from the males by the presence of cloacal valves in the former. (Fig. 34, female; Fig. 35, male) The adult female toads will weigh twice as much as the males. A toad of body weight 200 gm or more will always be female. Adult males will seldom exceed 100 gm in weight.

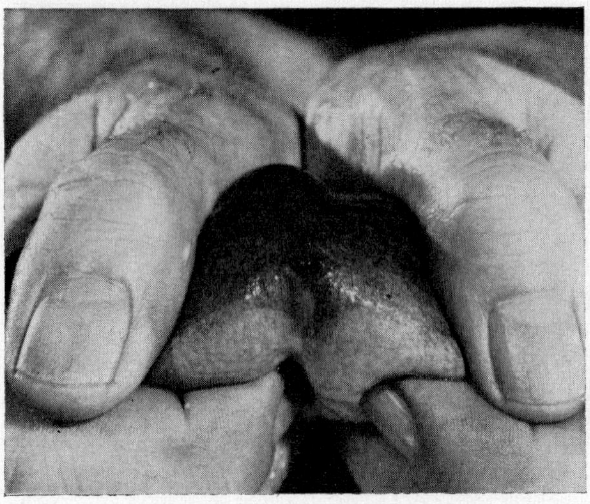

Fig. 34

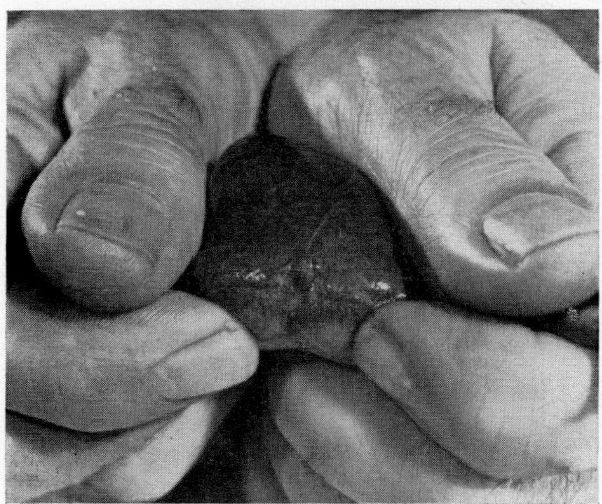

Fig. 35

Mouse

Much practice is needed to be able to sex young mice correctly. Fig. 36 is of day-old mice; the male is on the right and the female on the left. The anogenital distance in the male is twice that seen in the female. Bend the tail back slightly when sexing young mice.

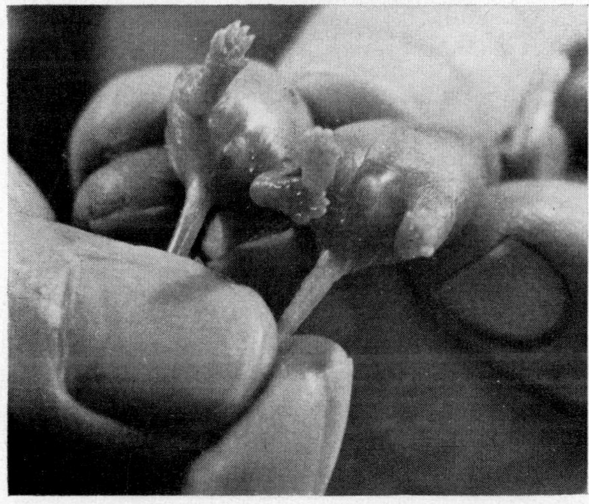

Fig. 36

FIG. 37

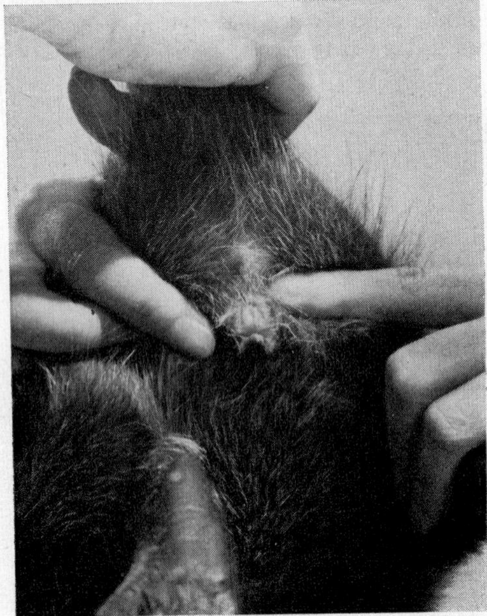

FIG. 38

Coypu

Put a wire grid on the floor and hold it in position with one foot. Stand the coypu on the grid, hold it by the tail, and when the animal clings to the grid with its front claws lift the tail so that the anogenital distance can be seen. The difference is most marked, the anogenital distance in the male being more than double that in the female (Fig. 37, male; Fig. 38, female).

PALPATION OF ANIMALS FOR PREGNANCY

Pregnancy may be confirmed in most laboratory animals at a third of the way through the gestation period by palpation of the foetuses through the abdominal wall. The examination must be made carefully and gently so that the animal sustains no injury, and every effort must be made to keep the animal calm and relaxed.

It is necessary to be able to distinguish between foetus (embryo) and faeces when palpating animals. In the common laboratory animals the faecal pellets are sausage-shaped, while the foetuses are bean- or kidney-shaped and lie diagonally, in the line of the uterus and not in the line of the rectum. The foetus is more mobile (can be moved more freely) than a faecal pellet, because the uterus is attached to the body wall by a broader ligament than is the rectum.

Rabbit

The rabbit should be placed on a flat, non-slip surface. Control the head with the left hand. The right hand is slipped under the animal and the first finger and thumb are used to locate the embryos in the uterus. Care must be taken to ensure that the embryo is not pinched.

Hamster

Hold the hamster by the loose skin at the back of the neck. The embryos may be felt with the thumb and first finger of the free hand. The embryos are the size of small peas at a third of the way through the gestation period.

Ferret

Hold as for sexing and use the thumb and first finger of the free hand to palpate the embryos.

Guinea pig

Hold as for sexing older animals and use the thumb and first finger of the free hand to palpate the embryos. Return the guinea pig to its cage and put all four of its feet on the floor before removing support. About ten days before parturition the pubic symphysis begins to open (the pubic symphysis is the anterior junction of the two bones of the pelvic girdle). At first the edges of the bones may be felt as two distinct ridges. The opening widens gradually until, immediately before parturition, the space between the edges of the bones will admit the width of the human thumb. With practice, accurate prediction of parturition dates may be made from this examination.

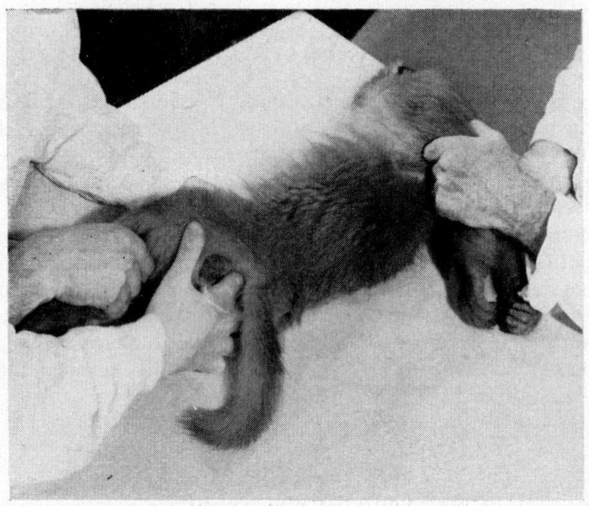

FIG. 39

Rat

Hold the animal with one hand and use the thumb and first finger of the free hand to palpate the embryos. They will feel like small beads on a string at a third of the way through pregnancy.

Monkey

Pregnancy may be confirmed by abdominal palpation from the 6th to 8th week of pregnancy. One person should control the arms and one the legs. The animal is placed on its back, and a third person carefully palpates the abdomen and locates the foetus.

Diagnosis may be made at the third week of pregnancy by rectal palpation. The monkey is laid on her right side, with one person controlling the head and arms and another the legs. A third person palpates the uterus through the rectal wall using the forefinger of the right hand while the left hand is placed on the abdomen to support the viscera and help locate and steady the uterus

(Fig. 39). The signs of pregnancy at rectal palpation of the Rhesus monkey are as follows:

PHYSIOLOGICAL STATE OF THE MONKEY	UTERUS	CERVIX
Normal, non-pregnant	$\frac{1}{2}$–$\frac{3}{4}$ in. long, hard texture	Firm texture, not distinguishable from the uterus
3rd–4th week of gestation	1 in. long, soft texture, ball-like	Soft texture, easily distinguished from the uterus.
6th week of gestation	$1\frac{1}{2}$ in. long, soft texture, ball-like or may be slightly ovoid	Soft, as above
8th week of gestation	Larger ($1\frac{3}{4}$ in.+) and firmer than at 6 weeks	Remains soft

The animal should be rewarded when it is returned to the cage.

Coypu

One person controls the coypu, and a second person places both thumbs on the animal's back above the pelvis and uses his fingers to palpate the embryos (Fig. 40).

Fig. 40

CHAPTER SIX

Methods of Identification

When it is necessary to identify individual animals some method of marking them has to be adopted. All methods should be simple, permanent, quickly applied, easily deciphered, and harmless to the animal. Some of the methods in general use are described below.

STAINING

The following stains, which are soluble in alcohol at a temperature of 78·8°F (26°C), are generally used:

COLOUR	STAIN
Yellow	Saturated picric acid or chrysoidin
Red	Fuchsin (acid, basic or carbol)
Violet	Methyl violet (gentian violet)
Green	Brilliant green, ethyl green, or malachite green
Blue	Trypan blue

Histological dyes should be used and made up as 3–5 per cent solutions in 70 per cent alcohol, except the yellow dyes, which are made up as saturated solutions. Stock solutions should be made up in bulk, as the efficiency of the stains increases with keeping.

Before applying the stains to the animal, consideration should be given to the side effects of staining on the coat and skin of the animal, and compatibility with the experiment for which the animal is being used.

A small area deeply stained is preferable to a large area lightly stained; reapplication at regular intervals is necessary. There is on the market a proprietary stain ('Darafur') specially prepared for animal marking. This preparation gives good durable results.

EAR PUNCHES

Surgical instrument manufacturers supply ear-marking pliers of various designs for use on large animals. For some small animals a chicken toe-punch may be used; these may be obtained from poultry-equipment suppliers. The ear-marking pliers and the chicken toe-punch are designed to cut holes right through the ears or toes, or in a series of notches round the edges of the ears.

Before marking the animals by this method, a well-defined code should be drawn up, and it is advisable to display a code chart wherever animals are being handled. By this method it is possible to mark individual animals in a colony. The method cannot be applied to animals, such as the rabbit, having large marginal ear veins.

EAR-MARKING STUDS

These studs are designed on the 'batchelor button' principle, and are made of brass or aluminium. The studs are lettered, numbered, or codified to requirement, and are supplied with a combined punch and stud forceps which pierces a hole in the ear and closes the stud in position.

TATTOOING

Surgical-instrument manufacturers supply tattooing outfits in several sizes and with a variety of interchangeable numerals and letters—see table for size of numeral recommended for each species. The colouring medium should be black for non-pigmented animals and green or red for heavily pigmented animals. The skin should be cleaned with spirit before it is tattooed.

A surgical, triangular cutting needle and marking ink may also be used for tattooing and can sometimes be less laborious than using forceps, which involve the continual changing of numerals. An electro-vibro tattoo may also be used.

The inner surface of the ear is the usual site for tattooing.

RING AND LEG-BANDS

For the identification of birds and poultry two types of leg bands are used—plastic split-rings, of different size and colour, and light metal, adjustable leg bands numbered or codified to requirement.

Codified or numbered metal leg-rings of a different size for each strain of rabbit are available. (See table for size of band and age at which rabbits may be ringed.)

WING-BANDS AND WING-CLIPS

Wing-bands are adjustable, and they are numbered or coded to requirement. These bands are clipped through the wing above the radial feathers in a manner unlikely to impede the bird's movements.

The wing-clip in general use resembles a small gilt safety pin. These bands and clips may be obtained from any poultry equipment supplier.

COLLARS, CHAINS, AND NECK-BANDS

Collars or neck-chains with alloy discs (numbered or coded) may be used on dogs and cats. Puppy collars may be used for cats. Monkeys may strangle themselves with a neck-band, therefore any chain or collar used on a monkey should be fixed around the waist-line.

TAMPERPROOF WING TAG (*'Ketchum tags'*)

This is a wing tag made of a light alloy, numbered or coded to requirement. It consists of a thin metal band doubled over and pointed at one end, and specially designed pliers are required for clamping the tags in position. This tag may be used for marking all species of birds; it is also a convenient method of ear-marking rabbits and guinea pigs, providing the tags are applied close to the head. These tags are also available in sizes suitable for farm animals.

ANIMALS

SPECIES	MARKING EQUIPMENT	METHODS OF APPLICATION
Day-old chicks, ducklings, and fowls	Wing-band	Clipped round the wing, above the radial, close to the body; must not impede movement
,,	Wing-clip	Clipped through the skin and neatly closed, so that no discomfort is caused to the bird
,,	Leg-band	Fit closely but comfortably round the leg
,,	Leg-ring	Fit closely but comfortably round the leg
Pigeons	Leg-band	Fit closely but comfortably round the leg
,,	Leg-ring	Fit closely but comfortably round the leg

ALL LEG-BANDS AND RINGS MUST BE INSPECTED REGULARLY FOR TIGHTNESS

Cats	Chain or collar and disc	Collars or chains should fit comfortably
Dogs	Tattoo	Tattoo the ears with forceps and numerals $\frac{3}{4}$ in. × $\frac{1}{4}$ in.
,,	Chain or collar and disc	To fit comfortably round the neck
Ferrets	Stain	On the back or head
,,	Tattoo	Tattoo the ears with forceps and numerals $\frac{3}{16}$ in. × $\frac{1}{4}$ in., or with triangular needle and marking ink
Frogs	Fine nylon braided suture thread. Small coloured glass beads. Sewing needle to take thread	The coloured beads are sewn to a fold of the skin covering the dorsal sac. A code of colour combinations can be used
Goats	Tattoo	Tattoo the ears with forceps and numerals $\frac{7}{16}$ in. × $\frac{1}{4}$ in.
,,	Ear punch	Use a well-defined code and cut a series of notches or holes in the ears
,,	Collar and disc	Fit comfortably round the neck
Guinea pigs	Stain	On the back or head
,,	Natural coat colour	Have, on the cage label, an outline drawing of the animal and mark on it the distribution of natural colouring
,,	'Ketchum Tag'	The sharp point is pushed through the ear close to the head, and is locked in position with special pliers
Hamsters	Tattoo	Tattoo the ears with forceps and numerals $\frac{3}{16}$ in. × $\frac{1}{4}$ in., or with a triangular needle and marking ink
Mice	Stain	On the back
,,	Tattoo	Use special mouse-ear tattooing forceps
,,	Chicken toe-punch	Use a well-defined code and cut a series of notches or holes in the ears (Figs. 1, 2, and 3)

ANIMALS (continued)

SPECIES	MARKING EQUIPMENT	METHOD OF APPLICATION
Monkeys	Tattoo	Tattoo the chest, upper lip, or forehead by using a triangular needle and marking ink, or an electro-vibro tattoo. *Monkeys should be anaesthetized for tattooing the lip*
,,	Chain and disc	Thin chains with discs securely fixed round the waist
Pigs	Tattoo	Tattoo the ears with forceps and numerals $\frac{7}{16}$ in. × $\frac{1}{4}$ in.
,,	Ear punch	Cut a series of notches or holes in the ears
,,	Ear studs	Pierce a hole in the ear into which the stud is clamped
Rats	Stain	On the back
,,	Tattoo	Tattoo the ears with forceps and numerals $\frac{3}{16}$ in. × $\frac{1}{4}$ in.
,,	Chicken toe-punch	Use a well-defined code and cut a series of notches or holes in the ears (Figs. 1, 2, and 3)
Rabbits	Tattoo	Tattoo the ears with forceps and numerals $\frac{3}{8}$ in. × $\frac{1}{4}$ in.
,,	Leg-ring	Choose the size of ring to suit the breed. Place the ring above the hock on a hind leg. INSPECT LEG RINGS AT FREQUENT INTERVALS
,,	Ear stud	Aluminium studs should be used for rabbits (see method described under pigs)
,,	Stain	On the back
,,	'Ketchum Tag'	As described for guinea pigs

Rabbit leg-rings

Rabbit rings are placed above the hock joint on a hind-leg. Small breeds may be ringed at six to seven weeks of age; larger breeds at nine to twelve weeks of age. Though it is easy to slip a ring on a young rabbit, the hock joint soon grows, so that the ring cannot be removed unless it is cut off. Provided the correct size of ring is used for each breed, the rabbits suffer no discomfort from the rings.

Recommended ring sizes are:

Dutch and Himalayan	Internal	diameter	$\frac{18}{32}$ in.
Chinchilla	,,	,,	$\frac{20}{32}$ in.
Half Lop	,,	,,	$\frac{26}{32}$ in.
New Zealand White	,,	,,	$\frac{28}{32}$ in.

Sheep	Tattoo	Tattoo the ears with forceps and numerals $\frac{7}{16}$ in. × $\frac{1}{4}$ in.
,,	Ear punch	Use a well-defined code and cut a series of notches or holes in the ears
,,	Ear studs	As described for pigs

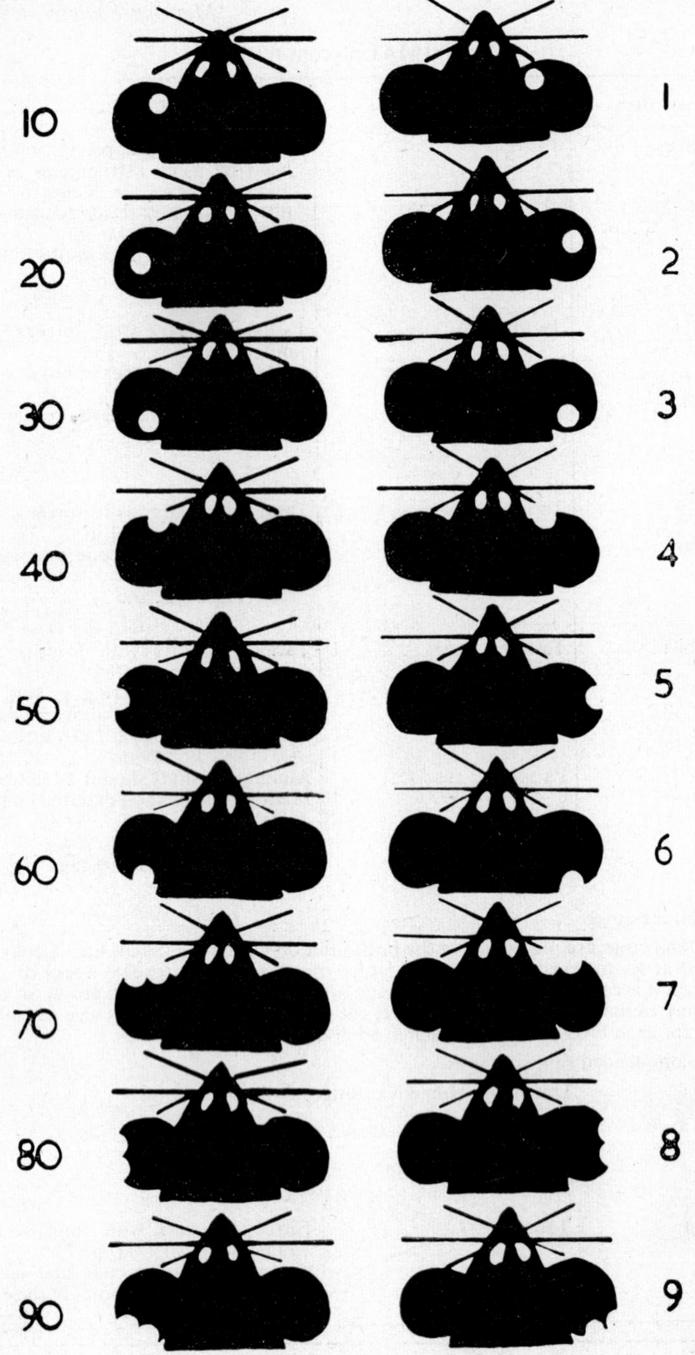

FIG. 1

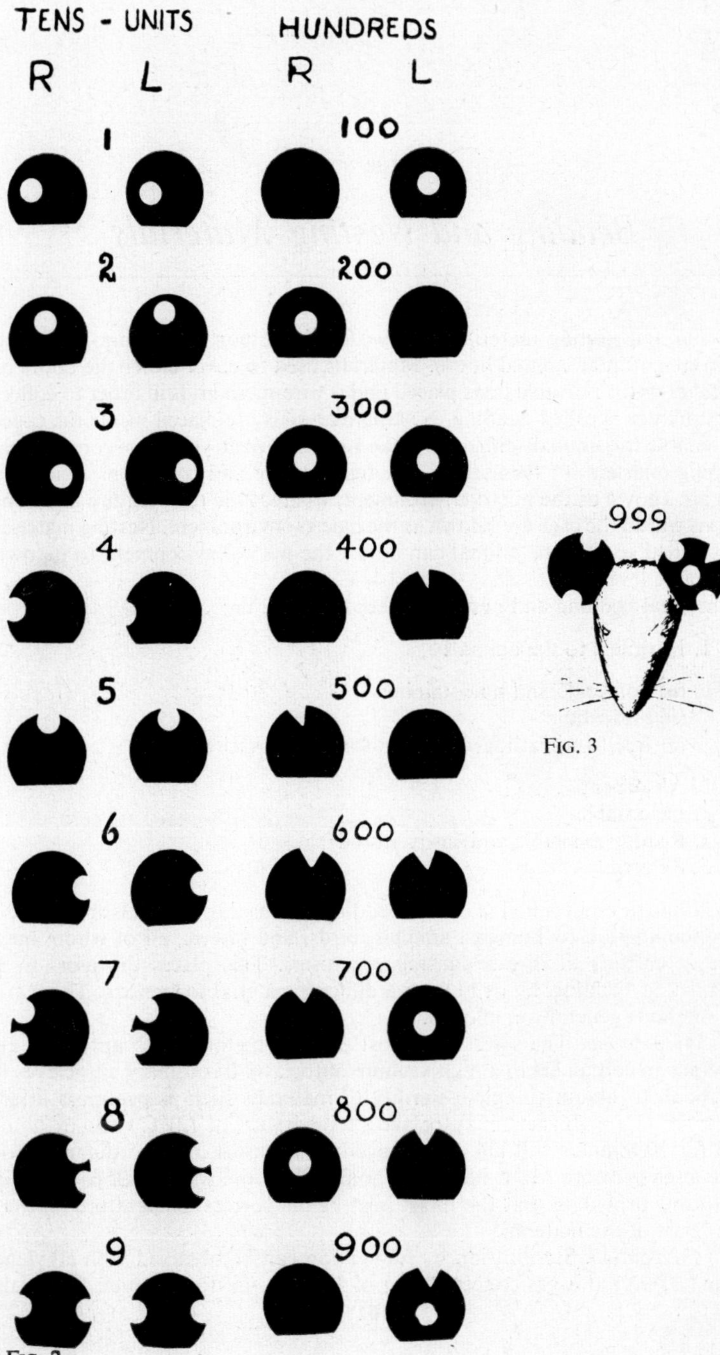
TENS - UNITS
HUNDREDS
R
L
R
L
1
100
2
200
3
300
4
400
5
500
6
600
7
700
8
800
9
900
999

FIG. 2

FIG. 3

CHAPTER SEVEN

Bedding and Nesting Materials

Bedding and nesting materials are two of the important factors governing environment in an animal house. Materials used to cover either the floors of the cages or the portable trays placed under wire-mesh grids in order to collect faecal matter is called bedding. Nesting materials are placed inside the cages and enable the animals either to make nests for their young or comfortable sleeping quarters for themselves. The temperature and conditions inside the nest are known as the micro-environment, whereas the temperature and conditions within the cage are known as the macro-environment. Nesting material is provided so that the animal can adjust the micro-environment to its own acceptable level.

The ideal bedding and nesting materials should be:

1. Harmless to the animals

   (*a*) non-toxic and non-staining
   (*b*) non-edible
   (*c*) free from pathogenic organisms and parasites

2. Absorbent
3. Disposable
4. Readily available and easily stored
5. Relatively cheap

It should be appreciated that most bedding and nesting materials are exposed to contamination by humans, animals, birds, and insects, all of whom may serve as carriers of disease-causing organisms. This places the users in a dilemma, as bedding, by its bulk, is a difficult material to sterilize. There are two methods generally employed.

(1) *Autoclaving*. The safest and most efficient method is to autoclave all materials in containers in a high-vacuum autoclave. In ordinary autoclaves it is difficult to obtain complete sterility of materials such as peatmoss litter. However, in a high vacuum autoclave a temperature of 121°C vacuum 29 in. held for 10 minutes will kill all pathogenic organisms. In an ordinary autoclave a temperature 121°C held for 1 hour at 15 lb/in. will kill all pathogenic organisms providing that the moist heat at the correct temperature reaches every part of the material.

(2) *Fumigation*. Sterilization by fumigation can be obtained with ethylene oxide (ETOX) at a gas concentration of 1,500 ml/m^3 for 6 hours. Materials

sterilized by ETOX must be well aerated at a temperature of not less than 10°C for several days before being used. These materials may have harmful effects on the animals if not properly aerated or if stored in a room where the temperature is under 10°C.

Bedding and nesting materials may also be sterilized by gamma irradiation, but at present the cost is a prohibitive factor.

BEDDING MATERIALS

Sawdust

Sawdust is most commonly used, but it must be from white softwood and obtained from a reliable source where the sacks are filled directly from the sawbench through the sawdust extractor chute, because it is less likely to be contaminated by cats, dogs, or other wild rodents than when it is obtained from a mill dump. Wet, dirty sawdust should never be used, because it is a possible source of infection.

Sawdust from resinous hardwoods, such as teak and mahogany, may contain phenolic substances that could be harmful to the animals. It is equally important to ensure that the sacks are not contaminated, and where possible they should be sterilized before being sent for refilling. Multiple paper sacks are recommended.

Trays below wire grids should be covered with a layer of sawdust; for small rodents $\frac{1}{2}$ in. thick, for larger animals $\frac{3}{4}$ in. thick.

For animals kept on solid floors, cover the floor of the cage up to $1\frac{3}{4}$ in. thick. For cats place up to 2 quarts in one corner or in the dirt tray.

Sawdust has to be changed frequently, otherwise the odours from the decomposition of food particles and the putrefaction of faecal matter becomes objectionable.

Softwood shavings or woodchips

These cannot legitimately be claimed as bedding, because they are generally used in conjunction with sawdust in cages with solid floors, where they act as a nesting material. This material can be successfully sterilized by either high-vacuum autoclave or by fumigation with ethylene oxide. It is practically non-absorbent and therefore not entirely satisfactory as a bedding and nesting material for rats. It is, however, satisfactory as a nesting material if the floor of the cage is given a liberal coating of either sawdust or peatmoss litter. Softwood shavings may be used as a bedding and nesting material for mice, hamsters, cotton rats, mastomys, or other small mammals. The release of ammonia is less rapid than with sawdust, nevertheless an unpleasant odour permeates the animal room unless this material is changed frequently. The shavings or chips should be either baled or bagged in multiple paper sacks at the mill and delivered direct to the animal house. The outer wrappings should be removed before the shavings or chips are taken into the bedding store.

Peatmoss litter

The initial cost is high, but since peatmoss requires changing less frequently than sawdust, it is labour saving, thereby reducing the overall cost of animal husbandry. Peatmoss has a high acid content, which delays the decomposition of faecal matter and the release of ammonia. It is obtainable in two forms—compressed, in bales, or granulated, in sacks—the latter is easier to handle, but the former is impervious to rodent infestation. Peatmoss litter can be rendered sterile in a high-vacuum autoclave at 121°C for 10 minutes. It is used in the same way as sawdust.

Sterolit

This is a waxy, greyish-green clay that after considerable processing yields the mineral attapulgite—a hydrated magnesium silicate. It is an American product and is claimed by the manufacturers to neutralize both acidity and alkalinity and to be completely inorganic and non-toxic. It is usually sterile on receipt and remains sterile for some considerable time in open bins in an animal room. As it is possible for the outer wrapping to become contaminated *en route*, sterilization is recommended. This may be done by laying the material out on trays which are passed through a high-vacuum autoclave at 121°C for 1 minute or through a normal autoclave at 120°C for 1 hour.

Trays below wire grids should be covered with a thin layer of Sterolit; for small rodents ¼ in. thick and for larger animals ½ in. thick. For animals kept on solid floors, cover the floor of the cage up to 1 in. thick. For cats place up to 2 quarts of Sterolit in a corner of the cage or in the dirt tray.

Sterolit reduces labour, because trays require cleaning less frequently. The trays should be removed at regular intervals and shaken gently in order to mix the dry material with the soiled. Trays are cleaned out when the Sterolit has lost its absorbency or when the odour level has increased.

Sterolit will not burn, therefore *it cannot be incinerated; nor can it be spread on farm land.* It must be buried or dumped. The cost is high.

NESTING MATERIALS

Cellulose wadding

This is a highly absorbent material. It is delivered in good-quality wrapping that protects it *en route* from heavy contamination. Nevertheless, it is not sterile, and should be sterilized by either high-vacuum autoclave or by fumigation with ethylene oxide. It is a relatively cheap nesting material.

Cut chaff and cavings (oat husks)

These materials are sometimes used on top of sawdust or peat moss litter for rabbits and guinea pigs on solid floors. There is the ever-present danger of introducing infection into the colony unless this form of bedding is sterilized, and sterilization presents many difficulties. It is therefore advisable not to use these materials unless absolutely necessary. The cost is low.

Meadow hay

Generally used as a fodder, it does, however, make a good bed for breeding rabbits, especially during cold weather, as soft hay makes a close and warm nest for the young. Hay can be sterilized by autoclaving or irradiation.

Straw (chopped)

Straw is seldom used as a bedding material for small animals, but for guinea pigs in floor pens, chopped straw with an underlay of sawdust makes a suitable bedding. Wheat or oat straw should be used.

Shredded paper

Clean shredded paper is highly recommended for nesting material. Shredded paper is obtainable from printers at a very reasonable rate. A special clean shredded paper as used for food packing is also obtainable, although slightly more expensive. The former should be sterilized if possible, the latter can be used without sterilizing.

BEDDING AND NESTING MATERIALS FOR EXPERIMENTAL ANIMALS

Paper, cotton-wool, and certain other materials can be used for experimental animals.

Post-operative animals may be given a liberal supply of sterile cotton-wool or sterile shredded paper. The sterile materials minimize the danger of introducing infection.

Animals on nutritional experiments may be deprived of any form of bedding, but they should be placed in a sterile cage with a wire grid with a mesh of the correct size for the species concerned.

Storing

All bedding and nesting materials must be stored in a vermin-proof building, either in racks or in large bins. Surplus bedding once taken out of the store should never be returned; undetected contamination may have occurred. Brooms and other equipment necessary for cleaning should be retained for use in the bedding store only.

Sterilization

As an added precaution, and where facilities are available, all material should be sterilized directly it is received, or at least before it is placed in the store.

Disposal

Sterilization of all infected bedding along with the cage must be regarded as a necessary routine. After sterilization the soiled bedding should be removed and conveyed to the incinerator by vacuum or a closed bin. Uninfected bedding may not generally be regarded as a source of danger, it should nevertheless be handled carefully and burned in the manner described above. All bins and utensils used in the process of cage cleaning should be sterilized after use.

SUMMARY

1. Procure bedding from a reliable source.
2. Sterilize all bedding before use.

3. Use sawdust or shavings from softwoods only.
4. Peat moss litter or Sterolit may be used in place of sawdust.
5. Cellulose wadding, fine woodwool, and shredded paper are recommended as a nesting material for rodents.
6. Sterile cotton-wool or shredded paper is recommended as a nesting material for rodents during experiments.
7. Store bedding in rodent-proof building.
8. Sterilize all infected bedding before handling and burning.

CHAPTER EIGHT

Routine Care of Laboratory Animals

A clean and tidy animal house is the hall-mark of a good animal technician. Much time and energy is spent on everyday chores in an animal house, but the effort is well worth while when it results in a healthy stock of animals maintained in surroundings in which it is a pleasure to work.

No hard-and-fast rules can be given for the conduct of any animal house; each must work out its own routine according to the demands made upon it. But whatever scheme is devised there are some aspects of the routine care of stock animals which are inescapable, and it is these points which are discussed here.

Environmental conditions

Room temperature and relative humidity are checked and recorded first thing each morning. The setting of time switches for lights should be checked. Female rats kept in windowless rooms and exposed to more than 15 hours artificial light in every 24 hours come into continuous oestrus and will mate, but fail to conceive.

Litter and bedding

Sawdust, wood shavings, and peat moss are the commonest litter materials. They are put in trays or pens to absorb the excreta in which the animals would otherwise run. Generally speaking, litter is renewed twice, if not three times, each week—usually on Mondays, Wednesdays, and Fridays. Animals which produce a large volume of urine (monkeys, dogs, cats) require fresh litter daily; rabbits infected with coccidia must be given fresh litter daily if the level of infection is to be held to a minimum. When peat moss is used the litter may need renewing only once or twice weekly, depending on the number of animals in the boxes or cages. Though peat moss is more absorbent than sawdust, it should be remembered that animals running on it do not excrete less urine or faeces or fewer infective organisms than they do when running on sawdust. Also, peat moss litter is as good a breeding place for flies and fleas as is sawdust.

Bedding (e.g. woodwool) is supplied to animals for nesting. It is renewed when it is dirty and/or damp. The frequency with which fresh bedding must be given is determined by the number and age of the animals. However, bedding should be renewed once each week whether or not it appears soiled.

The relative merits of different litters or bedding are discussed in the chapter on 'Bedding'.

Customarily, soiled litter is scraped from the trays into a waste bin. If a metal scraper is used the galvanizing of the trays is damaged (thus accelerating the rusting and ultimate decay of the trays), and the process is a noisy one. Wooden or plastic scrapers do not damage galvanizing and are quieter in use, but they are more troublesome to sterilize. The same comments apply to galvanized and plastic waste bins.

Disposable paper bags can be used instead of waste bins, and are especially useful for infected or contaminated waste. Triple paper bags, having considerable wet-strength, may be hung from portable frames which are designed to hold open the necks of the bags.

Litter can be removed from trays and boxes by a specially designed vacuum-cleaner-like apparatus (Charles *et al.*, 1962). Waste is drawn off through pipes into a sealed container (which may be situated outside the animal building), from which it is fed, automatically, into an incinerator. The nozzle of the vacuum pipe is fitted with baffles which admit the passage of litter and excreta, but not animals. This is an important feature when mouse boxes are cleaned by vacuum. The advantages of this system are obvious—waste bins, scrapers, brushes, brooms, floor sweeping, and, above all, dust are eliminated. Vacuum cleaning is widely used in the United States, and has been installed in at least one large Institute in this country, where the cost of the complete apparatus, fitted with six vacuum pipes which may be used simultaneously, was about £2,000 in 1962.

Cages

Animals should be given clean cages as often as is practicable; the frequency is usually determined by the number of spare, clean cages available and the speed with which dirty cages may be sterilized. The aim should be to give clean cages at least once a month. It is desirable to give a clean floor grid once a week. Ferrets kept on mesh are best given clean floor grids each day because of the sticky nature of their faecal matter. Rabbit cage floor grids can become clogged with trampled faeces, and if a rabbit is left on such a grid the animal will certainly develop sore hocks.

When cages are changed the tray, food container, and water bottle should also be changed. The rack on which the cage stands and the wall behind the rack should be washed thoroughly with hot, soapy water. Any 'disinfectant' used in this washing process must be chosen with care. All disinfectants are toxic to a greater or lesser degree (see chapter on 'Sterilization and Disinfection').

Cages may be moved, so it is useless to rely on cage position as a guide to which batch of cages is due to be changed. When clean cages are put up they should be marked to indicate the date. The mark should be a semi-permanent one, and should be visible when the cage is in position. Suitable marking can be achieved by the application of strips of coloured adhesive tape; different colours and/or positions indicating the month and week of the month.

For long-term experiments it is advantageous to use removable cage labels and to transfer them, with the animals, to clean cages, thus obviating the need for new adhesive labels. Removable labels must be fitted securely to the cages; the term 'removable' meaning 'removable by humans' and not 'removable by animals'.

Each new experiment (and each animal counts as one experiment) must be set up in a clean cage. No cage may be used twice for two different experimental animals. When an experiment ends the cage must be sterilized, even if it has been in use for only ten minutes.

It is customary to combine cage changing with routine cage cleaning. Thus, Mondays, Wednesdays, and Fridays become the days when the heaviest work is done.

Water bottles and food containers

All water bottles should be emptied and the bottles and fittings washed and sterilized once each week. Three or four days later in the week every bottle should be emptied, rinsed, and refilled. On the remaining five days the bottles may be topped up with water. Water bottles must be filled from a mains water supply, as infections may be introduced through contaminated water from storage tanks.

Water bottles for monkey cages must be fitted with metal drinking spouts which cannot be bitten through. The bottle must be fixed securely to a solid metal part of the cage so that it is: (i) out of the monkey's reach, and (ii) cannot be dislodged by the monkey pushing against the spout which protrudes into the cage.

It is important that a bottle which has been only rinsed or topped up is returned to the cage from which it came. Care in this matter eliminates one of the routes for cross-infection in an animal house.

Water bottles should be wide-necked and without sharp shoulders and corners which are difficult to clean. Round, wide-necked bottles may be quickly and efficiently cleaned by using a bottle-washing machine. The drinking spouts can be tiresome to clean if they are allowed to get very dirty and greasy. Where there are no facilities for sterilizing by heat, bottles and spouts may be immersed for 30 minutes in hot water containing 1 per cent Tego. *Pseudomonas pyocyanea* may occur in tap water, and it multiplies rapidly in dirty bottles. This organism is not normally pathogenic, but it can, when present in large numbers, prove toxic to animals.

Food baskets and hoppers containing pelleted diets should not be topped up daily unless that is necessary to supply sufficient diet to satisfy the animals' needs. Rather, these food containers should be allowed to run low before refilling them with fresh diet. Baskets and hoppers do not have to be filled to the brim, and are best filled only to a level which will supply sufficient food for three or four days. Pelleted diets which are left uneaten in baskets and hoppers constitute an excellent medium for mould growth, which can proceed rapidly in the warm, damp atmosphere of an animal house. Not all food baskets are designed so that the animals can eat only from the lower layers of pellets, and mice and rats will, if given the opportunity, climb the sides of food baskets and contaminate the diet with excreta.

Some animals cannot eat from food baskets or are fed diet which cannot be pelleted. Moistened diets must be freshly prepared each day, including the week-ends. They should not be prepared in bulk and stored, even in a cold room. Uneaten food must be discarded and not offered again the next day, when it will be sour and, consequently, refused by the animals.

Farm animals, dogs, cats, and ferrets cannot drink from bottles, and must be given open dishes of water. Guinea pigs can and will drink from water bottles, though it is customary to offer them water in open dishes or in constant-level water troughs. When food and water is supplied in open dishes these dishes must be washed daily, for they always become fouled with excreta and litter unless they can be placed where the animals can reach them but not walk over them.

Dishes, baskets, and hoppers always become greasy in the places where the animals rub against them, so they have to be scrubbed with hot, soapy water to clean them.

Open-air exercise runs for larger animals

Animals in exercise runs should at all times have access to fresh drinking-water and to shelter from strong sunlight and rain. Solid rubbish (e.g. fallen leaves, faeces, sawdust) which may choke drains must be removed before the runs are hosed down daily. The wire netting and fencing runs should be inspected regularly for rusted or weak places which animals could break through.

The care of rooms, equipment, and fittings

WASTE BINS must be kept covered when not in use, and all waste should be incinerated as soon as this can conveniently be done. Bins should be sterilized and washed out once each week. Waste bins are easier and quieter to move if they are fitted with rubber castors, but if the castors are to have a long, useful life they must be carefully cleaned and oiled when the bins are cleaned.

TROLLEYS should be dismantled and cleaned at weekly intervals, special attention being given to the cleaning and oiling of the wheels.

ANIMAL BALANCES and the weights should be washed thoroughly each week, care being taken to prevent water penetrating to the mechanism. The balance pan should be cleaned immediately after each use. When a balance is moved it must be set level in the new position. Most balances are fitted with spirit levels so that this may be easily accomplished.

DUST must not be allowed to accumulate in an animal house, where it would form an excellent breeding place for pests. Dust is best washed away, or at least removed with a damp cloth. Dry dusting tends only to move dust from place to place. Besides 'dusting' obvious places, such as tables, chairs, and window ledges, dust and dirt should also be removed from service pipes, light fittings, doors (including the handles), underneath sinks and fixed benches, racking, and solid-topped cages.

The washing of walls and ceilings is a tedious task, but it should be done as often as possible, and certainly not less often than twice per year. The floors of animal rooms and corridors have to be washed on at least six days of each week. On one day of the week hot water and a disinfectant should be used, and on this occasion all drain covers should be removed and the drains and gulleys cleaned. Cresols must be used with great caution. On other days floors may be washed with hot water only. Floors may be dried off by the use of a squeegee or a drying machine. The latter is particularly useful in drying dog and cat pens.

FOOD BINS should be kept in a cool place and not in an animal room. The lids should be kept closed when the bins are not in use. Empty bins should be washed out and sterilized before they are refilled. On no account should fresh diet be put into a bin containing old diet crumbs and dust.

BROOMS, brushes and dustpans, pails, mops and swabs used in the business of cleaning, themselves become dirty, and should be washed at least once each week.

PROTECTIVE CLOTHING AND FOOTWEAR must be worn at all times in an animal house. When not in use these garments are kept at the entrance to the animal house. High-necked gowns which fasten at the back afford more protection to the wearer than coats. Short gum boots are more comfortable to wear for long periods than are full-length Wellington boots. It is possible that within a few years all colonies will be of Specified-Pathogen-Free animals, in which case it will be necessary for all animal technicians to strip, bathe, and dress in clean clothing before entering the animal house.

FOOTBATHS at the doors to isolation quarters must be large enough and so placed that they cannot be stepped across or walked round. A large sponge rubber mat soaked with disinfectant solution may be substituted for a footbath. Fresh disinfectant solution must be used daily. Footbaths must be lined with non-slip mats. These devices are used to prevent the carriage of infective organisms on footwear, but they have the additional advantage of detering unnecessary traffic.

INSPECTION OF ANIMALS. Technicians must observe animals in addition to cleaning, feeding, watering, and breeding them. Animals housed in captivity require regular inspection if they are to be kept in perfect condition and free from minor ailments. Animals can only be helped by technicians who care *about*, as well as care *for* them.

THE RABBIT

Examinations should proceed in logical order so that nothing is missed. Start at the head and work down to the tail. The rabbit should be placed on a flat, non-slippery surface, as this animal becomes panic-stricken if it loses its foothold. The eyes should be bright and alert; if the rabbit looks as if it had been crying the fact should be reported. In most cases bathing with 2 per cent boric acid solution or the application of penicillin eye ointment will effect a cure.

Any discharge from the nose should be reported at once. The animal should be isolated immediately if the ailment is 'snuffles', as this disease is infectious.

The teeth should be inspected carefully. The front teeth of a rabbit are of an interesting construction. The two front upper teeth (incisors) are grooved, giving the impression that there are four, rather than two teeth. Immediately *behind* the main teeth are two small incisors, only about $\frac{2}{5}$ in. of which are visible. When, through accidental displacement, the lower incisors do not meet they grind against the small upper incisors: the lower incisors continue to grow like tusks (preventing the animal from eating properly), and may even grow up through the mouth (see Fig. 1). When this occurs it is necessary to cut the tooth off to its correct length with bone forceps, but continued attention is necessary, as the tooth will continue to grow wild.

FIG. 1

The ears should be examined inside as well as outside, for it is inside the ear that ear canker (caused by a mite) first appears. It is important that rabbits' ears are inspected regularly, as ear canker, although easy to cure, can be both troublesome and painful if not treated in the early stages. Any vegetable oil will help to soften the scabs, but the application of 2 per cent of phenol in liquid paraffin or of benzyl benzoate is recommended. Great care should be taken if the canker is bad, as treatment can be very painful if carried out clumsily. The following technique is recommended:

If possible, get a second person to hold the rabbit. If the task has to be done single-handed place a duster flat on a table, place the rabbit on the duster, tie the two front corners of the duster round the rabbit's neck (taking care its front feet are inside) and tie the other two corners on the top of the rabbit's back (seeing that both hind feet are inside the duster). Have ready some cotton wool, a pair of blunt-ended forceps, and a supply of medicant.

Pour a little of the oily substance into each ear and leave for at least 5 minutes so that the scabs will loosen. Then, with forceps and using great care, lift off the scabs and place them in a burnable bag. This operation must be done very gently, for much pain can be caused by rough treatment. When all the scabs have been removed wipe the ears out with a pad of cotton wool soaked in either benzyl benzoate or phenol in liquid paraffin. This treatment should be repeated after ten days.

The nails of the front feet should be inspected, and if necessary cut with a small pair of bone forceps. It is on the inside of the front legs that evidence of snuffles can be found, because when the rabbit sneezes it wipes its nose on its front legs.

Examine the length of the nails of the hind feet. Rabbits living in cages, and having no opportunity for digging, often grow long nails. Place the rabbit on its back and examine the pads of its feet. If there is any soreness (sore hocks) remove the floor grid from the cage and place the rabbit on a bedding of sterilized wood wool or hay and paint the sore hocks with iodine solution of the following formula:

Iodine (resublimed)	10 gm
Potassium iodide	6 gm
Distilled water	10 ml
Ethyl alcohol (industrial)	100 ml

When this condition is found in rabbits the cage grids must be sterilized and scrubbed at least once a week. If the skin on the foot is broken the fact must be reported at once. If the lesion becomes infected the animal will lose condition rapidly.

A healthy rabbit's coat shines and looks attractive; a sick rabbit has a dull and unattractive appearance. Run a hand down from head to the tail of the rabbit, feeling the skin for lumps or scratches. The backbone of a healthy rabbit is smooth and firm to the touch. The bones of the spine are palpable only when the muscles of the back have wasted.

The importance of good hygiene in the rabbitry cannot be overstressed, and the cleaning routine should on no account be allowed to lapse. This is illustrated by the fact that the life cycle of the protozoa, coccidia, may be broken by removing all faeces from the cage each day.

Coprophagy (eating of faeces) in rabbits

Coprophagy in rabbits has been attributed to a lack of B-vitamins in the diet. Blount (1957) studied this nocturnal habit and concluded that the swallowing of specially formed faecal pellets is a normal feature in rabbits. It does not represent a depraved appetite, but is, in fact, a specialized process. Several theories, none of which is conclusive, have been advanced to explain this phenomenon.

It should be understood that the stomach of a rabbit does not empty after every meal. Observation has shown that the stomach of the rabbit, like that of the horse, has no great powers of contraction except at its exit—the pylorus. No matter when a healthy rabbit is killed, its stomach will always be found to be more than half full of food. By nature the rabbit eats little and often, but never takes a large feed. If a rabbit becomes hungry but cannot, or does not, eat, bulk must be provided to push onward the stomach contents, and coprophagy is resorted to.

The points of interest are:

(i) Coprophagy is a normal occurrence in rabbits of all ages and both sexes, except in baby rabbits receiving only milk.

(ii) The formation of the special faecal pellets takes place about six hours after the last meal, usually at night.

(iii) These pellets are swallowed voluntarily an hour or so later, and sufficient are taken to fill about one-third of the stomach. They are taken direct from the anus and are not picked from the floor.

(iv) These pellets remain intact in the stomach for a number of hours, after which they undergo a gradual softening and disintegration.

(v) The formation of these pellets is not stopped by feeding large quantities of a vitamin B complex-rich foodstuff (Marmite) or by adding to the diet of the rabbit a proportion of normal rabbit or bovine (dairy cow) faeces daily.

(vi) It is suggested that coprophagy is developed as a compensatory habit during the natural life of the rabbit, when it remains in the burrow for long periods to avoid its enemies.

THE GUINEA PIG

In good health the smooth-coated guinea pig should feel compact and firm to handle, with the flesh evenly distributed. Any suggestion of 'lightness' when the animal is handled should be regarded with suspicion. The coat should be shiny, dense, and smooth; the eyes bright; the nose clean; and respiration regular.

Evidence of diarrhoea should be reported at once, as it may be the first sign of a salmonella infection in the colony. Slobbering from the mouth is unusual, and indicates either trouble with the teeth or a more serious complaint which must be investigated. Any rise above the normal death rate of a guinea pig colony should be carefully investigated by the veterinary officer, and, routinely, post-mortem examination should be made of all animals found dead or moribund, and adequate records of the causes of death should be kept. If necessary, laboratory reports should also be filed. These steps will ensure that serious, epidemic diseases are checked before they obtain a hold on the colony. This will call for close co-operation between the chief animal technician and the veterinary officer.

Guinea pigs are most susceptible to changes in temperature or humidity, and draughts. The actual temperature at which guinea pigs are kept is not so important as the maintenance of an even environmental temperature. As a guide, 62–65°F (17–18°C) room temperature and 50–60 per cent relative humidity is suggested.

Guinea pigs will eat and do well on a pelleted diet provided that vitamin C (ascorbic acid) is present in the diet, either in the form of a green food supplement or added to the pellets when they are made. Neither route is free from hazard, as ascorbic acid decomposes quickly in a pelleted diet unless great care is taken with the compounding and greenfoods may carry diseases.

Signs of vitamin C deficiency (Scurvy) can be detected after eleven to fourteen days, when the guinea pigs will exhibit difficulty in walking, steady loss of weight, staring coat and, later, swollen joints and bleeding from the gums around loosened teeth.

Some guinea pig colonies have suffered a high mortality rate among pregnant sows after the colonies had been transferred from pens to grid-floored batteries, and this embarrassment has been overcome by adjustment (usually by adding) of the calcium, phosphorus and magnesium content of the diet. This should be checked with the miller if breeding stocks are housed in batteries with grid floors.

The causative organism of cervical adenitis (abcesses in the neck region) in guinea pigs is *Streptobacillus moniliformis*. This bacillus also causes 'infective arthritis' in mice and 'middle ear disease' in rats; therefore rats, mice and guinea pigs should never be housed in the same room.

It is difficult to breed and rear guinea pigs for long periods without the addition of top-quality meadow hay to the diet. The exact reason for this has yet to be determined. Clean water is also necessary.

It is important to examine the hair and skin of guinea pigs from time to time, as they are parasitized by two species of lice. Treat the animals with a pyrethrin aerosol or sevin or malathion dust. The treatment should be repeated after 10 days to kill newly emerging larvae.

For further reading the *UFAW Handbook* is recommended, and also see Paterson (1957).

THE RAT

Paratyphoid disease (Salmonellosis) is the only serious, common, specific infection of all laboratory rats. However, about 75 per cent of all adult rats are affected with chronic, pulmonary disease, and middle-ear disease is quite common in some stocks of rats.

The first signs of paratyphoid disease in rats are acute diarrhoea and loss of condition. Prompt and ruthless action by the technicians and veterinary officer is necessary if the stock is to be saved (see chapter on Animal Diseases). Great attention must be paid to hygiene. Precautions against the contamination of food by wild rats and mice must be strictly enforced. One source of paratyphoid infection may be a 'carrier', that is, an animal which carries and transmits a disease without exhibiting signs of infection. Bacteriological screening should be instituted to detect carriers. It may be necessary to obtain a fresh colony with rats from a pathogen-free stock (see chapter on Breeding).

Sniffling at the nose (*Murine pneumonitis*) is present in most stocks of rats. The disease is transmitted from adults to offspring during the first few days of life. Childs and Rees (1958) recommend controlling the infection with sulphonamides, but to eliminate the infection it is necessary to take rats by caesarian section and raise them in a specific pathogen-free colony (see chapter on Breeding).

Fifty per cent of all rats, wild or domestic, carry *Streptobacillus moniliformis*, the causal organism of 'infective arthritis' in mice and of cervical adenitis in guinea pigs. Cross infections will occur if these three species are kept in the same room.

Ringtail sometimes occurs in newborn rats bred in an environment having a relative humidity of less than 50 per cent. The use of solid-bottomed, rather than mesh-bottomed, litter cages greatly reduces the risk of exposure for the pup, and colonies kept in such conditions rarely exhibit ringtail.

Scabies of the ear and tail is caused by mites (*Chorioptes cuniculi* or *Psoroptes cuniculi*). Such infestation causes dark, scaly lumps on the tail and 'cauliflower' ears. Affected areas should be treated with benzyl benzoate (a specific cure for scabies), the bedding burnt, and the cages sterilized.

THE MOUSE

Laboratory mice are subject to many bacterial and virus diseases, most of which can be avoided by strict attention to routine care and inspection by the technician in charge. Mice brought in from an outside colony should not be mixed with an existing colony until after they have been quarantined for at least four weeks and have been subjected to bacteriological screening.

Mice, rats and guinea pigs should be kept separately. Most rats carry *Streptobacillus moniliformis*, which can be transmitted to mice. The feet and tails of mice should be examined at least once a week. Swollen feet might indicate infection with *Streptobacillus moniliformis*, or, more serious still, *Ectromelia* (Mouse Pox). *Streptobacillus moniliformis* is also the causative organism of cervical adenitis in guinea pigs.

Infestation with mites should not be tolerated; apart from lowering the animals' condition and vitality, certain mites carry diseases which they can transmit from mouse to mouse and stock to stock, e.g. *Eperythrozoon coccoides*. Overcrowding of conventional stocks of mice is dangerous and may contribute to an outbreak of Tyzzer's disease.

Bald patches on mice may be caused by Ringworm fungus, which can be transmitted to humans. Encourage all technicians to wash their hands after touching mice, and to report immediately any unusual skin conditions appearing on either themselves or the mice.

Chloroform must not be used in the mouse room, because male mice are particularly susceptible to its vapour, which often causes infertility or even death. Recommended environmental conditions are room temperature 72–74°F (22–23°C) and a relative humidity of 50–60 per cent.

THE HAMSTER

This animal requires sympathetic treatment. Lack of care and attention can lead to the production of unsatisfactory and even sick animals. Gentle but firm and regular handling is essential (see section on Handling).

In spite of their usefulness in the study of experimentally produced diseases, golden hamsters are generally free from spontaneous disease. Both tapeworm and roundworm may infest a colony, and routine checks should be made for these parasites. Mites and fleas may be the intermediate hosts for worms, so it is essential that the hamster be kept free from ecto-parasites. These measures together with an adequate diet are the best means for disease control.

Hamsters dislike temperature variation and draughts. A sudden drop in room temperature in winter-time will send hamsters into hibernation. This is not an ailment but a natural phenomenon common to many animals. The hamster curls up tightly in its nest and is quite cold and rigid to the touch. Breathing is very slow and shallow, and the heart beats faintly. This condition can be mistaken for death. No attempt should be made to waken a hibernating hamster; it should be given gradual warmth until signs of life return.

Running eyes, high-pitched bronchial wheezes, and sniffling are all indications of a cold which, if not checked, may lead to a fatal pneumonia. Rooms should be controlled at a temperature of 68–70°F (20–21°C) and kept at a humidity of 50–60 per cent. Signs of diarrhoea in hamsters must always be treated seriously, as the commonest cause of death is a form of enteritis or 'wet-tail disease'. Diarrhoea is also connected with a bacterial infection associated with the infestation of the worm *Hymenolepis nana.*

All cases of diarrhoea should be reported at once.

Overgrown claws are seen occasionally in hamsters. These should be clipped with a pair of sharp scissors or small bone forceps. Hold the claw up to the light, notice the red blood-supplied portion of the nail and clip $\frac{1}{8}$ in. distal from this point.

Sometimes the teeth grow inwards spontaneously, and natural wear is prevented. Overgrown teeth can be a problem because there is no cure except to keep cutting the teeth with bone forceps.

Hamsters will eat any cubed diet suitable for rats and mice, and do not require any supplement other than fresh, well-washed greenfood.

THE CAT

Cats suffer from at least two virus diseases, one of which, infectious feline enteritis (Panleucopenia) may be fatal. This disease can be controlled by a vaccine which only provides protection for a few months (see section on Diseases).

Early diagnosis, isolation, and good nursing are very important to the recovery of cats from diseases. Particularly is this so in the case of feline pneumonitis, a disease caused by a virus of the psittacosis group. This disease manifests itself in both mild and severe forms (Scott, 1952), and kittens, both weaned and unweaned, are very susceptible. In the mild form the kitten has a discharge from the eyes together with inflammation of the conjunctiva. It is vital that this condition should be detected as early as possible, and the technician must be prepared to stop all other work to attend to the kitten. The eyes should be bathed with boracic solution and penicillin eye ointment smeared on the eyes. Scott (1953) has suggested that aureomycin is the most suitable antibiotic to administer. Short and Lamotte (1958) found aureomycin capsules given orally as the most effective remedy. The technician should make certain that sufficient liquid is given to the patient during illness. In most cases the kitten can be returned to the litter or group of kittens within a few days. It has also been suggested that injections of combined penicillin and chloromycetin should be given when the animal is suffering from the severe form of this disease.

Tapeworms and roundworms are common in cats. Healthy adult cats appear to tolerate worm infections quite well, but kittens and pregnant and lactating cats are adversely affected by the presence of worms. Three species of tape worm infect the cat, one of which, *Diplidium caninum*, can be transmitted to man. Strict care with personal hygiene should be taken after handling dogs or cats and their bedding.

The roundworm requires no intermediate host, as the eggs, if ingested, mature and reproduce in the duodenum of the cat. The contamination of rooms, food, water, bedding, and equipment by faeces from infested animals should be prevented.

Proprietary tablets are available for the treatment of round and tapeworm infections in the cat. The treatment is effective, and cats do not have to be starved prior to administration of the tablets.

The ears of cats and kittens should be inspected regularly for the presence of mites. Treatment is discussed on the chapters on Diseases and on Pests.

Ringworm can be troublesome in cats. La Touche (1957) reported that numbers of cats, kittens, and dogs and puppies examined in Leeds were found to be infected. Almost all these animals were from homes inhabited by people suffering from ringworm caused by *M. Canis*, the so-called animal type microsporum, presumably transmitted to them by their pets.

The most frequently affected areas are the bridge of the nose, and inside the ears and the back, although all parts of the hair-bearing surface may be affected.

It is essential that imported cats are carefully screened for ringworm and internal and external parasites before they are admitted to an existing colony. Maintain a high standard of personal hygiene.

THE DOG

Most laboratories rely on dealers and breeding kennels for supplies of dogs. It may be assumed that such dogs, on reaching the laboratory, have already been exposed to various infections and changes in diet. They may be infested with endo- and ecto-parasites, and will certainly be suffering from considerable nervous strain (see chapter on Receiving Animals). It is important that a temperature chart be kept for each dog. A variation of more than 2°F. out of normal may indicate the incubation of a serious disease and should be reported immediately. Running or gummed-up eyes may also be an indication of disease.

Where dogs have to be exercised with other dogs it is important that the group is watched to see that all the animals behave well towards one another. Persistently quarrelsome dogs should be isolated before other dogs contract the habit of quarrelling.

THE FOWL

The outward signs of ill health are loss of appetite, dullness of plumage, paleness of comb and wattle, running nose and paralysis of either a leg or wing. (Any of these signs of illness should be reported.) Handling the birds will give a good general indication of their health, and will also enable the technician to examine them for external parasites, which can be troublesome in cage-kept birds.

There are several species of lice which parasitize poultry; some attack the head and neck, some the wings, and others the body, especially the abdomen and tail. The body louse is the commonest species. It is about $\frac{1}{12}$-in. long, and is pale yellow with dark spots. When the feathers are parted the lice may be seen running over the surface of the skin. The eggs (or 'nits') are laid in clusters at the base of the feathers.

The wing louse establishes itself in the primary and secondary feathers of the wing. The head louse is most injurious to young chickens.

For the treatment of lice on poultry, 5 per cent Sevin dust or 5 per cent Malathion dust should be used. The dust should be applied thoroughly, particularly in the regions of the vent, and treatment should be repeated after 10 days. For the treatment of poultry red mite, the birds should be dusted with 5 per cent Sevin dust and the poultry accommodation thoroughly sprayed with a 0·25 per cent Sevin spray. Poultry red mite is the most serious ectoparasite of poultry, and the adult and nymphal stages both feed on blood. Feeding takes place mainly during the hours of darkness, and during the day the mites remain in cracks and crevices in the house, where the female lays its eggs. Thorough treatment of the poultry accommodation will eradicate the infestation within 2–3 days, as the mites emerge during the night and move across the treated walls and perches. As Sevin is effective at very low dilutions the deposit is active for many weeks.

The hen flea should never be allowed to establish itself, because it is a difficult insect to eradicate. It is necessary to thoroughly treat the birds and cages to control this parasite. A 5 per cent Sevin dust or a 5 per cent Malathion dust should be used.

N.B. For references, Harrison, I. R. (1960). The Control of Poultry Red Mite with 1-Naphthyl-*N*-methylcarbamate, *Vet. Rec.*, **72**, No. 16, 298–300. Harrison, I. R. (1961). Sevin for the Control of Poultry Red Mite and other Poultry Parasites, *Agric. & Vet. Chemicals*, 2–4.

THE MONKEY

Monkeys are an accepted laboratory animal, and may be kept in cages for several years provided that the cages, if they have mesh floors, also have a wood or metal seat, so that the animal can sit or stand on a solid surface and is not forced on to the grid. Failure to provide this simple seat can result in the monkey becoming paralysed in one or both hind-legs.

Monkeys do well on the cubed diets which are satisfactory for breeding mice and rats (Short and Parkes, 1949), but a daily supplement of greenfood (or some other source of vitamin C) must also be given. Monkeys deprived of vitamin C develop scurvy. It is known that the young monkey requires a *minimum* of 2 mg vitamin C daily.

Most monkeys bite, so it is important to learn how to handle them. Newly imported monkeys must be handled with special care.

The Rhesus monkey (*Macaca Mulatta*) is a fairly hardy creature and may be housed at a room temperature of 68–72°F (20–22°C). Monkeys from Africa and the Far East require much higher temperatures, 78–85°F (25·5–29·5°C). The relative humidity should be 55–65 per cent; at lower humidities the animals may develop chest troubles. It is important that these temperatures and degrees of humidity are maintained in the monkey rooms.

Newly imported monkeys are usually infested with lice. The insects usually congregate on the chest of the monkey and can be eradicated by a thorough dusting with 5 per cent Sevin or Malathion dusts.

The technician should inspect the animals each day and report any suspicious signs immediately to the veterinary officer. For further reading about monkeys:

The *UFAW Handbook*, 2nd Edition (1957).
VAN WAGENEN, C., 'The Monkey' in FARRIS, E. J., *Care and Breeding of Laboratory Animals*. Chapman and Hall Ltd. (1950).
RUCH, T. C., *Disease of Laboratory Primates*. W. B. Saunders Co. (1959).
'Care and Disease of the Research Monkey', *Annals of the New York Academy of Science*, **85** (May 1960).

REFERENCES

BLOUNT, W. P., *Rabbit Ailments*. Published by Fur & Feather, Bradford (1957).
CHILDS, R. T. and REECE, O., *Nature*, **181**, 1213 (1958).
LA TOUCHE, C. T., *UFAW Handbook*, p. 526. Baillière, Tindall & Cox (1957).
PATERSON, J. S., *L.A.C. Collected Papers*, **5**, 58 (1957).
SABIN, A. F. and WRIGHT, A. M., *Journal of Experimental Medicine*, **59**, 115–16 (1934).
SCOTT, P., *Journal of Physiology*, No. 118, 35–36 (1952).
SCOTT, P., Unpublished Observations (1953).
SHORT, D. J. and PARKES, A. S., *Journal of Hygiene*, **47**, No. 2, 209–12 (1949).
TUFFERY, A., *Journal of Animal Technicians Association*, **11**, 3 (1960).
CHARLES, R. T., POPPLETON, W. R. A., and STEVENSON, D. E., *Journal of Animal Technicians Association*, **13**, 1 (1962).
SHORT, D. J. and LAMOTTE, J., *Journal of Animal Technicians Association*, **9**, 1 (1958).

CHAPTER NINE

Sterilization and Disinfection

It is most important to understand clearly the special terms used in connexion with this subject. Much confusion has arisen through incorrect, loose usage of terms because they are of old derivation. For example, the terms 'antiseptic' and 'disinfectant' were first introduced some two hundred years ago, and, as is not uncommon, their meanings have been modified with the passage of time.

Sterilization

In its proper sense, sterilization is an absolute term, meaning the complete destruction or removal of all forms of life. The number of agents capable of achieving this is limited, and is confined to high temperatures (including fine saturated steam under pressure) and certain types of filters. A few of the many chemicals employed, and possibly one or two of the radiation treatments, suitably applied, are true sterilizing agents.

The term sterilization is often erroneously applied when disinfection is really meant. This should be noted in the medical field, where 'sterilization' is used to mean the destruction of micro-organisms undesirable under a particular set of circumstances. This use of the term has often led to confusion in the past, and illustrates the need for taking care always to use the correct terminology.

Disinfection

This may be defined as the process of eliminating or destroying infection; it is accomplished by the use of a *Disinfectant*. The term was introduced before the establishment of the germ theory of infection, and so because disease was always associated with foul odours, it tended to imply, primarily, the destruction or masking of these odours, although often the killing of bacteria attended. On this account the term *Disinfectant* is still frequently confined to the strong-smelling coal tar fluids, whereas, in fact, it has a much wider application and meaning. Several authorities, with some justification, prefer to confine the use of the word to the treatment of inanimate (lifeless) objects, and this is the generally accepted meaning.

Sanitization

This is an awkward word of recent vogue in the United States, and means the process of rendering sanitary or of promoting health. It is akin to disinfection, but carries with it the inference of cleansing as well as the removal of infection. It is not a term used to any extent in Great Britain.

Antiseptic

Another much misunderstood word, antiseptic, literally interpreted from its Greek origin, means 'against putrefaction', but it has now been extended to include activity against bacterial sepsis or infection. By inference, the word conveys a meaning similar to that of 'disinfection', and there is a tendency to use the term specifically of preparations for application to living tissues, especially in surgery and hygiene. It can also be used to denote a property of inhibiting or preventing the growth of micro-organisms under prescribed conditions of usage.

Bactericide

A bactericide kills bacteria, but not necessarily bacterial spores, while a *Bacteriostat*, or *Bacteriostatic Agent* prevents the growth of bacteria and so gives rise to a state of *Bacteriostatis.* Similarly, a *Fungicide* kills fungi and a *Viricide* kills viruses. A *Germicide* kills *all* micro-organisms. The suffixes -stat and -stasis are not used in conjunction with this term.

By way of general explanation it may be said that the suffix -cide always applies to any agent producing a killing effect on the micro-organisms concerned, whereas -stat means that the agent simply prevents or inhibits growth. *Stasis* is the state of suspended animation or inhibition produced by the latter type of agent.

METHODS OF STERILIZATION AND DISINFECTION

Physical

Heat: (*a*) Wet
(*b*) Dry

Irradiation

Chemical

(*a*) Liquid
(*b*) Gas

Heat

An important agent in the artificial destruction of micro-organisms, the effect of heat is to coagulate and denature cell proteins. In general, among bacteria which are parasites of mammalian animals the non-sporing forms in a moist state cannot withstand temperatures above 113°F (45°C) for any length of time. Bacteria are more susceptible to moist heat (e.g. in a steam sterilizer) than to dry heat (e.g. in a hot-air oven).

Bacterial spores

Some species of bacteria, those of the genus Bacillus and Clostridium, develop *Spores*—a highly resistant resting stage. The spore is not a reproductive structure, but it can survive unfavourable external conditions.

(A) HEAT (WET)

(i) Autoclave: sterilization by steam under pressure.
(ii) Free steaming: sterilization by steam not under pressure.
(iii) Boiling.
(iv) Washing machines.

(i) Autoclave

Autoclave is the use of steam under pressure in a specially constructed apparatus. The principle on which the autoclave depends is that water boils when the vapour pressure is equal to the pressure of the surrounding atmosphere. If therefore the pressure is increased inside a closed vessel the temperature at which water boils will rise above 212°F (100°C); the exact temperature depending on the pressure employed. The pressure generally employed is 15 lb per square inch. At this pressure water boils at 249·8°F (121°C), and 30 minutes exposure at this temperature kills all forms of organisms, *including spores*.

The point of using pressure is to raise the temperature, and it is the heat which does the sterilizing.

TEMPERATURES OF SATURATED STEAM UNDER PRESSURE

lb pressure on autoclave	°C	°F
0	100·0	212·0
5	108·4	227·1
10	115·2	239·4
15	121·0	249·8
20	126·0	258·0

It has been recognized for many years that rapid penetration of *Steam* to all parts of material in the autoclave is essential if complete sterilization is to be achieved. One of the traditional methods of removing air from steam sterilizers has been to use a steam ejector to draw out part of the air contained in the autoclave before steam was admitted. This method has proved inefficient; often it removed only about one-third of the air. Residual air prevents the penetration of steam and, thus, the attainment of the required sterilizing temperature. Recent attention has been drawn to high-vacuum and infinitely variable sterilizers for the processing of animal food, bedding, cages, add equipment. The degree of vacuum which these machines can be set to draw will vary from atmospheric pressure to 10 mm of mercury. The maximum and minimum pressures at which these units will operate are 15 in. of mercury absolute, equivalent to a temperature of approximately 80°C and to 32 lb/in., equivalent to 136°C.

These autoclaves may be equipped to perform fully automatic or manually operated cycles as follows:

High-vacuum steam sterilization—this cycle is indicated where penetration of the steam into the load is required.

Downward-displacement steam sterilization—this cycle is used where penetration of steam into the load is not required. Such items as fluids in bottles or animal cages or utensils of metal or plastic material.

Sub-atmospheric steam sterilization cycle with or without formaldehyde injections—this cycle is useful for the processing of heat-sensitive materials which cannot be subjected to temperatures greater than 80–90°C.

Ethylene oxide sterilization—this cycle utilizes ethylene oxide as the sterilizing agent. The ethylene oxide may be mixed with CO_2 or Freon to provide a non-flammable mixture. There are several ethylene oxide cycles available, depending on the type of material and packaging to be processed. Usually ethylene oxide would be used only where the materials to be sterilized are unable to withstand the temperatures associated with steam sterilizing.

The special, electrically operated door designed by sterilizing engineers has, by eliminating the restrictions of conventional radial locking-arm devices, enabled any size of chamber to be used and, for the first time, it is possible to design a chamber around a load module. The push-button, power-closing of the door gives ease of closure combined with a constant degree of locking force.

These newly designed high-vacuum autoclaves speed up the amount of work done by six times a day.

Glick, Gremillion, and Bodmer[1] (1961) have shown that a nesting-type of cage (designed to save space during storage and sterilizing) when stacked six high in an upright position and autoclaved for four hours at 15 lb per square inch pressure still contained active test organisms. When the cages were autoclaved lying on their sides 30 minutes autoclaving at 20 lb per square inch was sufficient to kill the test organism and its spores. The authors concluded that with the cages in an upright position pockets of air were trapped at the bottom of each cage, which excluded the steam necessary to raise the temperature.

These workers also studied the sterilization of infected carcases. Twenty dead guinea pigs were packed in fibre-board containers and autoclaved for various periods up to 16 hours. It was found that a period in excess of 8 hours was necessary for the centre of the load to reach sterilizing temperatures. For practical purposes the load was autoclaved for sufficient time to ensure sterility of the outside surfaces of the containers, and the contents were considered infectious during transport to the incinerator.

(ii) Sterilization by steam not under pressure

This method is the use of steam at atmospheric pressure in steamers or tanks. The maximum temperature which can be reached by this method is 212°F (100°C).

The Cage Sterilization Sub-Committee of the Public Health Laboratory Service[2] investigated the rate at which vegetative organisms in contaminated cage litter were destroyed by exposure to steam at atmospheric pressure applied in cabinets made of galvanized sheet iron. The committee concluded that pathogenic organisms such as *Salm. typhimurium* and tubercule bacilli were invariably killed in litter by steaming for 10 minutes. The exposure of contaminated litter, even when 4 in. deep, to steam at atmospheric pressure for 10 minutes is sufficient to destroy vegetative forms of pathogenic organisms. This committee recommended that, to ensure a wide margin of safety, steaming should be continued for 30 minutes.

It is important to note that the times given for sterilizing are AFTER the maximum temperature of 212°F (100°C) has been reached on all parts of the cages and litter. This method will not kill all organisms, and is not to be recommended if spore-forming bacteria are present.

(iii) Boiling

This method may be used for sterilizing animal cages in a large tank of water, heated by an immersion heater, or gas or steam pipes. Great care is necessary to ensure that the maximum temperatures and time of sterilizing (the same as for the previous method) are obtained and maintained. The cages are cleaned as they are sterilized. This method is not to be recommended in a modern animal house.

(iv) Washing machines

These machines can be used for *washing* and *sanitizing* cages, water bottles, and other items of equipment which can stand the rigours of this treatment. If used correctly, such machines are labour saving as large numbers of cages can be cleaned in a short time; but it must be remembered that few machines can exceed a temperature of 180°F (82°C). There are some expensive, specially designed machines which attain a temperature of 212°F (100°C) and can therefore be used to *sterilize* equipment.

(B) DRY HEAT

It should be emphasized that sterilization by dry heat depends upon the penetration of adequate heat to all parts of the article. It is therefore possible to sterilize apparatus already assembled and pre-sealed in a container, whereas steam and gases (such as ethylene oxide) can be relied upon only to kill organisms with which the steam or gas comes into direct contact. Another advantage of this method is that objects which are damaged by water or steam (e.g. food, bedding) can be sterilized, provided the heat penetrates to all parts of the substance.

The disadvantages of the method are, first, some equipment, such as hot air ovens, take a considerable time to reach sterilizing temperature; dry sterilizing requires a higher temperature and a longer exposure time than wet sterilizing; objects (e.g. metal ones) may become oxidized at high temperatures or may not withstand the temperatures.

Wentworth Cumming[3] (1962) reported that all materials, except food, taken into a Specific Pathogen-Free Unit were sterilized in a large electric oven at 250°F (121°C), for 2 hours *after* a 30-minute warm-up period. Darmady and Brock[4] (1954) suggested that it was important to check each hot air oven carefully, as a number tested did not reach a uniform sterilizable temperature, and they showed that there might be a variation of 86–104°F (30–40°C) between different parts of the oven. They also showed that ovens fitted with a fan required a shorter time to reach sterilizing temperature than still-air ovens. The forced circulation of air reduced temperature variations to a minimum. It was also found that unless the objects in the oven were loosely packed, there would be a delay in heat penetration to the centre of the load, even in ovens fitted with a fan.

(C) RADIATION (*vide* Sykes,[5] 1958)

The range of radiations which have been used for killing micro-organisms fall into two groups: (*a*) the ionizing radiations comprising X-rays, gamma rays, cathode rays, beta rays, and the heavy particles, neutrons, protons, etc.; and (*b*) the longer electro-magnetic radiations comprising ultra-violet rays, infra-red rays, and radio-frequency radiations; and ultrasonic waves of high frequency. The terms 'radiation sterilization' and 'electronic sterilization' (terms which have become quite commonplace in present-day parlance) apply exclusively to the various forms of ionizing radiations; they do not include the radiations of group (*b*) above.

Darmady *et al.*[6] (1961) investigated sterilization by radiation of disposable medical items and found that 2·5M rad (Mega) can be recommended to give a high degree of sterility. This was best achieved, in practice, by using either high-energy electrons from a machine such as a linear accelerator or gamma radiation from an isotope source such as Cobalt-60. Both types of radiation have been used commercially for sterilization, and they have similar bactericidal properties; but they differ in dose, rate, and degree of penetration.

Radiation sterilizing is not, however, the answer to all sterilizing problems, because certain products (e.g. rubber, plastic) could be damaged, and the capital cost of installation is high. Their powers of penetration, and the absence of any thermal effect, marks the high-energy ionizing radiations of particular potential value in the sterilization of materials which are difficult to treat by the more orthodox heating methods. Thus, any material which has low heat conductivity or is adversely affected by heat is specially suited to treatment by radiation, provided there are no adverse side effects. For this reason, investigations have been largely centred on the preservation of food and the sterilization of pharmaceutical products.

The outstanding advantage of the treatment is that it is virtually devoid of any thermal action, the maximum rise in temperature being only 3° or 4°. The treatment could be useful in the sterilization of animal bedding.

A table giving the radiation stability of various materials (Radioisotopes Review Sheet G1) is obtainable from the United Kingdom Atomic Energy Authority (RLV_1), Wantage Research Laboratory, Wantage, Berks.

CHEMICALS

The ideal chemical disinfectant should satisfy the following demands:

1. It should kill all pathogens which may infect any or all laboratory animals.
2. It should be bactericidal to a broad spectrum of bacteria, including Mycobacterium, tuberculosis, and spores.
3. It should be viricidal.
4. It should be mycoplasmacidal.
5. It should be fungicidal.
6. It should be oöcystacidal.
7. It should not be toxic to animals.
8. It should not be dangerous for the operator.
9. It should not be corrosive to the cages, racking, or rooms.
10. It should not be easily inactivated by dirt normally encountered in an animal house.
11. It should have high surface activity and penetration, easily removing infected dirt and leaving a protective film when dry.

12. It should not give rise to resistant strains of bacteria, viruses, or fungi.
13. It should be easy and safe to apply.

It is doubtful if any disinfectant will satisfy all these demands, for although there are many disinfectants having good biocidal activity, as shown by laboratory tests, their practical unsuitability eliminates many.

Phenol and phenol derivatives

Although phenol has occupied a prominent place in the field of disinfection since its discovery as a potential germicide by Lister in 1867, interest is now centred on the derivatives of phenol as practical disinfectants. Probably more is known about the anti-microbial properties of phenol than any other substance, and it is often used as a model for examining certain aspects of the theory of disinfection. Today its applications are virtually limited to that of the standard against which the germicides are compared and to that of a bacteriostatic for use in preparations administered by injection.

All the phenols can act either bactericidally or bacteriostatically, depending upon their concentration. Generally speaking, the phenols are not sporicidal, but they are effective against tubercle bacilli, and they exhibit anti-fungal activities, some being more active against fungi than against bacteria. They are not particularly viricidal. Small changes in concentration give rise to relatively large differences in killing rates, but the phenols remain bacteriostatic over fairly wide ranges of concentration.

The phenols become more effective with increase in temperature. The warmer the surfaces to be disinfected and the warmer the phenol solution, the better the result. Alkaline solutions are always less active than acid ones.

Mode of action

The phenols owe their anti-bacterial properties to their ability to combine with and denature proteins. This phenomenon has long been understood, but more recently other aspects of the mode of action of phenols have been studied; particularly the absorption of phenols by bacterial cells.

Effect of organic matter

The bactericidal power of phenols is always reduced in the presence of other organic matter (e.g. faeces). The organic matter may cover the organism and prevent the penetration of the disinfectant; it may neutralize the disinfecting action by combining with the disinfectant; or it may serve as an absorbing surface to reduce the amount of active disinfectant present.

Liquid synthetic phenolic disinfectants

A group of synthetic phenols is of recent advent, although the soap-based pine oil disinfectants were known forty years ago. The pine oil disinfectants are not phenolic in constitution, but are composed mainly of terpenes; however, it is useful to consider them in the present context, because many synthetic phenol disinfectants have pine oil fractions added to them. Pine oil disinfectants, if properly constituted with the right terpene fractions and a suitable soap base, are reasonably active against many bacterial types.

A correctly formulated pine oil disinfectant contains at least 60 per cent of pine oil, but many preparations on the market contain lower proportions than this of superior oils, and consequently merit being classed as little more than deodorants.

Coal tar disinfectants

The coal tar disinfectants are also phenolic in nature. They are much cruder than the pure synthetic phenols and are more toxic and irritating to the skin. Lysol, for instance, cannot be comfortably tolerated by normal skins at a concentration greater than 1 per cent.* Coal tar disinfectants should be used only for disinfecting inanimate objects.

The active constituents of this type of disinfectant are derived from coal tar distillation fractions and have only low solubility in water. Hence they have to be emulsified in order to obtain an adequate concentration for disinfectant purposes.

The germicidal activities of these disinfectants are determined by the quality of the phenol fraction used, the ratio of phenol to emulzent used and the nature of the emulzent. The mixed cresols which are the main constituents of the fraction boiling in the range 383–401°F (195–205°C) are used for making Lysol or solutions of Cresol in Soap B.P.

Lysol of the B.P. formulation contains about 50 per cent of cresol and 22 per cent of linseed oil soap. The higher the molecular weight of the compound, and the higher its boiling point, the more divergent are its activities against the gram-positive and gram-negative organisms. Thus, the various Lysols made with the boiling cresols have equal or slightly lower phenol coefficients against *Staph. aureus* than against *Salm. typhimurium.*

Too much reliance should not be placed on the phenol coefficient of a disinfectant (see Rideal Walker and Chick Martin tests). The response to these fluids differs with different organisms, and, in practice, there is also the problem of the presence of other organic matter. Generally, disinfectants of this type are used when organic matter of one sort or another is present, and the Chick Martin test was devised to meet these conditions.

Broadly speaking, the low Rideal Walker coefficient fluids are less affected by the presence of organic matter than the high ones. Thus, Lysol with a Rideal Walker coefficient of 2·5 has a Chick Martin coefficient of about 2·0, but a fluid with a Rideal Walker value of 20 may only have a Chick Martin value of 3 or 4.

Phenol coefficient test—Rideal Walker

The primary purpose of a phenol coefficient test is to compare the efficiency of phenolic coal tar disinfectants against a standard phenol solution. Rideal and Walker (1903) were the first people to standardize the conditions of testing; it included a reference germicide, phenol, and it used a culture of vegetative organisms grown in broth.

The Rideal Walker test

The Rideal Walker test is not only of historical importance, it is used extensively today, and therefore justifies consideration in some detail. In its original

* A solution of 3–5 per cent is recommended for most uses. Lysol is toxic for most animals, especially for cats, dogs, monkeys, and mice.

form, as published by Rideal and Walker (*ibid.*), the test represented the first real attempt at putting the assay of disinfectants on a quantitative basis; in its present form it is a valuable tool in the routine assessment of standard disinfectants. The test is over fifty years old and is today the same in principle as it was originally, although a number of modifications have been made, all with the object of improving the precision of the test. When first produced, it represented an entirely new departure in testing techniques in that it specified standardized cultural and other conditions and, most important, it used a standard substance, pure phenol, against which the disinfectant under examination was compared.

The test specified the species and age of the culture, the culture medium employed, the amount of inoculum used, the temperature and time of medication, the period of incubation of the subcultures, and the resistance of the test organism in terms of the phenol control. The organism used in the original test was the Rawlings strain of *Salm. typhimurium*, chosen because it is a typical representative of the entire group of the intestinal organisms and also because it was believed to possess a degree of constancy in growth characteristics superior to other organisms. For the same reason, Liebig's meat extract and Witte's peptone were specified, on the assumption that they would ensure reproducibility in nutrient properties of the culture medium, and therefore, in the growth characteristics of the organism.

In subsequent years, the authors made various modifications which they later published as an 'approved technique' (Rideal and Walker,[8] 1921). The preface contained the admonition that to avoid discrepancies 'strict observance of the conditions laid down by the authors cannot be too strongly emphasized'. The new technique included modifications to the volumes of disinfectant dilutions and of culture medium employed, the time of medication, the test temperature, and it substituted Witte's peptone by Allen and Hanbury's Eupeptone.

Calculation of the coefficient

The Rideal Walker coefficient is calculated by dividing the dilution of disinfectant which shows life after 2½ and 5 minutes, but not after 7½ and 10 minutes, by the dilution of phenol which shows the same end-point.

A typical test result is as follows:

DISINFECTANT	DILUTION	TIME (MIN) CULTURES EXPOSED TO ACTION OF DISINFECTANT			
		2½	5	7½	10
A	1 in 1,000	–	–	–	–
A	1 in 1,100	+	–	–	–
A	1 in 1,200	+	+	–	–
A	1 in 1,300	+	+	+	–
Control phenol	1 in 105	+	+	–	–

(+ = growth; – = no growth)

Rideal Walker coefficient = $\frac{1200}{105}$ = 11·4 (approximately).

Surface active compounds

Some substances when dissolved in water lower the surface tension and thus increase the 'wetting' capacity. The synthetic, surface-active substances which have this effect may be classified in three groups according to the electric charge (negative, positive, or both negative and positive) of the lipophilic portion of the molecule. They are referred to as anionic, cationic, or ampholytic surface-active compounds. The *anionic* surface-acting compounds possess high detergent, but only very limited bactericidal, properties. Modern synthetic soap-substitutes fall into this class.

The *cationic* surface-active compounds, as represented by the quaternary ammonium compounds, possess limited detergent, but high bactericidal and bacteriostatic properties.

Quaternary ammonium compounds ('quats'), e.g. Cetrimide, Benzalkonium Chloride, Vantoc B., and C. L. Cetavlon

The quaternaries are more active against the gram-positive organism than the gram-negative. Bacteria may readily adapt themselves to become resistant to the quaternaries in the same way as they do to the antibiotics and sulphonamides.

Sporicidal activity

Sykes (1958) reports that in spite of claims to the contrary, the quaternaries cannot be considered to be effectively sporicidal.

Antiviral activity

Sykes (*ibid.*) says that the viruses, in general, are rather more resistant than the bacteria and fungi. Because of this, quaternary ammonium compounds have been used in the preparation of vaccines and suspensions of organisms free from viable bacteria. The concentration of some compounds necessary to inactivate the influenza virus in a few minutes is of the order of 1 in 500 to 1 in 8,000, but not all quaternary compounds show such activity.

Toxicity

At the concentrations used for the purposes of disinfection the quaternaries are virtually non-toxic. At high concentrations they can cause severe skin inflammation and oedema, and when administered in large doses in the diet of animals they have proved fatal.

The uses of quaternaries

The quaternaries have been found most useful in the field of surgery (where they are used principally for pre-operative skin disinfection) and in the food industry (where they are used to disinfect food utensils, drinking glasses, and dairy equipment). The quaternaries are not recommended for sterilization of surgical instruments, because they are not active against bacterial spores.

In the food industry the particular virtues of these compounds are that they are colourless, non-staining, practically tasteless, and rapid in action.

The concentrations generally used range between 1 in 5,000 and 1 in 2,000.

One other advantage claimed for the quaternaries is that their solutions do not become exhausted as rapidly as do solutions of hypochlorites.

Ampholytic surface-active agents, e.g. Tego

Whereas surface-active compounds of the anionic, cationic, and non-ionic types have been rapidly developed in appropriate fields of use, the characteristics of the *ampholytic* surface-active compounds were relatively late to be recognized. *Ampholytic* surface-active compounds, as their name implies, combine the detergent properties of anionic compounds with the bactericidal properties of cationic compounds, and they remain highly active in the presence of protein.

The ampholytic surface-active agents

$$H_2N{\cdot}R{\cdot}COO^- \quad H_3N{\cdot}^+R{\cdot}COOH \quad H_3N{\cdot}^+R{\cdot}COO^-$$

Anion *Cation* *Amphoteric ion*

The Tego series of compounds are nitrogenous compounds of high molecular weight. Perkins and Short[9] (1957) investigated the bactericidal property of *Tego MHG* by two *in vitro* methods. The results of their investigations are shown in Table 1, from which it was clear that 1 per cent *Tego MHG* was an effective bactericide.

TABLE 1. KILLING TIME IN MINUTES OF 1 PER CENT *TEGO MHG* AT 71·6°F (22°C)

ORGANISM	KILLING TIME MINUTES
Staphylococcus pyogenes (*penicillin resistant*)	<1
Staphylococcus aureus	<1
Staphylococcus albus	<1
Staphylococcus citreus	<1
Streptococcus pyogenes	<1
Streptococcus faecalis	<1
Streptococcus viridans	<1
Streptococcus haemolyticus, Group D	<1
Escherichia coli	>1 <2
Salmonella enteritidis	>1 <2
Salmonella typhi	>1 <2
Shigella flexneri Type 4b	>1 <2
Corynebacterium diphtheria	>1 <2
Pseudomonas aeruginosa	>2 <5
Pseudomonas pyocyanea	>2 <5
Proteus vulgaris	>2 <5
Clostridium welchii	>2 <5
Epidermophyton rubrum	>5 <10
Epidermophyton interdigitale	>5 <10
Mycobacterium tuberculosis H 37RV	>20 <30

The second method was a modification of the 'Use Dilution Confirmation Test' in use in the United States (Stuart *et al.*,[10] 1953). The results of the tests are given in Table 2, which shows the bactericidal nature of the surface-active agent.

With this 'Use Dilution Confirmation Test' technique it was found that a spore suspension of *Bacillus cereus* NCTC 9689, which withstood boiling water for 90 minutes was killed by being brought to the boil in 1 per cent *Tego MHG*, a procedure which took, in all, 45 seconds.

Observations on the bactericidal property of *Tego MHG* thus described indicated that the use of the technique should prove effective in the sterilization of animal rooms, racks and cages.

TABLE 2. TIME TAKEN TO STERILIZE INFECTED GLASS CYLINDERS WITH 1 PER CENT *TEGO MHG* AT 96·8°F (37°C)

ORGANISM	TIME TO STERILIZE, MINUTES
Staphylococcus pyogenes (*penicillin resistant*) *NCTC* 8178	<1
**Staphylococcus aureus* (*penicillin resistant*)	<1
**Staphylococcus aureus* (*penicillin resistant*)	<1
**Staphylococcus aureus* (*penicillin resistant*)	<1
Staphylococcus aureus NCTC 7447	<1
Streptococcus pyogenes NCTC 8312	<1
Bacillus cereus (*Vegetative cells*) *NCTC* 9689	>1 <2
Bacterium coli NCTC 8009	>1 <2
Bacterium coli NCTC 86	>1 <2
Salmonella enteritidis NCTC 5694	>1 <2
Shigella flexneri, Type 4*b NCTC* 8336	>1 <2

* These three strains of penicillin-resistant staphylococcus aureus were recently isolated from human infections.

In the control of viruses field studies have been more successful than laboratory work. Reuss, Wagener, Bingel, and Kliewe have shown that regular use of Tego M.H.G. in chicken runs and pigsties prevents the spread of fowl pest, swine fever, and swine influenza, especially when a new litter or batch is introduced into old animal quarters. Perkins, Darlow, and Short put forward the following theories to explain the difference in experience between the laboratory and the field station. In the laboratory cell-culture techniques are extremely sensitive in detecting any viable virus particle, whereas the infection of an animal involves a minimum infection dose, usually many virus particles, reaching a particular site for multiplication. They suggest from the practical standpoint that it is far more meaningful to know that cross infection is prevented rather than a cell culture has detected a few virus particles.

Toxicity

In comparisons with quaternary ammonium compounds the ampholytic compounds are ten times less toxic. For the quats, a fatal dose administered orally to rats is around 0·2–0·5 gm per kilogram of body weight. The corresponding value for the *Tego* compounds is between 3 and 4 gm per kilogram of body weight.

The ampholytic compounds approximate to the pH of the skin, but in practice they are singularly non-irritant, even when in constant use for hand washing by persons with sensitive skins. Frisby[11] (1959) reported that the *Tego* series of compounds were inactivated by soap and synthetic detergents.

Evidence suggests that Tego is sporicidal, and Perkins and Westwood report that Tego is able to kill all mycoplasma.

It is important to have an easy and efficient method of applying a disinfectant to all surfaces of rooms, cages, and racking, with the chemical delivered in the right proportion. Fig. 1 illustrates a portable method and Figs. 2, 3, and 4 a central system for using Tego. All these machines automatically dispense 1 per cent of the disinfectant.

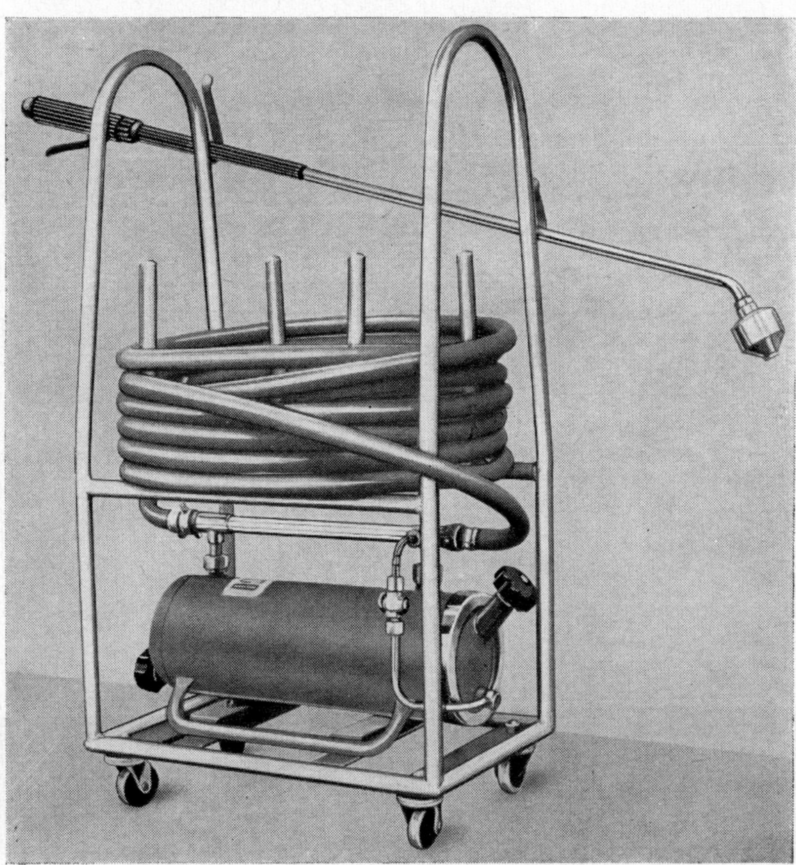

FIG. 1

HALOGENS

This group consists of the four elements fluorine, chlorine, bromine, and iodine, which have closely related and graded chemical and physical properties.

Chlorine

During the early part of the last century chlorine was found useful for disinfecting water and treating sewage, purposes for which it is still used today.

Chlorine is much used for the disinfection of dairy equipment, it has application in medicine for the treatment of wounds, in the domestic sphere for personal hygiene, and in the animal house it has many uses.

Hypochlorites

The hypochlorites were the first compounds of active chlorine to be made. Calcium hypochlorite is made by allowing chlorine gas to react with slaked

FIG. 2

lime in either the solid state or in an aqueous suspension. The trade name for this is lime bleach liquor. When freshly prepared chlorinated lime (bleaching powder) may contain up to 39 per cent of available chlorine, but it is unstable under normal storage conditions, and is most sensitive to moisture and heat. Chlorinated Lime B.P. contains not less than 30 per cent of available chlorine. Sodium hypochlorite is made for a variety of industrial purposes in solutions containing up to 20 per cent of available chlorine, but other preparations are sold under numerous trade names for general disinfection purposes and personal hygiene.

The disinfecting action of chlorine

The intense reactiveness of chlorine with organic compounds generally is undoubtedly the reason for its being a rapid and effective germicide, even at quite high dilutions. The hypochlorites are the most reactive, therefore they

FIG. 3

are the most rapid in their germicidal action. Chlorine is a bactericidal agent and possesses little, if any, bacteriostatic activity. The lethal action is the result of the direct action of the chlorine on some vital part of the cell, such as its protoplasm or an enzyme system. Chlorine and its compounds are much

K

more effective in acid solution than alkaline solution; this effect is illustrated in the following table by Charlton and Levine[12] (1935).

TABLE 3. THE EFFECT OF pH ON LETHAL ACTION OF HYPOCHLORITE AGAINST *B. METIENS* SPORES

Available chlorine, p.p.m.	pH	Time to kill 99 per cent of the spores
1,000	11·3	70 minutes
100	10·4	64 minutes
20	8·2	5 minutes
1,000	7·3	Less than 20 seconds

Effect of temperature

Studies on the effect of temperature on the bactericidal activities of chlorine compounds show that there is always an increase in activity with an increase in temperature. It has been found that about twice the concentration of hypo-

FIG. 4

chlorite was required to kill *Salm. typhimurium* at 35·6°F (2°C) than at 104°F (40°C).

Effect of organic matter

The bactericidal efficiency of chlorine falls in the presence of organic matter, and even quite small amounts of such material can exert a significant effect. When treating animal cages and cage trays it is important to see that all organic matter is removed before using chlorine or its compounds.

Iodine

Iodine is an element with a high chemical reactivity, and it is this reactivity which makes it an effective germicide. Among its outstanding characteristics are: (i) its lack of selectivity against different bacteria, all types being killed at about the same level of concentration; (ii) its exclusive bactericidal rather than bacteriostatic action.

The reactivity of iodine is similar to that of chlorine, but unlike chlorine, its disinfecting action is the result of the direct intervention of free iodine molecules which combine with the protein substances of the cell.

Iodine is effective against bacterial spores, but organic matter has a depressant effect on the activity of iodine.

Iodophors

The iodophors are a comparatively new development in the application of iodine for disinfection purposes, and have proved particularly useful in dairy hygiene. Strictly speaking, they are not compounds of iodine, but simply mixtures of iodine with surface-acting agents of all types which act as vehicles for the iodine.

The solutions have all the characteristics of iodine as a germicide, but display a complete lack of odour and of staining power and low irritant properties. Cationic, anionic, and non-ionic detergents can be used with equal success, but generally the more stable preparations are found with the non-ionic group.

Some of the latest iodine compounds have 'built-in' colour molecules which show whether or not the disinfectant is still active.

GASES AND VAPOURS

The sterilization of equipment, bedding, and food is usually thought of in terms of heat treatment, but at least one other method of sterilizing is available, i.e. by bactericidal gases or vapours. This treatment can, of course, effect only surface sterilization, but in dealing with solids this is practically all that is necessary. When the material to be sterilized is porous, or when it is in fine crystalline or powder form, most gases can be used for sterilizing.

The particular advantages of this form of treatment are that it can be carried out at normal or only slightly elevated temperatures, and the gas can be removed completely from the treated material after sterilizing. One important thing to be considered apart from the bactericidal efficiency of such agents is the reactivity of the chosen gas or vapour. No gaseous disinfectant is chemically inert—it would not be germicidal if it were so—but certain of

TABLE 4. COMMON

CLASS OR TYPE	INDUSTRIAL NAME (U.K.)	MANUFACTURED BY	CONCENTRATION NORMALLY USED
Halogens			
Chlorine	—	—	—
Iodine	—	—	—
Hypochlorites	Chloros	ICI	1–5%
	Sod. Hypochlorite (10% ww Chlorine)	Hopkin and Williams	3–5 (liquid)
Phenols and related compounds			
Cresol	Lysol B.P.	Prince Regent Tar Co. Ltd	Not exceeding 5%
Cresol and chlor-xylenol	Jeypine and other brands	Jeyes	Not exceeding 5%
Quaternary ammonium compounds			
	Nonidet 32G	Shell	0·5%
	Cetavlon	ICI	1·0%
	Cetrimide B.B.	ICI	1·0%
Ampholytic compounds			
	Tego MHG	Hough, Hoseason & Co. Ltd., Manchester	1%

them, such as sulphur dioxide and chlorine, are much too reactive for practical consideration.

Of the substances which are more acceptable, formaldehyde, ethylene oxide, and beta-propiolactone (B.P.L.) have received most attention. But even these have their limitations; they are all toxic to humans above certain concentration and they exhibit other unpleasant or undesirable side effects.

Formaldehyde

Under optimum conditions formaldehyde is lethal to bacteria, bacterial spores, viruses, fungi, and yeasts, and has been employed as a fumigant with varying degrees of success for the past fifty years.

CHEMICAL DISINFECTANTS

EFFECTIVE AGAINST	NOTES
Most bacteria	Quick acting
Most bacteria	All bacteria killed at same level of concentration
Most bacteria and viruses	Cheap; effective, somewhat corrosive to metal. Neutralized by soil, faeces, etc.
Most bacteria and viruses	Cheap, effective, somewhat corrosive to metal. Neutralized by soil, faeces, etc.
Most bacteria, including *Myco tuberculosis*, not spores or *Ps. pyocyaneus*	50 per cent cresol in soap solution. Good general disinfectant
Most bacteria, including *Myco tuberculosis*, not spores or *Ps. pyocyaneus*	General disinfectant. Compatible with soap
Most bacteria, including *Myco tuberculosis*, not spores or *Ps. pyocyaneus*	A bactericidal detergent for general use. Not to be mixed with other materials, e.g. soap
Most bacteria and viruses, including *Myco tuberculosis*, not spores or *Ps. pyocyaneus*	A bactericidal detergent for general use. Not to be mixed with other materials, e.g. soap. Low toxicity; useful for water bottles
Most bacteria and viruses, including *Myco tuberculosis*, not spores or *Ps. pyocyaneus*	A bactericidal detergent for general use. Not to be mixed with other materials, e.g. soap. Low toxicity; useful for water bottles
Most bacteria and some viruses	One of the most useful of the modern disinfectants. Low toxicity to animals

Formaldehyde has the advantage of being cheap, easy to handle, not injurious to fabrics, paints, or metals, and of being rapidly neutralized by ammonia. Its pungent and irritating odour gives adequate warning of its presence at concentrations (5 p.p.m.) too low to be toxic.

Formaldehyde gas can be generated by heating a solution of Formaldehyde B.P. (popularly known as formalin), which is a 37 per cent solution of formaldehyde in water stabilized by the addition of a small amount of methyl alcohol.

Another method employed is to mix about two parts of formalin with one part of potassium permanganate and allow the heat generated by the oxidative reaction to volatilize the remainder of the formaldehyde gas. The bactericidal efficiency of formaldehyde vapour is a direct function of the concentration, relative humidity, and temperature.

Concentration

Sykes (1958) found that bacterial cultures dried on cotton threads required several hours in a gaseous concentration of 1 or even 2 mg of formaldehyde per litre of air before a complete kill was obtained. Nordgren[13] (1939), using bacteria dried on metal and glass, found that organisms died in an atmosphere containing 1 mg of formaldehyde per litre of air within 20–50 minutes. Glick, Gremillion, and Bodmer[14] (1959) advise the vaporization of 1 ml of formalin or formalin–methanol solution for each cubic foot of space in treating rooms or buildings.

Temperature

An increase in the temperature influences disinfection in three ways: by increasing the amount of vapour, by reducing the loss of formaldehyde due to polymerization (the union of two or more molecules of the same compound to form larger molecules), and by reducing adsorption on to fabrics if they are present. A temperature of not less than 64°F (18°C) is recommended.

Relative humidity

At low humidities disinfection is slow, the rate of killing increases with rising humidity, reaching a maximum at 80–90 per cent. Above this level efficiency falls sharply, particularly if the objects are grossly wet.

Ethylene oxide

Kelsey[15] (1959) reports that experience has shown that some of the early claims made for ethylene oxide as the perfect sterilizing agent were not justified. Attempts had been made to define a safe process in terms of concentration, time, temperature, and humidity analogous to accepted standards for steam sterilization, but none could be accepted with complete confidence. Glick, Gremillion, and Bodmer report the extensive use of ethylene oxide in the form of a low pressure mixture with two of the chloro-fluorohydrocarbons (Freons) in a disposable can. Any steam autoclave can be inexpensively converted for applying the ethylene oxide–freons mixture without interfering with the conventional use of the autoclave. This conversion enables complete mechanical equipment and heat-labile substances (substances prone to undergo changes with heat) to be sterilized using the same piece of equipment.

Chemically, ethylene oxide is a powerful alkylating agent, and its bactericidal effect is probably due to this action. It is lethal to bacteria and viruses. It forms an explosive mixture when more than 3 per cent of it is present in air. Its toxicity may be compared with that of ammonia.

If mixed with carbon dioxide or freons in proportions of not more than 12 per cent by weight the mixture is not inflammable (Kelsey, *ibid.*).

A definite drawback to the use of ethylene oxide is the required exposure time. In concentrations practical for use (11 gm per cubic foot of air) a minimum of 6 hours is required to sterilize materials contaminated with bacterial spores.

Ethylene oxide does not readily penetrate thick layers of dirt, grease, or oil. It is active over a wide range of humidities.

Beta-propiolactone

The use of beta-propiolactone (B.P.L.) as a vapour disinfectant is relatively new. However, it has been found to be effective in the decontamination of rooms, buildings, and closed chambers in which airflow can be held to a minimum. The temperature and humidity requirement are the same as for formaldehyde.

Kelsey (personal communication, 1962) writes that beta-propiolactone is a carcinogen and, as such, this substance should be treated with the utmost respect.

Beta-propiolactone has the following advantages over formaldehyde as a vapour disinfectant:

1. It does not polymerize as readily on surfaces, so there is little or no residue.
2. It acts more rapidly. However, in the liquid state B.P.L. is more toxic than formaldehyde, and in handling care must be taken to see that it does not contact the skin.

Glick, Gremillion, and Bodner (*ibid.*) report that a vaporizer suitable for the dissemination of B.P.L. and also formaldehyde has been used successfully, 1 gallon of the lactone being vaporized for each 12,000 cu ft of space.

As with formaldehyde, the room or building need not be hermetically sealed and, after a waiting period of 2–3 hours, doors and windows can be opened and forced ventilation resumed. At this point entry into treated areas should be made with protective clothing and respiratory protection. After good airing for a further 3 hours normal entry can be made.

REFERENCES

1. GLICK, C. A., GREMILLION, C. G., and BODMER, G. A., *Proceedings of Animal Care Panel* (February 1961).
2. MONTHLY BULLETIN M.O.H. and P.H.L.S., *The Sterilization of Animal Cages by Steam*, **47**, 16, 64.
3. WENTWORTH CUMMING, C. N., 'The Commercial Production of Rats under S.P.F. Conditions', *Journal of Animal Technicians Association*, **12**, No. 4 (March 1962).
4. DARMADY, E. M. and BROCK, R. P., *Journal of Clinical Pathology*, **7**, 29 (1954).
5. SYKES, G., *Disinfection and Sterilization*. Spon (1958).
6. DARMADY, E. M., HUGHES, K. E. A., BURT, M. M., FREEMAN, B. M., and POWELL, D. B., 'Radiation Sterilization', *Journal of Clinical Pathology*, **14**, 55 (1961).
7. RIDEAL, S. and WALKER, J. T. A., *Journal of the Royal Sanitary Institute*, **24**, 424 (1903).
8. RIDEAL, S. and WALKER, J. T. A., *An Approved Technique of the Rideal Walker Test*. H. K. Lewis, London.
9. PERKINS, F. T. and SHORT, D. J., 'A New Technique in the Sterilization of Animal Houses, Racking, and Cages', *Journal of Animal Technicians Association*, **8**, No. 1 (1957).

10. STUART, ORTENZIS, and FRIEDL, *Journal of the Association of Agricultural Chemists*, Washington, **36**, 466 (1953).
11. FRISBY, B. R., 'Tego Compounds in Hospital Practice', *Lancet*, pp. 57–58 (July 1959).
12. CHARLTON, D. B. and LEVINE, M., *Journal of Bacteriology*, **30**, 163 (1935).
13. NORDGREN, G., *Acta Pathologica Microbiologica Scandinavica*, Suppl. 40 (1959).
14. GLICK, C. A., GREMILLION, C. G., and BODMER, G. A., *Proceedings of Animal Care Panel* (February 1961).
15. KELSEY, J. C., 'Sterilization by Ethylene Oxide', *Journal of Clinical Pathology*, **14**, 59 (1961).
16. REUSS, U. Berl. u. Munch. *Tierartzliche Wochenschrift*, **24** (1954).
17. WAGENER, K. Special Report, 5th January 1962.
18. BINGEL, K. F. Special Report, 14th February 1962.
19. KLIEWE, H., Special Report, 4th May 1959.
20. PERKINS, F. T., DARLOW, H. M. and SHORT, D. J., *J. Inst. Anim. Techn.* **18**, No. 2 (1967).

CHAPTER TEN

Transport of Laboratory Animals

PART ONE

THE CARRIAGE OF LIVE ANIMALS

Animals are, on the whole, adaptable and travel well, providing they are properly packed and protected from extremes of heat and cold. The commonest accidents during transit are suffocation of the animals because the ventilation holes in crates become blocked, overheating, the crushing of fragile containers, and escape.

Travelling containers

Nocturnal animals prefer darkness, and all species deserve privacy. Therefore an all-wire cage is unsuitable unless it can contain enough bedding for the animals to cover themselves completely.

Well-designed and reinforced cardboard boxes are not easily crushed, and may be used for sending mice, rats, rabbits, and guinea pigs on journeys of not more than 12 hours duration within this country.

Metal boxes offer poor insulation against extremes of temperature unless they are well filled with bedding. Metal boxes are commonly used for air transport, as their light weight keeps down freight costs.

Wooden (or cardboard lined with wire mesh or metal) boxes may be used for rodents destined for long journeys. Such containers are also acceptable to airlines. Unlined cardboard boxes are not acceptable.

Ventilation

At least 10 per cent of the top and the same proportion of the sides of the container (or the equivalent on two or more sides) should be for ventilation by means of wire-mesh windows. The mesh prevents the animals from escaping and protects handlers from bites. Spacing bars fitted to the top and sides of the box prevent the ventilation areas being blocked during stowage. The risk of blocked ventilation panels is minimized if the containers are drum-shaped or are rectangular with inclined sides.

Food and water

All animals should have access to food and water as late as possible before dispatch.

FIG. 1

Top

Left: Collapsible cardboard rabbit box—18 in. × 10 in. × 10 in. ($14\frac{1}{2}$ in. with handle) high—and light-weight, no-nail, wooden rat or mouse box—$12\frac{1}{2}$ in. × $5\frac{1}{2}$ in. × 5 in. high.

Right: Cardboard boxes for mice—12 in. × 5 in. × 5 in. high and 18 in. × 11 in. × $5\frac{1}{2}$ in. high—and for guinea-pigs—24 in. × 14 in × $8\frac{1}{2}$ in. high.

Below

Stapled cardboard box suitable for rabbit, in use on the Continent. (Note sloped sides to keep air vents free.)

Food in the form of grains, pellets, or cubes, sufficient for the journey, should be placed in the travelling box.

Moisture must be provided, but preferably not in the form of a wet mash, which quickly becomes soiled and unpalatable. Rats, mice, hamsters, and guinea pigs will travel well if moisture is supplied by succulents, e.g. apples, beetroot, raw potatoes, raw carrots, or grapes; in this case water bottles or other containers are unnecessary.

Water bottles invariably leak because of jolting in transit. Water bottles are suitable for use only during resting stages in the journey, and they should be provided by the sender, together with full instructions for their use.

However, for journeys of not more than five days the majority of rodents will obtain all their water requirements from succulents packed with them, and the need for watering will not arise. If the consignor wishes his animals to be watered on the journey he must provide a suitable utensil—bottle, dish, etc.—*and full instructions on the container*. Otherwise the label should state specifically that no food/water is required and should not be given.

Bedding

The floors of all types of containers for rodents must be covered with fine softwood sawdust, granulated peat moss, or fine silver sand to a depth of 1 in.

Rabbits, guinea pigs, and hamsters should be provided with a liberal supply of hay or shredded paper. Travelling boxes containing rodents should be almost filled with bedding through which the animals can burrow and remain dispersed throughout the box.

Adequate bedding must be provided for the following reasons:

1. For the comfort of the animal.
2. To absorb moisture.
3. To protect against shock, which is unavoidable even with the best handling.
4. To insulate against temperature fluctuations.
5. To keep the animals dispersed throughout the container.

Labels

All boxes must be well and clearly marked with the following information:

(*a*) Name, address, and telephone number of the consignor.
(*b*) Name, address, and telephone number of the consignee.
(*c*) The word LIVESTOCK in bold red letters.
(*d*) Label—'Box contains rodents: *Please do not open*'.
(*e*) Instructions for feeding and watering, if necessary, or instructions 'Do *not* feed or water'.
(*f*) Instructions on the care of the animals in the event of delays *en route*.
(*g*) Label marked 'To be called for'.

Birds

LIVE POULTRY. The British Transport Commission rigidly enforces the following conditions for the packing and care of live poultry:

1. Crates must be substantially constructed and of a height and size reasonably sufficient for the type and number of birds. Maximum floor area of the crates must not exceed 24 sq ft, and any one compartment must not exceed 10 sq ft. The crates must be constructed in a manner which will prevent head, legs, and wings protruding through the top, bottom, or sides. Baskets covered with wire netting are permitted under the above conditions. Flimsy cardboard boxes are not acceptable, but the recognized pullet boxes are. If more than one species of bird is being dispatched in one crate the species must be in separate compartments.

2. If sent by passenger train delivery must be ensured within 30 hours.

3. *Feeding and watering.* Birds will suffer no hardship if they are deprived of food for up to 48 hours. Water must be given at intervals not exceeding 12 hours. Instructions to this effect should be printed on the crate. *Birds will not eat or drink in the dark.*

4. *Day-old chicks* should be packed in recognized day-old-chick containers. Chicks generate much heat, therefore adequate ventilation is necessary. However, they must not be left in draughts. They are liable to suffocate if the packages are stacked too high or too tightly. They can survive for 48 hours without food or water.

During cold weather transport should be in heated and well-ventilated vans.

Cats

Wicker hampers and wooden crates serve well as containers for cats. Crates should have solid floors, and the sides and top should be of 2-in. slats, $\frac{1}{2}$ in. thick and set $\frac{1}{2}$ in. apart. These crates may be divided into three separate compartments by using two movable wooden partitions. Drinking-vessels should be securely fixed in each compartment. Dimensions of crate or hamper 3 ft $\times$ 16 in. $\times$ 1 ft 3 in. high, as recommended by the RSPCA.

Dogs

Dogs may travel loose if they are provided with stout collars and chains and are securely muzzled. The labels must be attached to the collars beyond the reach of the animals.

Dogs may also be sent in well-ventilated crates with solid floors of the type described for cats. Drinking-vessels must be provided, because dogs must be watered every few hours.

Ferrets

Ferrets must travel in strong boxes or hampers from which they cannot escape. The lids must be firmly secured to prevent opening *en route*. Adequate ventilation must be provided. Ferrets will not be accepted for transport in Britain if the journey is to exceed 12 hours.

Frogs and Toads

A cage intended for the carriage of frogs and toads should be constructed of wood, hardboard, or plywood. The cage or box should be well ventilated and all ventilation holes covered by wire gauze or perforated metal sheet. The number per box or compartment should be kept to a minimum, because these animals pile on top of each other. Bedding: damp, rough, leaf-mould, in which the animals can bury themselves.

Goats

Goats may travel loose, under the same conditions as dogs. Kids should be sent in crates.

Guinea pigs

For short journeys stiff cardboard boxes (reinforced, if necessary) are suitable. These should have solid floors and tops and ventilation holes, 1 in. in diameter, on all sides.

For long journeys boxes should be constructed of hardboard, on a framework of 1-in. square timber, with solid floors and lids. The lids should be hinged and secured by a catch or a spring-loaded hook. Ventilation should be by 1-in. diameter holes on all sides, and a bar should be fixed around the box to prevent blocking of the ventilation holes; light metal boxes may also be used, but not in hot climates.

Hamsters

These may be packed in strong wooden boxes or metal tins with adequate ventilation (as for guinea pigs). The floors and lids should be solid.

Mice

For short journeys, stiff cardboard boxes with perforated metal windows inserted on all sides for ventilation are satisfactory.

For long journeys (over 6 hours) light timber boxes are preferable. The whole of one side of the box should be of perforated metal or wire scrim. Light metal boxes (e.g. half-size biscuit tins) with adequate ventilation holes on all sides are ideal for sending small numbers of mice by air.

Rabbits

Boxes should be constructed on a framework of 1-in. square wood. The floor, lid, and lower third of every side should be of hardboard or plywood, the remainder being covered with wire netting. The lid must be hinged and secured by a catch or lock-nut.

Rats

Pack as mice, but consult density chart.

Reptiles

Reptiles travel most comfortably at a temperature not lower than 60°F (15°C), and they should not be exposed to extremes of temperature. Boxes should be constructed of wood or plywood, and the ventilation holes should be covered with strong gauze or perforated metal sheet.

Snakes should be packed singly in strong jute cloth bags. These bags should be slung inside a well-constructed box in such a way that the bags cannot touch each other or the sides of the box.

FIG. 2. An efficient cardboard travelling box for rabbits for journeys up to 12 hours

Each bag should be clearly labelled, stating the species it contains and whether or not the snake is venomous.

DENSITY CHART

Species	Weight of animal	Maximum number per compartment	Sq in. per animal (sq cm)	Height of box, in. (cm)
Guinea pigs	170–280 gm	12	14 (90)	6 (15)
	280–420 gm	12	25 (160)	6 (15)
	over 420 gm	12	36 (230)	6 (15)
Hamsters	Young	12	5 (32)	5 (13)
Mice	15–20 gm	25	3 (20)	5 (13)
	20–35 gm	25	4 (26)	5 (13)
Rabbits	Under 2·5 kg	4	120 (770)	8 (20)
	2·5 kg	2	150–180 (970–1160)	10 (25)
	Over 5·0 kg	1	220 (1400)	12 (30)
Rats	35–50 gm	25	6 (40)	5 (13)
	50–150 gm	25	8 (52)	5 (13)
	Adult	12	16 (100)	5 (13)

For further information on the transport of reptiles see British Standards *Recommendations for the Carriage of Live Animals by Air*, Part 6.

EXPORT OF ANIMALS

The following documents are the minimum which will be required when exporting animals. Individual authorities (e.g. Government Departments) and countries may have various additional requirements which must be met and which will be specified when the order for livestock is placed.

1. Clearance certificates must be obtained from the consignee stating that the livestock will be admitted subject to local regulations.

2. Veterinary certificates of health must be provided in triplicate, each signed by a veterinary surgeon whose signature is acceptable to the Government of the receiving country. One copy of the veterinary certificate (clearly marked) must be affixed in a prominent position on the travelling box; two copies must be handed to the shipping agent or airport official. The certificates must have been signed within the 24 hours preceding the time of departure from the transport terminus. The following wording is recommended:

> I hereby certify that I have today examined (number and specie of animals) which were bred at (address) and have found them to be free from any visible sign of disease.
>
> (*Signed*) A. N. Other, M.R.C.V.S.
>
> (*Date*)

Transport arrangements should be made with one of the recognized shipping agents or with the commercial attaché of the receiving country. Livestock should be delivered to the port or the agent's depot. The recipient must be informed of the flight number, times of departure and arrival at destination (in G.M.T.) or such similar information as is appropriate, so that arrangements may be made for the reception of the animals. Air cargoes require a Waybill and three copies of the veterinary certificate. The Waybill may be completed at the departure airport (or air terminus), and the information required includes: the number of packages, and—of each package—method of packing (wooden box), nature and quantity of goods (13 male hooded rats), marks and numbers (box No. 1), dimensions or volume (18 in. diameter by 8 in. high), gross weight, full names and addresses of consignor and consignee.

Cargo space on specified flights should be booked in advance with the freight department of the airline. Choose a flight which is speedy (has few and brief intermediate stops) and arrives at a reasonable time of day from the recipient's point of view; thus the animals will suffer the minimum inconvenience. Animals should travel overnight to destinations in hot climates so that they arrive in the cool of early morning.

Airlines will not accept animals packed in cardboard boxes.

GENERAL CARE DURING TRAVEL, AT LOADING, AT INTERMEDIATE STOPS AND AT UNLOADING

1. Animals should be kept under cover at all times and should be protected from sudden and drastic changes of temperature, strong draughts, rain, strong sunlight, noise, and rough handling.

2. Animals should not be left on railway platforms, airport aprons, or in general merchandise goods sheds, but should be removed to a warm room (or a cool room, in the tropics). Animals should not be placed against hot pipes, open fires, or radiators.

3. Loading and unloading should be quick and efficient, and be done immediately the animals arrive.

Customs authorities and transport personnel are usually very helpful about livestock and will, if asked, give clearance priority to consignments of animals and permit the consignee's vehicle on to the airport tarmac to collect animals.

4. Livestock which suffers unexpected delay in the course of a journey (e.g. through the grounding of aircraft) should be handed over to a person competent to feed, water, and repack, if necessary, according to instructions on the box. Instructions relating to the consignment must be strictly adhered to, because some animals may bite or be dangerous to handle, and others will endeavour to escape.

The consignee should be fully informed of any delay and the new time of arrival.

Animals should not be reshipped before notifying a qualified person, who should either inspect them or give instructions for their care.

Summary

The successful transport of livestock is dependent on the following:

The consignor packing the animals well.

The consignee collecting the animals immediately on their arrival.

The transport personnel treating all livestock as urgent and valuable cargo.

APPENDIX A

Medical Research Council recommendations on the humane shipment of monkeys by air

I GENERAL REQUIREMENTS

1. The duration of time in transit should be as short as possible.
2. Factors causing stress to monkeys should be reduced as much as possible.
3. Monkeys under six months of age should not be transported by air (a guide is given in Appendix B on estimating the age of monkeys by weight).
4. Monkeys in the same cage should be approximately the same weight.
5. Pregnant monkeys should not be shipped except when specifically requested by the importer.
6. Pregnant monkeys and monkeys over 10 lb (4·5 kg) in weight should be shipped in cages specially approved for this purpose.
7. When possible, only monkeys of the same species should be shipped in the same aircraft. It is most desirable that no other species of animal or bird are carried at the same time.
8. At no time during transit should monkeys be left unattended. At least one animal handler should be present at all times when the plane is on the ground.

II TRAVELLING CAGES

1. The type of travelling cage recommended and the containers for food and water are shown in the diagram in Appendix A.
2. A cage of the type shown in the Appendix must not contain more than twelve monkeys.
3. The total weight of monkeys in one cage must not exceed 50 lb (23 kg).
4. No individual monkey in this cage shall weigh more than 10 lb (4·5 kg).

III CERTIFICATE OF FITNESS

A certificate of fitness, on the prescribed form (see Appendix C), must accompany each consignment of monkeys. In the absence of such a certificate the captain of the aircraft will not accept the consignment for shipment.

The certificate must be signed by a person whose qualifications and experience are acceptable to the Government of the exporting country, to the shippers, and to the consignee.

IV CARE IN FLIGHT

1. Ventilation, temperature, and light

(*a*) A minimum of twelve changes of air per hour is required.

(*b*) Draughts are to be avoided.

(*c*) There should be no 'dead' pockets of air.

(*d*) The temperature considered optimum is 75°F (23·9°C) (sixteen air changes per hour). Maximum 80°F (26·7°C) (twenty air changes per hour). Minimum 65°F (18·3°C) (twelve air changes per hour). The variation in temperature should be no greater than 1°F (0·6°C) every 5 minutes, and extremes of temperature should be avoided.

(*e*) Humidity should be kept as low as possible during shipment.

(*f*) Except when monkeys are being fed and watered, it is better that they should travel in semi-darkness. This will make the monkeys quieter and less inclined to fight and give them better opportunities for resting.

2. Food (Food container—see Appendix A)

A monkey can comfortably go without food for 24 hours if it has been fed regularly during the quarantine period before shipment.

If the period from the scheduled time of departure is longer than 24 hours the animals must be fed at the expiration of this period and thereafter at intervals of 12 hours. A sufficient stock of food must be available on the plane and at likely stopping-places for this purpose. Three ounces (85 gm) of food per monkey is required daily. The food should consist of dry cereal grain or gram. Fruit and vegetables should not be offered immediately before or during shipment, as experience has shown that this is liable to upset the monkeys.

3. Water (Water container—see Appendix A)

Monkeys must be watered not less than every 6 hours. The water supplied must be fit for human consumption and piped from tanks within the aircraft. Manifold taps from a pipe running along the roof of the aircraft must be suitably placed (e.g. at 10-ft intervals) so that a movable tap can be used to fill the water containers in the cages.

A minimum of ¼ pint (142 ml) of water must be allowed for each monkey daily, i.e. 50 gallons (approximately 227 litres) for 1,600 monkeys.

4. Sickness and injury of monkeys

Wherever feasible, injured and dead monkeys should be removed. Injured monkeys should be put into a spare cage kept for this purpose and dead ones into impervious disposable bags. Tongs should be provided for handling dead monkeys. Sick animals should not be removed from their travelling cages.

Experience has shown that the application of dressings and administration of medicines to individual monkeys during flight is of little value.

V HYGIENE

It is desirable that any aircraft carrying monkeys be kept as tidy as possible during the flight. Cage trays must be cleaned not less than once every 24 hours and the refuse put into the impervious disposable bags which must be provided.

VI LOADING, INTERMEDIATE STOPS, AND UNLOADING

The conditions required during the flight are required also after the aircraft has landed at intermediate stops and at destination. In addition, it is considered necessary that:

1. The monkeys should be under cover from the time of leaving the collecting unit.
2. Loading and unloading should be carried out quickly and efficiently. A roller conveyor with a canopy could be used for this purpose.
3. The stowing and lashing of cages should be supervised by the senior animal handler.
4. Every effort should be made to avoid subjecting the monkeys to extremes of temperature, draughts, etc.
5. On arrival at destination the Customs authorities should be asked to give clearance priority to shipments of monkeys.
6. Airport authorities should allow the consignee's vehicles on to the tarmac to collect monkeys on arrival.

VII ANIMAL HANDLERS

There should be an approved list of animal handlers. Facilities can be offered by the importing organization for training handlers in the proper care of monkeys. Every approved handler will sign a statement certifying that he has received a copy of the M.R.C. recommendations and has read them.

Medical attention

Prior to approval, animal handlers will be required to produce evidence of physical fitness and good health, including freedom from active tuberculosis. All animal handlers should receive six-monthly chest X-ray examination, and tuberculin-negative personnel should be retested for tuberculin conversion at six-monthly intervals. In the absence of anti-tuberculosis vaccination, tuberculin convertees will be laid off duty until further evidence of freedom from active tuberculosis has been produced. It is advisable that protection against tetanus and the enteric group of fevers should be offered to all animal handlers as a routine.

Note: 'Carriage of Live Animals by Air', 'Monkeys for Laboratory Use', **B.S. 3149:** Part 1, 1959 (Available from British Standards Institution, 2 Park Street, London, W.1. Price 4*s*.).

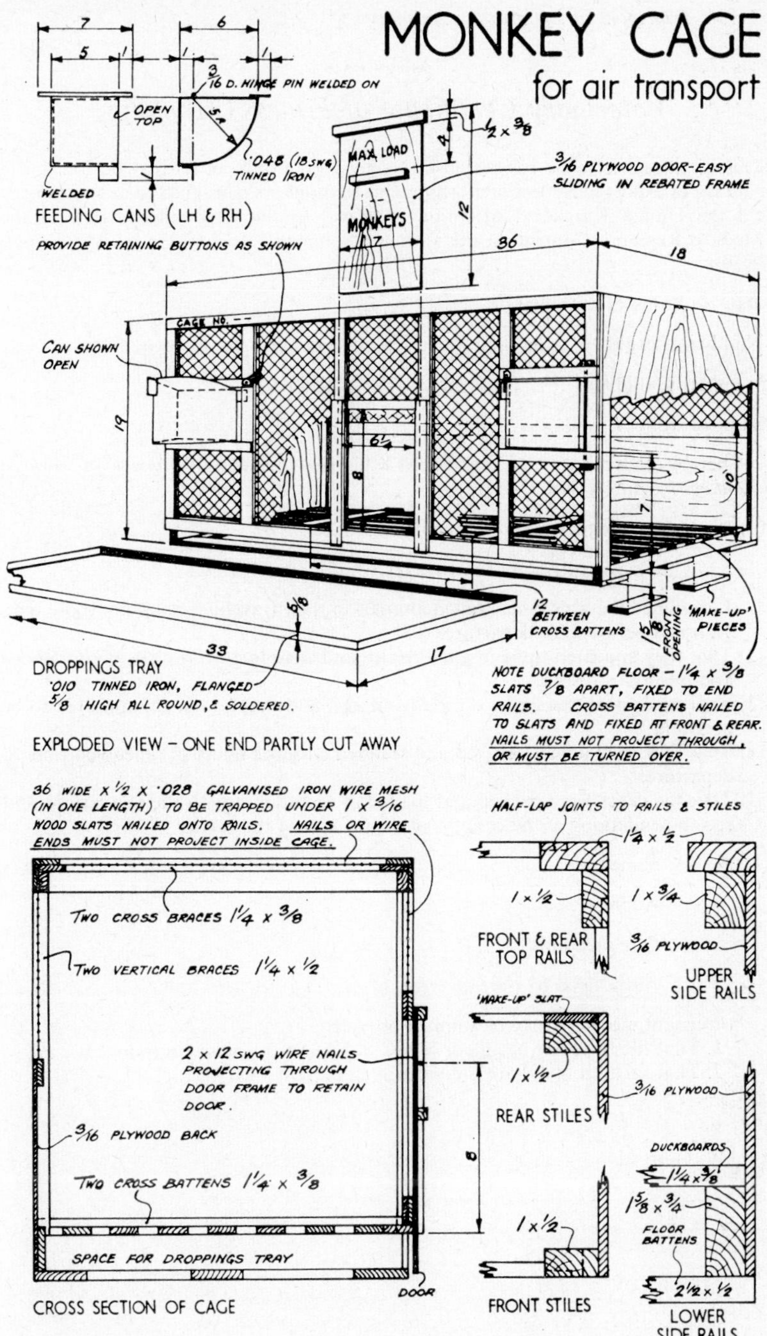
MONKEY CAGE
for air transport
FEEDING CANS (LH & RH)
3/16 D. HINGE PIN WELDED ON
OPEN TOP
WELDED
·048 (18 SWG) TINNED IRON
PROVIDE RETAINING BUTTONS AS SHOWN
MAX LOAD
MONKEYS
1/2 x 3/8
3/16 PLYWOOD DOOR-EASY SLIDING IN REBATED FRAME
CAGE NO. --
CAN SHOWN OPEN
12 BETWEEN CROSS BATTENS
FRONT OPENING
'MAKE-UP' PIECES
DROPPINGS TRAY
·010 TINNED IRON, FLANGED 5/8 HIGH ALL ROUND, & SOLDERED.
NOTE DUCKBOARD FLOOR – 1 1/4 x 3/8 SLATS 7/8 APART, FIXED TO END RAILS. CROSS BATTENS NAILED TO SLATS AND FIXED AT FRONT & REAR. NAILS MUST NOT PROJECT THROUGH, OR MUST BE TURNED OVER.
EXPLODED VIEW – ONE END PARTLY CUT AWAY
36 WIDE x 1/2 x ·028 GALVANISED IRON WIRE MESH (IN ONE LENGTH) TO BE TRAPPED UNDER 1 x 3/16 WOOD SLATS NAILED ONTO RAILS. NAILS OR WIRE ENDS MUST NOT PROJECT INSIDE CAGE.
TWO CROSS BRACES 1 1/4 x 3/8
TWO VERTICAL BRACES 1 1/4 x 1/2
2 x 12 SWG WIRE NAILS PROJECTING THROUGH DOOR FRAME TO RETAIN DOOR.
3/16 PLYWOOD BACK
TWO CROSS BATTENS 1 1/4 x 3/8
SPACE FOR DROPPINGS TRAY
DOOR
CROSS SECTION OF CAGE
HALF-LAP JOINTS TO RAILS & STILES
1 1/4 x 1/2
1 x 1/2
1 x 3/4
FRONT & REAR TOP RAILS
3/16 PLYWOOD
UPPER SIDE RAILS
'MAKE-UP' SLAT
1 x 1/2
REAR STILES
3/16 PLYWOOD
DUCKBOARDS
1 1/4 x 3/8
1 5/8 x 3/4
FLOOR BATTENS
1 x 1/2
2 1/2 x 1/2
FRONT STILES
LOWER SIDE RAILS

FIG. 3

APPENDIX B

Veterinary Certificate of Fitness (Monkeys)

This certificate is to be completed and signed by a duly authorized person in the country of export in respect of all shipments of monkeys consigned to or transported via the United Kingdom. Attention is drawn to the Recommendations of the Medical Research Council on the Humane Shipment of Monkeys by Air (revised 1959).

Date and time of examination ....

Number of cages .... Species ....

Number of animals .... Species ....

Shipped from .... To ....Via ....

I hereby certify that I have read the M.R.C. recommendation *Humane Shipment of Monkeys by Air* (revised 1959).

(1) That, at the request of (consignor) ....

....

I examined the above-mentioned animals in their travelling cages not more than 12 hours before their departure.

(2) That they appeared to be in good health and were free from signs of contagious and infectious disease.

(3) That no animal was under 6 months of age, and that no animal appeared to be pregnant.*

(4) That they were adequately fed and watered within 2 hours of scheduled time of departure.

(5) That the cages were packed and loaded in accordance with the regulations and recommendations of the country of export and of the carrier.

Signed ....

Address ....

Qualifications ....

Authorized by the Government of ....

* Pregnant monkeys may be shipped provided:

1. That the importer has made a special request for pregnant monkeys.
2. That such monkeys are individually caged.

APPENDIX C

Livestock Progress Report

Service No. .... Captain ....
(on arrival in London)

Aircraft Registration .... Animal handlers ....

Date of Departure: London........... Arrival: Singapore ....

Singapore .... London ....

Consignment No. (most important)	Cage No.	Number of animals or birds dead	Notes

Were the M.R.C. recommendations observed? If not, in what respects were they not observed? ....

Signature of Senior Handler...................................

NOTES TO ANIMAL HANDLERS:
(*to be completed by BOAC*)

PART TWO

THE RECEPTION OF ANIMALS FROM OUTSIDE SOURCES

It is most important to remember that animals received from outside sources have been under stress, and therefore cannot be regarded as normal animals until they have settled into their new environment. Animals sent on a journey of any length, 100 yards or 500 miles, are exposed to a variety of hazards—changes in personnel, surroundings, temperature, and diet; overcrowding; draughts; noise; physical agitation; exposure to infections. Ensure that animals suffer no unnecessary delay at the end of their journey, e.g. remember that there is no routine, afternoon goods delivery by British Railways, though special transport can be arranged for livestock, providing British Railways is given adequate notice of this need. Animals deserve some special and immediate attention on arrival at their destination. First, they should be inspected by a person competent to recognize any gross abnormality, such as a broken limb or an infectious disease, and then they should be housed, fed, and watered in the quarantine area and kept there for at least fourteen days.

Quarantine area

Isolated quarantine quarters for holding animals from outside sources should be clearly designated as quarantine space, and the interchange of animals, personnel, and equipment between this area and the main animal buildings must be strictly prohibited. Each room of the quarantine quarters should have its own hot- and cold-water supply, system of heat control, and a set of apparatus which is not removed from the room.

Hygiene

A strict routine, suitable to the species held, must be instituted and maintained for the protection of both animals and personnel. A minimum requirement for personnel is to use protective clothing and to wash their hands before and after handling quarantined animals.

Overcrowding

Animals are always liable to some cramping or overcrowding when boxed for travel, and should be released into more spacious quarters as soon as possible. Mice, rats, and cavies tend to crowd together, and even on top of one another, in the corners of cages, and small or weak animals may suffocate. The following table gives a guide to the numbers of animals which may safely be housed together:

Mice	15–20
Rats	6–10
Cavies	4–6
Rabbits	Singly
Cats	Singly
Monkeys	Singly
Dogs	Singly

Animals are nearly always thirsty on arrival, so it is advisable to double the usual number of water bottles to each cage for the first 24 hours.

Diet

Avoid making unnecessary changes in the animal's diet. A change of diet nearly always causes a pause in the growth of animals, but the body weights usually return to the normal growth curve within a few days. Animals which have travelled and suffered a change of diet will not, at first, have a normal body weight for their age. All animals eat when they are hungry, so it is not necessary to tempt newly received animals with tit-bits unless there is some special reason for doing so. Tempting foods must be given sparingly or the transfer to standard diet may be delayed.

Travelling boxes

All containers and the food and bedding therein must be regarded as contaminated and should be burned or sterilized without delay. If this cannot be done at once, the containers should be removed from the quarantine building and left in the open.

Source of animals

Animals from breeders are an advertisement for the breeder, who will take steps to ensure that his animals are in good condition when dispatched and are suitably packed for the journey.

Animals from dealers may have changed hands several times within a short space of time, and consequently their condition may be poor and their resistance to disease low. They will have had ample opportunity for exposure to infection.

CARE OF ANIMALS

General

During the quarantine period animals should be observed carefully for abnormalities (e.g. loss of weight or condition; respiratory distress; diarrhoea) and examined for internal and external parasites (see chapters on diseases and parasites).

Rabbit

Examine for sore hocks; overgrown or broken teeth or claws; ear canker; nasal discharge (suspect Snuffles); diarrhoea (suspect Coccidiosis).

Cat

In the absence of a well-authenticated history of vaccination, a prophylactic injection of infective feline enteritis antiserum should be given, followed by injections of the vaccine if the cat is to be kept for any length of time. Watch for respiratory infections.

Dog

Dogs enjoy the company of humans and will settle into a new life more readily if some trouble is taken to befriend them. Dogs may be tempted to eat by offering small amounts of highly flavoured foods (raw meat or liver) or mixing a warm solution of 'Bovril' or 'Marmite' into the biscuit. If the dog is to be kept for any length of time vaccinate it against distemper, contagious hepatitis, and leptospirosis, unless it is known that the animal is already so protected.

Monkey

Newly imported monkeys may be carrying B virus, which is transmissible to man, to whom it is usually fatal (see p. 189). Monkeys often arrive in poor condition, as they have suffered considerable stress during their capture and long journey.

Monkeys are caged individually during the quarantine period so that they may be observed more accurately and are saved from the fighting and bullying which commonly occurs when monkeys are first caged in pairs. Antibiotics are usually given to control apparent and latent infections (see p. 191). Monkeys are susceptible to both human and bovine tuberculosis, and newly imported animals should be screened for the presence of reactors to the Mantoux Test (see p. 189). Cage trays must be inspected carefully each day for signs of diarrhoea or dysentery.

Poor eaters may be encouraged to take food by offering small amounts of orange, banana, dates, brown bread, or whole grains in addition to the standard cubed diet. Monkeys require a dietary source of vitamin C.

Protection of personnel

The hazard from B virus is real. Newly imported monkeys must be quarantined for at least four weeks. During this time the handling of the animals must be kept to a minimum and must not be attempted unless the handler wears full protective clothing (gown, elbow-length gloves, vizor-type face mask, and boots) and the animal has been sedated. Mechanical devices (nets and boxes) must be used for catching the animals. These precautions must be continued until the monkey has been certified free from B virus, tuberculosis, and intestinal infections transmissible to man. Cuts, scratches, or abrasions on the hands or arms of handlers present a potential route of entry for infections.

All persons handling monkeys should be required to have chest X-rays every six months. This procedure checks that no person has contracted tuberculosis from a monkey and that no human carrier has the opportunity to transmit tuberculosis to the monkeys.

IMPORTS INTO THE UNITED KINGDOM

ITEM	IMPORT PROHIBITED FROM	OTHER RESTRICTIONS OR EXEMPTIONS
Horses, asses, and mules	Afghanistan, Cyprus, India, Israel, Iraq, Iran, Jordan, Lebanon, Pakistan, Saudi Arabia, Turkey, United Arab Republic, and any country in Africa	
Dogs and cats (canines and felines)		Six months quarantine
Live poultry and hatching eggs (domestic fowls, turkeys, geese, ducks of any species, pheasants, guinea-fowl, partridges)	All countries except Irish Republic, Northern Ireland, Channel Isles, and Isle of Man	Import permitted from Australia, New Zealand, Denmark, Norway, Sweden if in conformation with conditions set out in Diseases of Animals Act, 1950, and its Orders Licences may be obtained from Min. Agric. Fish., and Food to import from other countries. Such imported stock liable to quarantine and tests for disease
Parrots and similar birds	All countries except Jersey, Isle of Man, and Northern Ireland	
Hay, straw, and dried grass	All countries except Australia, Canada, Channel Isles, Denmark, Faroe Isles, Greenland, Falkland Isles, Finland, Iceland, Isle of Man, Netherlands, New Zealand, Northern Ireland, Norway, Irish Republic, Sweden, South Africa, and the U.S.A.	

CHAPTER ELEVEN

Common Diseases of Laboratory Animals

MICE

BACTERIAL

Salmonellosis (mouse typhoid)

CAUSAL AGENTS. Bacteria of the Salmonella genus: *Salmonella typhi-murium*, *Salmonella enteritidis*, and more rarely other members of this genus.

SYMPTOMS. The acute form is characterized by rapid loss of condition, sometimes followed by purulent conjunctivitis and/or diarrhoea. Death occurs within seven to fourteen days of exposure.

The chronic form probably follows a sub-acute stage which passes unnoticed. The animal progressively loses condition as infection spreads. Clinical recovery may occur without treatment, though the animal can continue to excrete Salmonella organisms in the faeces for a long period.

POST-MORTEM. In the acute stage the animal may only show a generalized inflammation of the internal tissues. The more chronic the infection, the greater is the degree of involvement of the body tissues. The following lesions occur—peritonitis with excess peritoneal fluid; enlargement of the mesenteric and colonic lymph glands; hepatic enlargement, congestion, and necrosis; liver surface studded with small microscopic foci; splenic enlargement with thickening of the capsule. Sometimes the lungs are consolidated and there is excessive pleural exudate. Inflammation of the gut and hypertrophy of Peyer's patches.

DIAGNOSIS. During life the causal organism can be cultured from the faeces by the methods described for Salmonellosis in the guinea pig. These methods are also used for the examination of post-mortem materials.

ROUTE OF INFECTION. Probably by the oral route, though it may be that the conjunctiva is also a portal of entry.

SOURCE OF INFECTION. Investigations over the past few years have revealed the ubiquitous nature of the Salmonella, and many sources of the infection are possible.

Food and bedding may be contaminated by the excreta of domestic and wild-animal carriers or actual cases of Salmonellosis. Outbreaks of Salmonellosis have also often followed fresh importations from infected carrier stocks.

CONTROL. The following points may be made:

(*a*) Mice for breeding purposes obtained from an outside source should never be introduced into the existing colonies without first undergoing a period of quarantine of at least fourteen days. During this time it is worthwhile, should the facilities exist, to test faecal pellets from each animal for the presence of Salmonella. If these organisms are isolated, then the whole consignment should be destroyed and the equipment and quarantine room efficiently sterilized.

(*b*) If the disease is detected in an established colony, then eradication measures will depend on the system of management. For example, if, of a number of well-isolated rooms, only one is affected, then it may be sufficient to sacrifice the animals in that room and carry out sterilization procedures. However, should overt cases and carriers disperse throughout the stocks, eradication of the disease may involve destruction of all animals. After slaughter, equipment, rooms, etc., must be sterilized. Salmonella eradication is possible provided the methods used are sufficient. No lasting benefit is achieved by vaccines, drugs, or antibiotics.

Tyzzer's disease

CAUSAL AGENT. This disease is usually attributed to a gram negative bacillus designated *Bacillus piliformis*. The taxonomic position of this organism and its exact role in the aetiology of Tyzzer's disease is obscure because the organism cannot be cultivated on laboratory media, nor will tissue suspensions containing it readily produce the disease.

SYMPTOMS. There is a loss of weight and condition and an associated diarrhoea; death is common.

POST-MORTEM. The liver is studded with 'target-like' white and yellow necrotic foci. These tend to remain discrete. The gut may be inflamed, particularly at the ileo-caecal junction.

DIAGNOSIS. The typical appearance of the hepatic foci is usually sufficient for diagnosis. The slender gram negative bacilli can readily be seen in stained sections of the liver and the affected areas of the gut. The macroscopic lesions can be confused with those caused by Salmonella or Corynebacterium, but can be differentiated by successful cultivation of causative bacteria in these cases.

ROUTE OF INFECTION. Probably oral; mice have also been artificially infected by the intravenous injection of infected material.

SOURCE OF INFECTION. This organism is apparently host-specific. Infection, when it occurs, is usually endemic, with occasional outbreaks of epidemic disease. Once introduced, faulty management, such as overcrowding, favours epidemics, while with good management few cases of overt disease appear. Possibly wild mice are sometimes responsible for fresh introductions of the disease.

CONTROL. Mice with diarrhoea should be killed and the liver and gut carefully examined. In-contacts should be sacrificed, equipment sterilized, and the bedding destroyed. The author has experience of a colony in which endemic infection disappeared after the introduction of sterilized bedding in place of bedding known to be liable to contamination with the faeces of wild mice.

Arthritis (mouse rheumatism).

CAUSAL AGENT. *Streptobacillus moniliformis* (*Actinomyces muris*): a gram negative bacillus.

SYMPTOMS. Acute: the animal loses weight and condition, there is an accompanying semi-purulent conjunctivitis and occlusion of the palpebral fissures. Death frequently ensues within a few days of infection.

Chronic: a more characteristic syndrome is observed. Weight and condition are progressively lost. There is conjunctivitis and keratitis. The legs and feet become oedematous, ankylosed, and arthritic, and they may ulcerate; the tail may be similarly affected. Subcutaneous nodules may be found, and there is sometimes enlargement of the inguinal and maxillary glands.

POST-MORTEM. The liver and spleen are enlarged and show areas of necrosis, which may become confluent. There is necrosis affecting the splenic pulp. Infection of the heart is not uncommon, e.g. pericarditis and endocarditis. The lymph glands are enlarged, and examination of the joints reveals disintegration with caseous exudation.

DIAGNOSIS. This is based on clinical symptoms, post-mortem findings, and the isolation of the causal organism by cultivation on a medium containing a high proportion of serum.

ROUTE OF INFECTION. Not definitely known; could possibly be by bites or through the conjunctiva. Feeding the organisms orally does not result in active disease.

SOURCE OF INFECTION. Other affected animals: primary introduction can be by the introduction of infected stock or even wild rodents, since the maintaining of rats in rooms with mice is a likely source of infection for the latter.

CONTROL. Mice injected with a heat-killed suspension of the causal organisms develop an immunity, but it is doubtful if this is a practical method of control. In a colony where an epidemic occurs it is usually necessary to sacrifice all the animals, sterilize equipment and rooms, and start afresh with non-infected animals.

Mouse septicaemia

CAUSAL AGENTS. Various bacterial agents may be responsible for sporadic cases of murine septicaemia and the following merit some consideration.

AGENT (1). *Corynebacterium kutscheri*. A gram-positive bacillus of the corynebacterium genus.

SYMPTOMS. Loss of weight and condition.

POST-MORTEM. Tubercular-like septic foci may be found in the lungs, the spleen, the kidneys, the cervical gland, and, less frequently, in the liver. The foci contain free or caseated pus, and may become so numerous as to coalesce over considerable areas.

DIAGNOSIS. The causal organisms are demonstrable in gram-stained smears of affected tissue and are recoverable by culture of the tissues on blood-agar plates.

ROUTE OF INFECTION. Probably by oral ingestion.

SOURCE OF INFECTION. Not known.

CONTROL. Sacrifice of immediate contacts.

AGENT (2). *Klebsiella pneumoniae*: a gram-negative bacillus.

SYMPTOMS. Dyspnoea, loss of condition, and finally death.

POST-MORTEM. Congestion and consolidation of the lungs and enlargement of the spleen are the chief lesions.

DIAGNOSIS. The organisms are demonstrable on gram-stained smears of affected tissue. They are heavily capsulated, best shown by preparations stained with a special capsule stain. The organism can be cultivated on routine laboratory media.

ROUTE OF INFECTION. Probably the respiratory route.

SOURCE OF INFECTION. Not established.

CONTROL. Sacrifice of in-contacts and sterilizing of cages, etc.

AGENT (3). *Erysipelothrix muriseptica*: a gram-positive bacillus.

SYMPTOMS. Loss of weight and condition, and there may be a conjunctivitis.

POST-MORTEM. Septic foci may be present in the liver, spleen, and lungs.

DIAGNOSIS. The causal organism can be cultured from infected material by inoculating blood-agar plates.

ROUTE OF INFECTION. Unknown.

SOURCE OF INFECTION. Unknown.

CONTROL. It is usually sufficient to kill in-contacts.

In addition to the above causal agents, there are numerous others which may be responsible for sporadic cases of murine septicaemia, e.g. *Streptococcus haemolyticus*, *Diplococcus pneumoniae*, *Bordetella bronchiseptica*, etc.

VIRAL

Ectromelia (mouse pox)

CAUSAL AGENT. The Ectromelia virus—a member of the pox group of viruses.

SYMPTOMS. Acute: the animal may show no obvious signs of illness before being found dead. The first sign of the disease is usually an inconspicuous oedematous skin lesion which develops a scab, but this often goes unnoticed.

Chronic: animals surviving the acute stage develop a secondary generalized skin eruption followed by scab formation about two weeks after infection. In chronic cases the tail, one or both hind feet, or even forefeet become oedematous and later gangrenous, with sloughing of the affected extremity.

POST-MORTEM. Acute: sometimes few characteristic lesions are seen. The spleen is usually somewhat enlarged and presents white areas of fibrosis. The liver may be yellow and show minute necrotic foci or more diffuse necrosis. Fluid is often present in the peritoneal cavity. In the author's experience, the disease may result in sporadic deaths in a colony and the only post-mortem finding a mild hepatitis or splenitis.

Chronic: the superficial skin lesions have been described. There is extensive hepatic necrosis and numerous small foci on the surface of the liver and the spleen; necrosis of the pancreas and mesenteric fat is characteristic.

DIAGNOSIS. Ectromelia should always be suspected when any lesions suggestive of the disease are seen and unexplained losses follow the pattern of an epidemic. Confirmation may be sought in the following way. The spleen should be ground in horse-serum saline to give a final 10 per cent W/V suspension. 0·1 ml of this is injected intraperitoneally into two mice and 0·05 ml into the sole of the hind feet of two other mice. These mice must be from ectromelia-free stock. If the virus is present the mice injected intraperitoneally die within a few days, with hepatitis and splenitis as prominent lesions at autopsy. Those injected in the foot develop localized oedema and gangrene of the foot, and may die if the infection becomes generalized. If sections are prepared from the oedematous skin the presence of large cytoplasmic inclusions in the epithelial cells is diagnostic. Diagnosis can also be confirmed in other ways. For example, mice recovered from infection with vaccinia are immune to ectromelia; the sera of mice recovered from ectromelia neutralizes the vaccinia virus and prevents the formation of pocks on the chorio-allantoic membrane of fertile eggs. The haemagglutinative inhibition test may also be used to identify the virus or to detect antibodies against the haemagglutinin.

A diagnostic method has been described using specific antiserum containing fluorescent tagged antibody. Suspensions containing suspect materials are inoculated on to monolayers of tissue cells. The virus, if present, invades the cells and multiplies within them. Its presence within the cells can be detected after treatment with a tagged specific antiserum followed by examination by the U.V. light microscope.

ROUTE OF INFECTION. Virus excreted by infected animals gains entry to a new host chiefly through skin abrasions of the tail and feet.

SOURCES OF INFECTION. Usually by direct contact with infected mice, both overt cases and carriers, or contact with material soiled by such mice.

CONTROL. Should ectromelia be present in a colony, the usual procedure is to sacrifice the colony and take the usual measures to destroy residual virus by adequate disinfection before a new stock of mice is introduced. Although mice develop solid immunity to ectromelia after inoculation with vaccinia virus, it is likely that a colony infected with ectromelia virus will continue to harbour this virus even after immunization and thus remain a source of infection for healthy mice. Vaccination, however, is of value under conditions where the introduction of ectromelia is a permanent unavoidable hazard. Furthermore, by making use of the fact that mice which have had ectromelia do not react to vaccination, the inoculation of vaccinia offers a test to ascertain that fresh intakes from outside sources are not infected. Thus, in the author's establishment it is the practice to test fresh intakes of foundation stock by the injection of 0·05 ml of dilute vaccinia intradermally into the tail. Non-immune mice show a local positive inflammatory response after seven days, and a mouse reacting in this way can be regarded as being free from infection and previous exposure to ectromelia virus. Ninety per cent positive 'takes' on any group of mice are regarded as evidence of freedom from infection.

Infantile diarrhoea

Two virus diseases of infant mice have been described by Kraft:

(*a*) Epidemic diarrhoea of infant mice (EDIM).
(*b*) Lethal intestinal virus of infant mice (LIVIM).

SYMPTOMS. This is a disease occurring in mice of less than fourteen days old. Whole litters are usually affected and show a yellow fluid diarrhoea which bespatters the litter. The disease is usually first noticed about ten days after birth. Mortality may be high, and survivors are often retarded in development. Infection may involve from 1 to 80 per cent of all litters and have a disastrous effect on output.

POST-MORTEM. The gut contents are semi-fluid and the animal dehydrated. There are no other microscopically demonstrable changes.

DIAGNOSIS. By the clinical symptoms.

ROUTE OF INFECTION. The route of the infection is probably oral, but the virus is disseminated in droplets and spreads rapidly over considerable distances. Presumably nursing mothers carrying the virus infect their offspring, and this is followed by widespread dissemination to neighbouring litters.

CONTROL. Owing to the great capacity of the infection to spread, eradication has only so far been possible under special conditions, known virus-free parents rearing their progeny in virus-proof isolators. This method has, as yet, been used only to study the disease, but could be applied to the foundation of 'infantile-diarrhoea-virus free' colonies. It seems probable that the toll exacted by the disease depends very much upon the management and genetic constitution of the mice, apart altogether from eradication of infection. With conventional methods of management success in control of the disease has been claimed by the provision of special diets and types of cages.

FUNGAL

Ringworm

CAUSAL AGENT. Ringworm fungus. *Trichophyton mentagrophytes.*

SYMPTOMS. Loss of hair; scaling and encrusting of lesions. The tail may be affected.

DIAGNOSIS. Skin scrapings and hair from infected areas are mounted in 10 per cent potassium hydroxide and examined microscopically for the presence of spore sheaths and mycelia invading infected hair roots. Cultures may be made on a suitable medium, such as Sabouraud's, to provide final identification of the causal fungus.

CONTROL. Infected mice and their in-contacts should be sacrificed and equipment, etc., sterilized. Treatment is not worthwhile; topical applications are of doubtful efficacy, and systemic fungicides, at present, are too expensive for routine use. *T. mentagrophytes* is pathogenic for man and an infected mouse is a hazard to the handler.

Note. Another Trichophyton—*T. Quickeanum* causes favus in mice. It is identified by the methods described above.

PROTOZOAL

Though the mouse gut is inhabited by several types of protozoa, e.g. coccidia flagellata, and entamoeba, they are seldom of pathogenic importance.

ECTO-PARASITES

Body mange

CAUSAL AGENT. A mite—*Myocoptes musculinus.*

SYMPTOMS. Infestation is characterized by a marked loss of hair and sometimes generalized inflammation of the underlying skin. Frequently the mites produce these symptoms only in the breeding females and youngsters; the adult male, though infested, shows no hair loss or skin inflammation.

DIAGNOSIS. Hair and skin scrapings are mounted in 10 per cent potassium hydroxide and examined microscopically. Mites in all stages of development from ova to adults are usually present in large numbers.

CONTROL. Both *M. musculinus* and *Myobia musculi* (see below) are difficult to control, and the continuous use of BHC and DDT and other hydrocarbon insecticides which are known to leave residues in animal fat are not recommended. Badly infested animals should be removed from the unit and dusted or dipped in Malathion. Frequent treatments with pyrethrin aerosols which can be used twice a week without any harm to the animal or the operator help to keep infestation at a low level. It is seldom possible to completely eradicate these mites.

Head and neck mange

CAUSAL AGENT. A mite—*Myobia musculi.*

SYMPTOMS. Inflammation of the above areas is followed by serous exudation and scab formation. Males are more frequently affected than females.

DIAGNOSIS. Skin and hair scrapings are made as described for *Mycoptes musculinus*. *Myobia musculi*, however, burrows deeply, and relatively deep scrapings may be necessary to demonstrate its presence.

CONTROL. As for *M. musculinus* (see above).

Ear and body mange

CAUSAL AGENT. A mite—*Psorergates simplex.*

SYMPTOMS. This mite is concerned in two forms of mange. In the first the ears of the infected animal are found almost completely occluded by a firmly adherent, white, waxy mass which consists mainly of adult mites and early developmental stages. The second form, body mange, is characterized by the mites burrowing under the skin and forming skin pouches—nidification.

DIAGNOSIS. Pieces of ear wax or material dissected from skin pouches are mounted in 10 per cent potassium hydroxide and examined microscopically.

CONTROL. For the treatment of ear mange, a large drop of undiluted dibutyl phthalate is instilled into the ear and the treatment repeated after a period of seven days. It is unnecessary and inadvisable to attempt to remove the adherent wax. Care should be taken to prevent the chemical from entering the eyes of either the mouse or the operator, where it can cause serious irritation. For the treatment of body mange, a 2 per cent aqueous solution of 15 per cent Aramide [(2-(*p*-tert-butyl phenoxy)isopropyl-2-chlorethyl sulphite)] is used twice as a topical application at intervals of a week.

Notes. A technique has been described for the detection of ecto-parasites in mouse stocks. A percentage of the mice are killed and immediately pinned out on black paper covered with transparent sticky tape. As the mouse slowly 'cools' the mites leave the body and are trapped on the paper, where they are detected, examined, and identified.

HELMINTHS

Roundworm infestations of the caecum and large colon

CAUSAL AGENTS. Two species of oxyuridae—*Aspicularis tetraptera* and *Syphacia obvelata.*

SYMPTOMS. Neither of these two worms cause recognized syndromes. Nevertheless, such heavy infestations as commonly occur, especially in the caecum, must be deleterious to well-being. It is also noteworthy that heavy infestations, in young weaned mice, of *Syphacia obvelata* appear to be related to the occurrence of rectal prolapses.

DIAGNOSIS. At post-mortem adult worms are found mainly in the caecum. The two species are distinguishable by the microscopic examination of gravid females.

CONTROL. Complete eradication is difficult in a large colony. However, reasonable control is achieved by the addition of piperazine acid citrate to the drinking-water (12 gm to 8 litres) for a period of seven days. This should be repeated on two occasions, with a rest period of seven days between each treatment. Individual mice can be treated by dosing orally with 50 mg of the drug contained in 0·2 ml of water. Eradication is difficult to achieve as

(*a*) piperazine, while highly active against the adult worms, is less efficient in the eradication of immature forms;

(*b*) the gravid females of *S. obvelata* can deposit their ova around the anal exterior, where they hatch and migrate back into the colon, and before this migration takes place may escape contact with the drug being used; and

(*c*) foodstuffs contaminated by wild-mouse faeces will almost certainly contain ova of these helminths. Re-introduction of the parasite quickly results from ingestion of foodstuffs contaminated with faeces from wild mice.

Cysticerosis

CAUSAL AGENT. The cystic stage, *Cysticercus fasciolaris*, of the cat tapeworm, *Taenia crassicollis.*

SYMPTOMS. Seldom symptomatic, and infestation is usually revealed only at autopsy.

POST-MORTEM. One or more cysts are found in the liver, and these contain the strobilocercus. Occasionally they may become infected by micro-organisms.

ROUTE OF INFECTION. Oral.

SOURCE OF INFECTION. Materials contaminated by the faeces of cats infested with *Taenia crassicollis*. Bedding, such as sawdust and wood shavings obtained from sawmills, is a common source of infestation.

CONTROL. Avoidance of possibly contaminated materials or the adequate sterilization of these before use in the colony.

CHEMICAL AGENTS

The volatile vapours of chloroform and carbon tetrachloride are unusually toxic to mice. When chloroform is used in animal houses to kill unwanted animals it sometimes happens that adult male mice housed nearby die and show diffuse tubular nephritis at autopsy.

RATS

The rat is less liable to serious diseases than most of the other laboratory animals, but it is prone to chronic respiratory infection.

BACTERIAL

Labyrinthitis (Middle ear disease—'circling')

CAUSAL AGENTS. At least two agents appear to be implicated, viz. Mycoplasma (pleuro-pneumonia organisms—P.P.L.O.) and a bacterium, *Streptobacillus moniliformis.*

SYMPTOMS. Middle ear disease is characterized by the affected animal tilting its head to one side and moving in a curve rather than a straight line. There may be difficulty in regaining balance should the animal fall over.

POST-MORTEM. There is infection of the middle ear. This may be due to an extension of infection from the outer ear or, alternatively, from the upper respiratory passages.

DIAGNOSIS. This is usually made by observing the clinical symptoms. *Streptobacillus moniliformis* can be isolated by inoculation of infected material on rich laboratory medium such as 30 per cent serum-agar plates. Mycoplasma can only be cultivated with regular success on the special media used for these organisms.

ROUTE OF INFECTION. Probably by either the oral route or the upper respiratory tract.

SOURCE OF INFECTION. Other rats.

CONTROL. Virtual elimination of the disease can be accomplished by frequent examination and destruction of animals with symptoms of labyrinthitis or upper respiratory tract distress.

Salmonellosis

CAUSAL AGENTS. *Salmonella typhi-murium* and *Salmonella enteritidis*.

SYMPTOMS. As described under mouse salmonellosis. In this country salmonellosis in rat colonies is not common. Post-mortem appearance, etc., is as described in mice.

VIRAL

Epidemic murine pneumonia (E.M.P.)

CAUSAL AGENT. Probably a virus.

SYMPTOMS. This disease, though probably the most common of all rat diseases, frequently causes little distress in spite of the presence, in individuals, of considerable lung damage. Close examination, particularly of old breeders, will often reveal dyspnoea, sometimes accompanied by a 'chattering noise'.

POST-MORTEM. Varying degrees of pulmonary involvement are noted, from discrete foci to widespread consolidation and hepatization.

DIAGNOSIS. In the past a large number of different bacteria have been thought responsible for rat pneumonia, but are now regarded as secondary invaders, and the true agent, a virus. Diagnosis is dependent on the examination of histological sections of lung tissue and the inoculation of suspensions of suspect lung into mice by the nasal route. The mouse inoculation test is carried out by preparing a 10 per cent W/V suspension of lung and dropping 0·05 ml on to the noses of anaesthetized mice; an infected suspension causes pneumonia within twenty-eight days.

ROUTE OF INFECTION. Respiratory.

SOURCE OF INFECTION. Adult females infect their offspring soon after birth.

CONTROL. Has not proved successful by any of the conventional methods used to combat disease. Every rat is at least a carrier of the virus, and a high proportion of adults will have demonstrable lung lesions. The disease impairs the development of rats and can seriously interfere with some kinds of research. Complete eradication of this disease has been achieved by the establishment of specified-pathogen-free colonies (see chapter on SPF animals).

PROTOZOAL

Coccidiosis

CAUSAL AGENTS. Three species of coccidia have been described: *Eimeria miyairii*, *Eimeria separata*, *Eimeria nieschulze*.

SYMPTOMS. Acute diarrhoea is the common symptom and usually occurs only in weanling rats.

POST-MORTEM. Haemorrhagic enteritis of the small intestine.

DIAGNOSIS. Large numbers of oöcysts are demonstrable microscopically in the gut contents and the faeces.

ROUTE OF INFECTION. Oral: by the ingestion of sporulated oöcysts present in faecal material from infected rats.

SOURCE OF INFECTION. Other rats.

CONTROL. It is probable that all conventionally managed rats are at some time sub-clinically infected with Eimeria spp. and that the infection only becomes evident when massive infection takes place. Provided the system of management ensures adequate nutrition and regular cleaning, signs of coccidiosis are seen rather rarely and only in young animals.

ECTO-PARASITES

Body mange

CAUSAL AGENT. The following species of mites can be responsible for body mange: *Myobia ratti*, *Myobia musculi*, *Bdellonyssus bacoti*, and *Notoedres spp.*

SYMPTOMS. Irritation of the skin and occasionally skin and tail lesions.

DIAGNOSIS. The particular agent responsible for the occurrence of mange can be demonstrated by microscopic examination of hair and skin scrapings after treatment with 10 per cent potassium hydroxide.

SOURCE OF INFECTION. Other infested rats.

CONTROL. As for murine body mange (see above).

HELMINTHS

Cysticercosis

CAUSAL AGENT. The cystic stage, *Cysticercus fasciolaris* of the cat tapeworm, *Taenia crassicollis*. Both the rat and mouse act as intermediate hosts to the tapeworm, and the information about this parasite given in the section on mice applies also to rats.

OTHER PARASITES

Though the rat may carry in its intestine, roundworms, tapeworms, protozoa, etc., these do not usually, apart from coccidia, give rise to recognizable disease.

PHYSIOLOGICAL

Ringtail

CAUSE. Attributed to low humidity in rat houses, viz. less than 50 per cent.

SYMPTOMS. Newborn rats develop swollen tails with distinctive corrugated ringing, and the tails may eventually fall off. The other extremities sometimes become affected.

NUTRITIONAL

Vitamin E deficiency

CAUSE. Diet containing insufficient vitamin E.

SYMPTOMS. Infertility or lowered fertility, resorption of foetuses, abortion, and testicular inactivity.

CONTROL. This deficiency has become uncommon, as it is now widely recognized that adequate amounts of vitamin E and antioxidants must be added to compounded diets.

GUINEA PIGS

BACTERIAL

Pseudotuberculosis

CAUSAL AGENT. *Pasteurella pseudotuberculosis*: a gram-negative bacillus.

SYMPTOMS. Usually chronic: the animal becoming emaciated and often dying, though recovery may take place.

POST-MORTEM FINDING. The primary lesion is usually found in either the mesenteric or colonic lymph nodes. The infection spreads to the gut wall, the liver, and the spleen. Infection may also be present in the genito-urinary tract and mammary tissues; there is seldom pulmonary involvement. The lesion is a white or yellow nodular necrotic focus containing free or caseated pus.

DIAGNOSIS. The causal organism is recoverable by culturing suspect material on routine laboratory media such as 5 per cent blood agar.

ROUTE OF INFECTION. Oral: by the ingestion of materials contaminated with the organism. Infection may occur rarely by other routes, e.g. through the peridontal tissues surrounding the incisors. When this occurs the primary lesions are found in the regional lymph glands.

SOURCE OF INFECTION. Primary infections are introduced from outside the colony, and secondary infections then occur by spread within the colony. Birds can be carriers of infection or overt cases. In the author's experience, pigeons are a frequent source of infection, contaminating growing green foods such as kale with excreta containing *P. pseudotuberculosis*.

CONTROL. (*a*) By exclusion from the colony of materials liable to be contaminated. (*b*) By the elimination of diseased animals: early non-clinical cases may be detected by palpation of the primary lesions in the gut glands.

Salmonellosis

CAUSAL AGENT. Bacteria of the Salmonella genus: *Salm. typhi-murium*, *Salm. enteritidis*, *Salm. dublin*, etc.

SYMPTOMS. Acute: animals may die without showing any symptoms: occasionally there may be an associated diarrhoea.

Chronic: wasting and emaciation.

POST-MORTEM. Acute: splenic and hepatic enlargement which may be accompanied by a haemorrhagic and necrotic invasion of the gut glands.

Chronic: enlargement of the liver and spleen. Infected foci containing free or caseated pus may be present in the liver, the lymph glands, and the walls of the gut: there may be a peritonitis.

DIAGNOSIS. The chronic form of the disease may resemble, at post-mortem, pseudotuberculosis. A preliminary distinction may, however, be made by staining material from affected tissue by Gram's method. Salmonella infected tissue usually contains large numbers of gram-negative bacilli: *P. pseudotuberculosis* is more difficult to detect. Final diagnosis is, however, dependent on the isolation of the causal organism on laboratory media, such as McConkey's agar, or more selective media, such as selenite F and desoxycholate citrate agar. Suspect isolates are identified by biochemical and serological techniques. Rarely, a condition similar to that described above can be caused by the bacterium, *Escherichia coli*. This organism is morphologically indistinguishable from members of the Salmonella genus, but it can be readily identified biochemically.

ROUTE OF INFECTION. Infection is believed to occur by the oral route, though some evidence suggests that infection of the eye may occur and spread by the local lymph glands to the bloodstream and other tissues.

SOURCE OF INFECTION. Several sources of infection have been incriminated. The contamination of feeding stuffs and bedding by wild rodents: flies can also spread contamination. It is now also recognized that certain of the materials used in compounded animal pellets can harbour Salmonellae, though it is probable that these are largely destroyed during processing. It is perhaps significant that the species of Salmonella usually responsible for the disease in laboratory animals are those commonly associated with Salmonellosis in cattle, though a direct association may be difficult to demonstrate.

CONTROL. (*a*) The removal and sacrifice of active cases and their contacts. (*b*) The sterilization of infected equipment and the destruction of bedding, etc. (*c*) It may be necessary in severe outbreaks to sterilize the animal house by physical and/or chemical methods. (*d*) If possible, the source of infection should be discovered and eliminated.

Streptococcal pneumonia

CAUSAL AGENT. *Streptococcus pyogenes*: a gram-positive coccus producing a soluble haemolysin and usually of Lancefield's Group C.

SYMPTOMS. Emaciation, respiratory distress, and nasal discharge: the urine may contain haemoglobin.

POST-MORTEM. The lung consolidates in varying extents. Infection may be found in the pericardium and myocardium, and the kidneys may show congestion.

DIAGNOSIS. The causal organism can be seen in gram-stained films made from infected tissue. It may be isolated on 5 per cent blood agar, which will also show the haemolytic nature of the organism.

ROUTE OF INFECTION. Probably by the respiratory route.

SOURCE OF INFECTION. Unknown.

CONTROL. (*a*) The elimination of infected animals and their contacts. (*b*) The sterilization of equipment and the destruction of bedding, etc.

Streptococcal lymphadenitis

CAUSAL AGENT. *Streptococcus pyogenes*: (Lancefield Group C—see streptococcal pneumonia).

SYMPTOMS. Acute: there is a fulminating septicaemia, the animal dying within three to four days of infection.

Chronic: the animal frequently remains in normal health. The cervical glands and other neck glands are obviously enlarged. Occasionally the infection may involve the axillary and inguinal glands, and there may be an associated arthritis and cellulitis. The disease in pregnant females sometimes results in abortion.

POST-MORTEM. Acute: some degree of lung hepatization is usually found. There may be a diffuse haemorrhagic peritonitis and congestion of the spleen and liver.

Chronic: free pus is found in the affected glands. There may be small abscesses present in the liver, lungs, peritoneum, and the gut wall.

DIAGNOSIS. Streptococci are seen in gram-stained smears made from infected pus or tissue. Cultures are made as in streptococcal pneumonia.

ROUTE OF INFECTION. The organism is considered to enter the body through abrasions in the buccal mucosa caused by ingestion of materials such as the thistles sometimes found in hay. In the acute form the lymph glands fail to localize the organisms.

SOURCE OF INFECTION. There is little definitely known, haemolytic streptococci are a common parasite of humans and animals, and either may be responsible for primary introduction.

CONTROL. Cases and their contacts should be sacrificed, equipment sterilized and bedding, etc., should be destroyed. Hay containing abrasive ingredients should not be used. Should infection become widespread, it may even be necessary to destroy the colony and restock from a clean nucleus.

Bronchiseptica pneumonia

CAUSAL AGENT. *Bordetella bronchiseptica*: a small gram-negative bacillus formerly known as either *Brucella bronchiseptica* or *Haemophilus bronchisepticus*.

SYMPTOMS. Affected animals may show signs of dyspnoea, though early cases may go undetected; there is frequently a tracheitis.

POST-MORTEM. The lungs show areas of consolidation which may contain a purulent exudate.

DIAGNOSIS. Recognition of the organism in gram-stained films of infected material is difficult. Cultures of infected tissue on blood-agar plates or bile-salt agar plates usually results in profuse pure growths of the organism. It is identifiable by its failure to ferment carbohydrates and its ability to produce ammonia from urea.

SOURCE OF INFECTION. Probably many normal guinea pigs harbour *B. bronchiseptica* in the respiratory tract, and some disturbance of the host–parasite relationship allows the organism to become pathogenic. Rabbits, rats, cats, and dogs also harbour this organism, and may occasionally be the source of infection for guinea pigs.

CONTROL. Usually only isolated cases of this disease occur, and no control measures required other than removal of the affected animal and close observation of in-contact animals. However, widespread epidemics can take place, and it may be advisable to sacrifice in-contacts and sterilize contaminated materials. An autogenous vaccine has been used with success in eliminating pneumonia in rats caused by *B. bronchiseptica* and guinea pigs can also develop a good immunity following vaccination. However, carriers persist and may infect non-vaccinated stock. Providing the source of continuing infection is removed, heat-killed autogenous vaccines can be successfully used to eradicate infection. The author has personal experience of an outbreak in a large colony probably caused by the housing of rabbits in close proximity to the guinea pig unit. Removal of the rabbits and all young guinea pigs at weaning accompanied by the vaccination of all breeding and reserve breeding stock resulted in the cessation of further cases. No respiratory tract carriers were found when 10 per cent of the breeders were examined at three, six and twelve months.

Cervical adenitis

CAUSAL AGENT. *Streptobacillus moniliformis*: a gram-negative bacillus.

SYMPTOMS. There is a lymphadenitis of the cervical glands, these become obviously swollen and may rupture through the skin. The general health of infected animals is not usually seriously impaired.

POST-MORTEM. The affected glands contain pus: infection sometimes extends to the trachea, the bronchi, and the lungs, where it manifests itself as pustular nodules. Infection of the axillary and inguinal glands has been observed but is uncommon.

DIAGNOSIS. The causal organism is not easily found in stained smears of pus: it is isolated by cultures of pus on media containing a high proportion of serum.

ROUTE OF INFECTION. Through abrasions in the buccal mucosa.

SOURCE OF INFECTION. The respiratory tract of most rats is infected with this organism and, in mice, it produces chronic arthritis. While little is definitely known about the source of infection for guinea pigs, rats constitute a dangerous reservoir of infection. Minor injuries to the mouth tissues may be the portal of entry.

VIRAL

So far, no well-defined virus diseases of major importance have been recognized in guinea pigs.

FUNGAL

Mucormycosis

CAUSAL AGENT. *Absidia ramosa*: a mucor mould.

SYMPTOMS. The infected animal is usually asymptomatic, though, as described below, rare cases are found where the infection becomes generalized and the animal dies. In the usual asymptomatic cases the primary lesion is a nodular lymphadenitis of the mesenteric lymph gland. This swelling is sufficiently large to be palpable, and may be confused with glandular enlargement caused by *Pasteurella pseudotuberculosis* or the Salmonella spp.

POST-MORTEM. Benign: the usual lesion is chronic enlargement of the mesenteric lymph node which contains free or caseated pus.

Generalized: infection is found spreading from the lymph gland to the kidneys, spleen, liver, and lungs. These organs are enlarged, congested, and necrotic. There is sometimes a marked excess of peritoneal fluid and a pleural effusion.

DIAGNOSIS. Typical non-septate hyphae are demonstrable microscopically in wet preparations of pus and infected tissue treated with 10 per cent potassium hydroxide. Cultures on blood-agar plates develop fine cotton-wool-like growth typical of mucor spp.

ROUTE OF INFECTION. Oral.

SOURCE OF INFECTION. Absidia sp. are widely distributed in nature. Heavy growths of *Absidia ramosa* may be found in hay, particularly when made during wet weather, and probably this represents the principal source of infection for guinea pigs.

CONTROL. The avoidance of poor-quality, mouldy hay.

Note. The importance of this disease rests in the likelihood of its being confused with the lymphadenitis caused by infections with Salmonella or *P. pseudotuberculosis*.

PROTOZOAL

Coccidiosis

CAUSAL AGENT. *Eimeria caviae*. A protozoan.

SYMPTOMS. Diarrhoea and emaciation.

POST-MORTEM. The disease is characterized by a gelatinous oedema of the large gut just beyond the caecal junction. The contents of the large gut are semi-fluid without the normal pellet formation.

DIAGNOSIS. This condition may be diagnosed during life by the microscopic detection in faecal material of large numbers of oöcysts.

ROUTE OF INFECTION. Oral: by the ingestion of materials contaminated with the faeces of guinea pigs, containing sporulated oöcysts.

SOURCE OF INFECTION. *Eimeria caviae*, like other coccidia, is host specific. As guinea pigs are not found in the wild state, infection must arise within the colony. Most guinea pigs harbour coccidia, and active disease is usually only found in animals from four to eight weeks of age, probably the period when maternal passive immunity has declined and active immunity not yet developed.

CONTROL. Oöcysts, after being shed in the faeces, take five to eight days to sporulate to the infective stage. Hence, regular cleaning and sterilization of equipment in shorter periods of time than five days will greatly minimize spread of infection. Similarly, any system of management which prevents gross faecal contamination will tend to lessen the liability to overt disease. Overt cases should be removed and sacrificed, and to prevent further spread it may be necessary to treat in-contacts with the coccidiostats used for poultry, such as sulphamezathine or sulphaquinoxaline. Bactericides are usually ineffective against oöcysts, which will, for example, sporulate in solutions of formalin and hypochlorites. Oöcysts are, however, killed by heat (steam or flame) and by 10 per cent ammonia solution.

Lice

CAUSAL AGENT. Three biting lice (Mallophaga) are commonly associated with guinea pigs. These are *Gyropus ovalis*, *Glirocola porcelli*, and *Trimenopon jenningsi*. A fourth type, *Menopon extraneum*, is only rarely found.

SYMPTOMS. Guinea pigs can be heavily infested without showing symptoms. However, in individual cases the animal may scratch and cause skin irritation.

DIAGNOSIS. Naked-eye examination will usually detect infestation, though a hand lens is of value. Identification of the type of louse present requires microscopic examination.

SOURCE OF INFECTION. Other guinea pigs.

CONTROL. As for murine body mange (see above).

HELMINTHS

Three worms have been described as causing pathological conditions in guinea pigs. They are: the fluke, *Fasciola hepatica*, the roundworm, *Paraspidodera uncinata*, and the larvae of *Trichinella spiralis*. It is not proposed to describe these in detail, as their occurrence in guinea pigs is rare. Briefly, *F.*

hepatica has been found in the liver and musculature; cysts containing *Tr. spiralis* larvae have been found in the limbs, the tongue, and the palate. *P. uncinata* is found in the caecum and colon of the guinea pig, and heavy infestations have been thought to cause loss of condition, emaciation, and diarrhoea.

PHYSIOLOGICAL

Stillbirth and inanimation

High losses may be experienced in guinea pigs from stillbirths and deaths shortly after birth. Certain colonies have experienced up to 30 per cent losses. These losses are a problem of management rather than disease control, and have been much reduced by replacement of breeding stock by the progeny of mothers with a satisfactory breeding record. Unsuitable diet and cages may also play some part in this problem.

Pregnancy complications

Dystocia is not uncommon and results from a number of causes. Common ones are dual presentation at full term; torsion of the uterus; the mating of females when too young; delaying mating until they are too old.

Intrauterine haemorrhage may also occur during the early part of the second half of the gestational period. The site of bleeding is found at the placental junctions. Careful examination of the animal at autopsy reveals the stomach to be displaced forwards and upwards and to the left when the animal is placed in its normal posture and viewed from behind. This gastric displacement results in partial or intermittent occlusion of the bile duct. The back pressure thus caused can give rise to marked and permanent damage to the liver. There may then be such a degree of hepatic disfunction as to prevent the formation of prothrombin factor VII, one of the substances required in the clotting of blood.

The placentas and their attachments are delicate and very vascular and particularly prone to massive haemorrhage should the blood coagulation mechanism be impaired.

The occurrence of the disease is directly related to the nutritional plane of the animal. Guinea pigs offered unpalatable hay refuse to eat it and consume more pelleted diet. This increased intake of richer diet results in a more rapid development of foetuses in the pregnant female, and the encroachment of the left horn of the uterus into the abdominal cavity causes gastric displacement and interference with the bile duct.

If there are no fatal consequences as a result of hepatic disfunction the stomach returns to its normal position when the foetuses are expelled at full term. However, the liver is probably permanently damaged. Histological examination of the livers of old discarded breeding sows show many of these to have periportal necrosis and fibrosis of the liver.

Pregnancy toxaemia also occurs, presumably due to faulty metabolism in the later stages of pregnancy.

Anal and vaginal fissures are sometimes found in heavily pregnant females. These are probably initiated in part by excessive urination, with consequent continuous wetting of the affected areas. Chronic inflammation may follow and produce oedema in the adjacent tissues.

Vitamin C deficiency

The recommended *daily* intake of ascorbic acid is 1·6 mgm per 100 gms body weight. When the vitamin is added to a pelleted diet the *final concentration must be at least 1 gramme per kilogramme.* In young pigs, the first sign of deficiency is the animal moving reluctantly with a mincing gait. Autopsy findings may include massive haemorrhages in the musculature round the knee; reddening, and, in extreme cases, fracture, of the long bones at the epiphyseal line; widespread petechial haemorrhages; and, in chronic, sub-acute cases, enlargement of the wrists and costochondral junctions. Untreated cases always die.

Hay deficiency

Guinea pigs eat large quantities of hay, and if deprived of it they show adverse effects, particularly in young stock, such as high mortality and bizarre dentition, which interferes with nutrition. The exact nature of the deficiency has not been determined.

Soft tissue calcification

This is a condition found occasionally in breeding boars, and more rarely in breeding sows. The affected animals are in poor condition and lose weight. At autopsy calcium deposits are visible on the external wall of the colon flexure, the heart, the kidneys, the external wall of the colon flexure, the heart, the kidneys, the external wall of the stomach, and the genito-urinary systems, particularly the seminal vesicles of the male. The condition has been produced experimentally by the feeding of diets deficient in magnesium and is probably caused by a dietary imbalance of calcium, phosphorus and magnesium.

Nephritis

Chronic nephritis is not uncommon in old breeding stock. It is thought to be due to the absorption, over long periods, of toxic substances which may be present in materials used for guinea pigs, such as the toxins present in wood shavings often used for bedding. Guinea pigs will, for example, eat large quantities of shavings if deprived of sufficient hay.

Exudative hepatitis and oedema (Paget's disease)

A number of deaths have been reported as occurring among young weaner guinea pigs, following the feeding of certain batches of pellets prepared according to the formula of the Bruce–Parkes Diet 18. The affected animals have staring coats and a characteristic dropsy of the abdomen. At autopsy the most marked feature is a subcutaneous oedema due to the collection of clear, watery fluid in the connective tissues of the abdomen, chest, and neck. There is sometimes pleural effusion, free ascitic fluid in the peritoneal cavity, and enlargement of the mesenteric lymph gland. The liver may be pale, but otherwise appears normal, though histological examination shows marked hepatic changes. There is dilatation of the liver-cell columns, cellular infiltration of the portal tracts and dilatation of the portal lymphatics. The mesenteric gland and spleen may show considerable hyperplasia of the sinusoidal reticulum cells, and there may be an increase in the number and size of the islets of the pancreas. Biochemical tests have shown serum protein values lower than those found in comparable normal guinea pigs.

This disease has been produced experimentally by feeding to guinea pigs diets containing groundnut meal contaminated with a toxin (aflatoxin) produced by the mould, *Asperigillus flavus.* As a result of these investigations it is now generally accepted that the outbreaks reported were caused by the feeding of compounded diets containing toxic groundnut meal. Soya-bean meal is now used as a replacement for groundnut meal in the majority of laboratory animal diets.

RABBITS

BACTERIAL

Snuffles (infectious nasal catarrh)

CAUSAL AGENTS. Believed to be bacteria: at least three gram-negative bacilli are regarded as associated with outbreaks of snuffles—*Pasteurella septica, Bordetella bronchiseptica,* and members of the Haemophilus spp. That so many agents have been considered to cause a disease syndrome raises the suspicion that the true cause may not yet have been discovered.

SYMPTOMS. Sero-purulent discharge from the nostrils, though symptoms may vary from occasional sneezing to marked dyspnoea. Rabbits with dyspnoea make snuffling noises, from which the disease has received its popular name. Some cases are mild and chronic, but others develop a severe pneumonia which often terminates fatally.

POST-MORTEM. Lesions may vary from mild inflammation of the nasal passages to severe rhinitis, sinusitis, tracheitis and pleuro-pneumonia.

DIAGNOSIS. This is usually dependent on the clinical symptoms. Bacteriological examination of exudates and infected tissues results in the isolation of either one or a combination of the bacteria mentioned above.

ROUTE OF INFECTION. Probably by the respiratory route.

SOURCE OF INFECTION. Other infected rabbits.

CONTROL. When an isolated case occurs in a rabbitry, sacrifice of the animal and sterilization of the cages, bedding, etc., may prevent further spread. However, the disease is often endemic in a colony, and then successful control is difficult. There is not yet any fully satisfactory curative treatment, although some success has been claimed by parenterally administered sulphonamides, water-fed antibiotics, and bactericidal inhalants. At least one bacterial species associated with snuffles may be found in the respiratory tract of normal rabbits, and their complete eradication is difficult. An adequate diet, sufficiently ventilated quarters, and the regular provision of cleaned and sterilized cages, etc., usually control the disease sufficiently to prevent serious outbreaks.

Pneumonia

CAUSAL AGENTS. The bacteria associated with snuffles.

SYMPTOMS. There is usually marked dyspnoea, but often this is overlooked, and the disease is only recognized after death.

POST-MORTEM. There is a pneumonia, though the amount of lung involved varies widely. Sometimes almost all the lung substance is consolidated and large cavities are filled with creamy pus.

DIAGNOSIS. Associated bacteria are recoverable by culture of infected material on blood plates, though richer medium, such as chocolate-agar plates, may be necessary for the isolation of organisms of the Haemophilus spp.

ROUTE OF INFECTION. Probably by the respiratory tract.

SOURCE OF INFECTION. Presumably other rabbits, but most of the organisms associated with the disease can be recovered from other species of laboratory animals.

CONTROL. Pneumonia may follow as a sequel to snuffles, or it may occur in a rabbit without involvement of the upper respiratory tract. Methods of treatment and control described under snuffles apply to pneumonia.

Pseudotuberculosis

CAUSAL AGENT. *Pasteurella pseudotuberculosis.*

Pseudotuberculosis in the rabbit is essentially the same as that described in guinea pigs. The disease is much less common in rabbits.

Rabbit syphilis (venereal disease).

CAUSAL AGENT. A bacterium, *Treponema cuniculi.*

SYMPTOMS. Ulcerative lesions are found on the penis and prepuce of the male, and the vulva and vagina of the female. There may be involvement of the anus, the eyelids, the lips, and the nose.

POST-MORTEM. Death from rabbit syphilis is not common, though it has been recorded after generalization of infection.

DIAGNOSIS. The causal organism is a spirochaete and is difficult to demonstrate by routine staining methods. Serous exudates require to be examined as wet preparations under the dark-ground illuminated microscope or in the dry state to be stained by a silver impregnation method.

ROUTE OF INFECTION AND SOURCE. By sexual intercourse with infected rabbits.

CONTROL. The disease is curable by a single injection of neosalvarsan or penicillin. Either of these agents should eradicate the disease from a closed colony. Before introduction of rabbits for breeding, they should be examined to ensure their freedom from this disease.

Listeriosis

CAUSAL AGENTS. *Listeria monocytogenes*: a gram-positive bacillus of the genus Listeria.

SYMPTOMS. The disease appears to be confined to young stock and pregnant does.

Young stock. Rapid loss of flesh, with death occurring suddenly.

Breeding does. Abortion of dead foetuses near term or delayed parturition with putrid foetuses.

POST-MORTEM. Young stock. Focal necrosis of liver, heart, and spleen, enlarged mesenteric glands.

Breeding does. Focal necrosis of liver, metritis, and associated peritonitis.

DIAGNOSIS. Isolation of *Listeria monocytogenes.*

ROUTE OF INFECTION. Not known.

SOURCE OF INFECTION. Not known.

CONTROL. The disease does not spread within a colony to any extent if management is good, but slaughter of all in-contact animals is recommended.

VIRAL

Myxomatosis

CAUSAL AGENT. Myxomatosis virus.

SYMPTOMS. The primary lesion of myxomatosis occurs on the eyelids, which are at first congested and swollen. Later a severe blepharo-conjunctivitis accompanied by a purulent exudate occurs. The nose and lips are swollen, and the genitalia may be similarly affected.

POST-MORTEM. Little change may be present in the internal organs. The spleen may be slightly enlarged, and subcutaneous tumours are sometimes found.

DIAGNOSIS. Domestic rabbits are highly susceptible to the myxomatosis virus. The clinical symptoms and very high mortality are sufficient to make a probable diagnosis. Confirmation can be obtained, by the histological examination of sections made from lesions, and by inoculation of suspected material into normal rabbits.

ROUTE OF INFECTION. The virus is introduced into the bloodstream by the bite of either mosquitoes or rabbit fleas which have previously fed on an infected rabbit.

SOURCE OF INFECTION. Fortunately, so far in this country, the mosquito has not been the vector responsible for many outbreaks of myxomatosis among wild rabbits, the rabbit flea has been the usual vector. The control measure necessary in Britain has been the exclusion from rabbitries of wild rabbits. However, where there is any likelihood that the virus may be spread by mosquitoes, steps must be taken to exclude them from buildings housing rabbits. A vaccine prepared from the Shope fibroma virus confers immunity against myxomatosis for about six months. This would be useful for immunization of contact rabbits if a case happened to appear in a colony of domestic rabbits.

Rabbit pox

CAUSAL AGENT. Rabbit pox virus.

SYMPTOMS. The most common and obvious features are blepharitis and keratitis, accompanied by a blood-stained mucopurulent discharge: there is a similar discharge from the nose. Close examination reveals glandular enlargement, particularly of the head and neck glands. Macular and papular rashes occur over the whole body, being most evident on the ears, exposed areas of skin, and the genitalia. The tissues of the mouth and tongue are swollen and necrosed. The disease is very contagious, and mortality rates are high.

POST-MORTEM. There is hepatic enlargement, and grey nodules are distributed throughout the liver substance. The spleen is enlarged, and areas of focal necrosis may be seen. The testicles of the male and the uterus and ovaries of the female usually show many small lesions. A few lesions may be present in the lungs. The lymph glands, particularly of the head and neck, are enlarged, haemorrhagic, and oedematous, and may show focal lesions. The nose and associated sinuses are haemorrhagic and contain a mucopurulent exudate.

DIAGNOSIS. Suspensions of infected material inoculated on to the chorio-allantoic membrane of suitable fertile eggs give rise to pocks, but such pocks are absent if the infectious material is incubated with immune vaccinia serum prior to inoculation of the eggs.

CONTROL. This is not a common disease in this country, though several outbreaks have been recorded from the United States. It is mentioned because this virus is sometimes used experimentally in laboratories and may easily spread to normal rabbits housed close by. Such an outbreak has been recorded in Britain. A solid immunity is conferred by vaccination with vaccinia virus.

FUNGAL

Ringworm

CAUSAL AGENT. Trichophyton and Microsporon spp.

SYMPTOMS. There is destruction of the hair follicles, resulting in areas of baldness which may exude serous fluid and become encrusted.

DIAGNOSIS. Scrapings made from infected areas and mounted in 10 per cent potassium hydroxide show fungal hyphae and spores when examined microscopically. Cultures are made by inoculating hair and skin scrapings on Sabouraud's medium.

ROUTE AND SOURCE OF INFECTION. Probably by contact with infected rabbits or other rodents.

CONTROL. It is doubtful if ringworm in rabbits is worth attempting to cure. Topical applications are of doubtful value, and oral treatments are expensive. Infected animals mostly recover without treatment, but since they are likely to infect others before recovering, it is probably more satisfactory to sacrifice the animal and effectively sterilize the cages, bedding, etc.

PROTOZOAL

Intestinal coccidiosis

CAUSAL AGENTS. Two species of Eimeria: *Eimeria perforans* and *Eimeria magna.*

SYMPTOMS. Diarrhoea accompanied by loss of condition, often terminating fatally.

POST-MORTEM. Small infected foci can be seen in the internal wall of the small gut. Often these lesions are accompanied by a more diffuse enteritis.

DIAGNOSIS. The oöcysts can be demonstrated microscopically during life by examination of faecal suspensions or at autopsy in scrapings from the gut mucosa.

ROUTE OF INFECTION. Oral: by the ingestion of material contaminated by the faeces of rabbits excreting oöcysts which have sporulated since being voided.

SOURCE OF INFECTION. Other rabbits: Eimeria spp. are host species specific.

CONTROL. Reasonable control depends upon removal of faeces before the contained oöcysts have sporulated. This requires a high standard of management to ensure regular cleaning and sterilizing of cages, etc. In addition, it has become a common practice to include sulphadimidine or sulphaquinoxaline in the drinking-water of breeding does immediately prior to littering and of growing stock for some days following weaning. Certain manufacturers now include the above-mentioned drugs and other coccidiostats in compounded feeding pellets. None of the treatments available are fully effective in either controlling or eradicating the disease.

Hepatic coccidiosis

CAUSAL AGENT. *Eimeria stiedae.*

SYMPTOMS. Mild cases may show no clinical signs of infection, but gross involvement of the liver results in faulty metabolism, loss of weight, and death.

POST-MORTEM. In early and mild cases the only finding may be of the presence of oöcysts, demonstrable microscopically in the bile of the gall bladder. As the diseases develop there is increasing invasion of the hepatic bile ducts, which thicken and show themselves as multiple white lesions throughout the liver substance. The lesions, when opened, are filled with greenish creamy fluid packed with oöcysts.

DIAGNOSIS. Microscopic examination of the bile or material from infected foci reveals the oöcysts. They are also voided in the faeces and require to be distinguished from intestinal coccidia.

ROUTE OF INFECTION. Oral: as in intestinal coccidiosis. It is said that the sporozoites liberated from sporulated oöcysts find their way to the liver via the portal vein.

CONTROL. As in intestinal coccidiosis; the so-called coccidiostats used in either the drinking-water or the feed appear to give a much higher degree of control of hepatic coccidiosis than they do of the intestinal disease.

ECTO-PARASITES

Ear canker

CAUSAL AGENT. Either of two species of mite may be involved. *Chorioptes cuniculi* or *Psoroptes communis*.

SYMPTOMS. Animals with severe ear canker usually show signs of discomfort, such as excessive scratching of the ears; mild cases often go undetected.

DIAGNOSIS. The ears contain yellow-white crust: a piece of the crust crushed in 10 per cent potassium hydroxide and examined microscopically will show the causal mites, nymphs, and eggs.

SOURCE OF INFECTION. Other infested rabbits.

CONTROL. The canker can be cured by a single installation into the ears of a diluted solution of tetraethyl-thurium monosulphide (Tetmosol, I.C.I.) one volume in nine volumes of warm water; it is unnecessary to remove the cankerous crust. Regular aural examination of rabbits should be practised. This can conveniently be done when the rabbits are set up for mating, at weaning, and when issued for experimental use. Such a routine eradicates ear canker or reduces the number of cases to a minimum.

Body canker

CAUSAL AGENT. *Notoedres spp.* and *Sarcoptes spp.* of mites.

SYMPTOMS. Irritation of the skin, followed by loss of hair and serious lesions.

DIAGNOSIS. The causal mites are demonstrable in hair and skin scrapings mounted in 10 per cent potassium hydroxide. Body mange is not common in domestic rabbits in this country. However, should it occur, it can be treated with gamma-benzene-hexachloride. A 0·05 per cent solution should be used, and the affected parts bathed twice with a ten-day interval between treatments.

HELMINTHS

Cysticercosis (bladder worms)

CAUSAL AGENT. *Cysticercus pisiformis* and *Cysticercus serialis*, the cystic stages of the dog tapeworms, *Taenia pisiformis* and *Taenia serialis*.

SYMPTOMS. Neither of these worms cause serious ill effects. The cysts of *C. pisiformis* are usually situated internally. The cysts of *C. serialis* develop in the subcutaneous tissue, causing large round swellings under the skin. The flank of the animal is a common site.

POST-MORTEM. *C. pisiformis* is found in the peritoneal cavity. It frequently occurs as clusters of multiple small cysts adherent to the pancreas, the liver, the stomach, and the mesentery.

C. serialis is found subcutaneously.

DIAGNOSIS. The presence of these typical cysts is diagnostic. When opened, immature tapeworms can be expelled from them.

ROUTE OF INFECTION. Oral.

SOURCE OF INFECTION. Foodstuffs, bedding, etc., which have been contaminated by the faeces of a dog harbouring the tapeworms. The larvae hatch from the ingested eggs and migrate to their final site by passing through the liver, their passage leaving fibrous tracts in that organ.

CONTROL. By avoiding the use of materials contaminated by dog faeces or sterilizing materials which may have been exposed to infection. The subcutaneous cysts of *C. serialis* can easily be removed, provided care is taken not to rupture them.

INJURIES

Fractures

Rabbits are easily frightened by noises, inexperienced handling, etc. Fracture of the spine may occur as a result of the animal jumping in fright and catching its back on some obstruction. Rabbits hind-legs are easily trapped in weld mesh, broken wooden boxes, etc., and a fracture results from the animals attempting to free the trapped limb. Injuries to the spine often pass unnoticed at the time of occurrence, and are detected later when signs of hind-quarter paralysis develop, resultant upon damage to nerves.

THE NON-SPECIFIC ENTERIC COMPLEX

Growing rabbits are prone to several ill-defined enteric conditions which are not yet fully understood. These conditions are mainly found in animals of from six to twelve weeks of age, the critical period of post-weaning adjustment. Many rabbits are ill and often die, showing at post-mortem examination either one or a combination of the following lesions.

Mucoid enteritis and/or mucoid typhlitis

The affected area of the gut contains large quantities of jelly-like mucus, which is sometimes excreted with the faeces.

Enteritis and/or typhlitis

There is inflammation of the gut, which may vary from a mild irritation to a severe haemorrhagic reaction.

Diarrhoea

A simple diarrhoea with fluid intestinal contents is seen. Evidence of pellet formation in the large gut is absent.

Impaction

The contents of the large gut present a solid impaction. There is gross distension of the gut wall, which is often without tone and is 'parchment-like' in appearance.

Fluid distention

This condition often follows the onset of impaction, the animal drinking excess water, causing distension of the stomach and intestines.

Although poorly understood, these conditions account for a large part of the losses in rabbitries. A mortality of 20–25 per cent of all litters born is not uncommon. Various causes have been suggested: faulty nutrition, inadequate housing, pathogenic bacteria, viruses, protozoa, etc.

Recent investigations have shown that viable bacteria are virtually absent from the gastro-intestinal tract of unweaned sucking rabbits. This apparently results from the production in the stomach of these animals of octoic and decoic acids; both of these substances possess active bactericidal properties. No appreciable amounts of either of these acids are produced when the young rabbit is fed the milk of other species. Enteric disease in the young rabbit frequently occurs at the age of six to seven weeks, and the onset of the disease is often accompanied by a large increase in numbers of viable bacteria present in the gut, particularly *E. coli.* It may be that the young weaned rabbit, during the period of post-weaning adjustment, when it is deprived of mother's milk and has lost any passive immunity it may have possessed, is prone to suffer from the ill-effects of largely unhindered bacterial growth in the gut and possibly die as a result of the absorption of toxic bacterial products. Though this may offer an explanation for the occurrence of post-weaning enteric disease, it does not account for the cases of mucoid enteritis and caecal impaction found in rabbits when they are about sixteen weeks old.

MONKEYS

The diseases described are those commonly affecting the *Macaca mulatta* species. These are still the most widely used monkeys in this country. However, should other species become used in large numbers, then other diseases may require to be described at some later date. For example, naturally occurring malaria is absent in the *M. mulatta* monkey, but is widespread in other species found in Africa, Asia, and South America.

BACTERIAL

Bacillary dysentery

CAUSAL AGENTS. Gram-negative bacilli of the Shigella genus.

SYMPTOMS. There is a diarrhoea which varies from a stool of semi-fluid consistency to a frank exudate of blood and mucus. The animal loses weight and condition, and in untreated cases the condition may terminate in death.

POST-MORTEM. The carcase is emaciated and dehydrated: the whole of the gut may show inflammatory and haemorrhagic changes and necrosis.

DIAGNOSIS. The presence of blood and mucus in the faeces is strongly diagnostic (microscopic examination of the exudate shows large numbers of pus cells). The causal organisms are recovered by inoculating the faecal exudate or gut contents on a selective medium such as desoxycholate citrate agar. Suspect isolates are examined biochemically and serologically. Commonly incriminated Shigella are *Shigella flexner* types 1, 2, 3, 4, 5, X, and Y; *Sh. sonnei* and *Sh. schmitzei.*

ROUTE OF INFECTION. Oral.

SOURCE OF INFECTION. Other infected monkeys and man.

CONTROL. Many monkeys imported into this country are either clinical cases or carriers of Shigella. Infection rates of between 30 and 40 per cent have been recorded in batches of these animals. Should the monkeys be penned together in large numbers after their arrival, outbreaks of dysentery may be severe. It is now common practice to cage monkeys in single units, and to treat them as a curative and prophylactic measure with broad spectrum antibiotics such as chloramphenical injected parenterally or terramycin given in the drinking-water. Resistant cases may be treated orally with a chemotherapeutic agent such as sulphaguanidine. Strict hygiene is also necessary; trays containing faeces should be removed, cleaned, and sterilized daily, and freshly sterilized cages regularly provided. Apart from the danger of spread among monkeys, Shigella are also human pathogens, and there are many recorded cases of handlers becoming accidentally infected. It is probable that most of these infections would have been avoided had adequate hand washing taken place after handling the animals or infected fomites.

Pneumonia

CAUSAL AGENTS. *Diplococcus pneumoniae*: a gram-positive diplococcus. *Klebsiella pneumoniae*: a gram-negative bacillus. Members of the Pasteurella group of gram-negative bacilli.

SYMPTOMS. Respiratory distress, inappetence, loss of condition and, if left untreated, the pneumonia often terminates in death.

POST-MORTEM. The lungs may be congested, consolidated, and hepatized.

DIAGNOSIS. During life there is usually apparent respiratory distress, which can be confirmed by stethoscopic examination and X-ray examination of the chest. At post-morten examination the particular causal organism involved can be demonstrated by plating infected material on laboratory medium such as blood agar. *Diplococcus pneumoniae* and *Klebsiella pneumoniae* are well-defined bacterial types and are easily identified. Members of the Pasteurella group may present more difficulties in classification.

ROUTE OF INFECTION. Respiratory tract.

SOURCE OF INFECTION. Other monkeys or humans.

CONTROL. The building containing monkeys should be warm, 70–75°F (21–24°C), dry, and adequately ventilated, the latter reducing opportunities for the occurrence of cross-infection. Diagnosed cases can usually be successfully treated with parenteral antibiotics.

Tuberculosis

CAUSAL AGENT. *Mycobacterium tuberculosis* var. *hominis* and var. *bovis*; more rarely *Mycobacterium avium*. (These are 'acid-fast' bacilli which require, for microscopic demonstration, to be stained by a technique such as the Ziehl Neelsen's method. After staining, the bacilli resist decolorisation by strong acid solutions and the application of a blue counterstain shows the organisms as red bacilli on a blue background.)

SYMPTOMS. The monkey becomes progressively emaciated and may, when the lungs are involved, show respiratory distress. Death is a common sequel to infection.

POST-MORTEM. Two main types of infection are usually encountered. The first is a generalized infection; whitish necrotic foci being distributed in all organs. Principal sites are the lymph nodes, the alimentary tract, the liver, and the spleen: infection may spread to the lungs. The second type of infection is primarily pulmonary, though there may be secondary spread. Lesions in the lungs can exist as large cavitations.

DIAGNOSIS. During life this is mainly by the use of the tuberculin test, which will be described below. At post-mortem material from infected areas can be stained to demonstrate the organisms. These may not be always easy to find, and a careful microscopic search is required. Cultures can be made on a selective medium, such as Lowenstein and Jensen's medium. Visible growth on culture medium is not present until after three to four weeks' incubation at 98·6°F (37°C).

ROUTE OF INFECTION. Probably orally or by the respiratory tract.

SOURCE OF INFECTION. Primary infection in wild monkeys is probably contracted from man, although the majority of infections in captive monkeys arise within the colony.

CONTROL. Freshly imported monkeys should be screened for the presence of reactors. Diluted mammalian tuberculin or its protein-purified derivative

(P.P.D.) is used. The injection is made into the eye-lid, which, within three days in a positive reactor, becomes reddened and oedematous, with the eye sometimes closing completely. Reactors should be killed. The test, while of value, possesses certain disadvantages. Very advanced cases of tuberculosis may not react positively. A number of positive reactors subsequently sacrificed are found to show no macroscopic evidence of infections. Infection may be confined to the lymph glands, which may show little if any change and require to have smears and sections made and stained to demonstrate the presence of acid-fast bacilli. Very early cases may fail to react positively. This last source of error can be partially corrected by testing a colony at frequent intervals. No attempt should be made to treat tubercular monkeys, as these animals represent a real hazard to human attendants. The sick animals would require to be kept in strict isolation and managed under conditions designed to ensure the safety of the attendants. These conditions can seldom be met, and there is rarely any justification for attempting treatment. As in other infectious diseases of monkeys, the spread of infection can be minimized by the caging of the animals in small units; by the provision of warmth and adequate ventilation, and the regular changing and sterilization of equipment. The use in monkey houses of bactericidal substances (e.g. hexyl resorcinol) which vaporize on heating to give continuous clouds of bactericidal particles are said to be of value in preventing the spread of disease.

VIRAL

B virus infection

CAUSAL AGENT. Herpes virus simiae.

SYMPTOMS. 'Herpes-like' lesions are found on the lips, hard and soft palate, buccal surface of cheeks, and the tongue. These sites may show considerable ulceration. The lesions start as small vesicles which ulcerate and scab. There is sometimes an accompanying nasal discharge and conjunctivitis. The animal usually shows little signs of systemic disturbance, and the lesions disappear within two weeks from the appearance of the vesicles: recovery is apparently complete.

POST-MORTEM. Pathological changes can only be demonstrated by histological examination of the central nervous system, liver, and kidneys.

DIAGNOSIS. By observing the typical lesions.

ROUTE OF INFECTION. It is thought that there are two methods of infection: by contact with food and water contaminated by the saliva of infected animals and by scratching and biting.

SOURCE OF INFECTION. Other monkeys.

CONTROL. Monkeys are often infected on arrival in this country. One series of observations disclosed that of 14,400 screened animals, more than 2 per cent had active lesions. It is doubtful, however, if the disease affects any monkeys adversely. The importance of the disease lies in that it is communicable to man by a bite or scratch from an infected animal, and the resulting B virus infection is usually fatal: of fifteen recorded cases, thirteen have terminated in death. Though many individuals are bitten by monkeys without apparently contracting this disease, the possibility that infection may occur is real, and precautions should be taken when handling these animals. They should, for example, be caged in such a way that they can be caught without undue handling. During any handling rubber gloves should be worn. Should, in spite of all precautions, a handler be bitten, then the wound should immediately be thoroughly cleansed and treated with an efficient antiseptic and the individual placed under medical supervision. Existing scratches and broken skin on the hands and arms of handlers also present a potential portal for virus entry.

It is considered that animals with active infection can transmit infection for about fourteen days after the appearance of the primary lesions and require to be in close contact with other monkeys before transmission can take place. It has become standard procedure in many laboratories to cage these animals individually when they are received from suppliers and to avoid all but essential handling for three to four weeks. During this time active cases, if any, should recover and become non-infective to other monkeys or the handlers. While this method may not be 100 per cent effective, it does substantially reduce the possibility of human infections taking place.

HELMINTHS

Oesophagostumum infestation

CAUSAL AGENT. A nematode: *Oesophagostumum apiostumum.*

SYMPTOMS. In severe infestations there may be a diarrhoea, with loss of weight and condition; the animal may die. It is more common, however, for an animal to be only slightly infected and show no signs of illness.

POST-MORTEM. Black rounded nodules are found in the wall of the large intestine, which when opened are found to contain developing worms measuring about 10 mm in length. The gut contents may be bloodstained and contain adult worms.

DIAGNOSIS. During life the adult worms may be found in the faeces and deposited ova from gravid females detected by a suitable flotation method.

ROUTE OF INFECTION. Oral.

SOURCE OF INFECTION. By the ingestion of material contaminated with the nematode larvae from the faeces of other infected monkeys.

CONTROL. By good hygienic management preventing the occurrence of opportunities for cross-infection. In known infected cases phenothiazine has been used to eliminate the worms from the gut, though immature forms encysted in the gut may resist such treatments.

ECTO-PARASITES

Pulmonary acariasis

CAUSAL AGENT. Members of the Pneumonyssus species of acari, e.g. *Pneumonyssus simicola.*

SYMPTOMS. Though these parasites have been said to cause pulmonary distress and other lung complications, opinions are varied as to their pathogenicity, and the condition is described here because infestation is widespread, and evidence of this is frequently seen at routine autopsy.

POST-MORTEM. The lungs show discrete foci which may vary in appearance from small white spots to greenish-yellow-coloured lesions measuring a few millimetres in diameter.

DIAGNOSIS. The pulmonary foci, when opened, contain the causal mites, which can be demonstrated microscopically. The finding of the typical foci containing mites is diagnostic.

SOURCE OF INFECTION. Other monkeys.

ROUTE OF INFECTION. This has not yet been definitely established. It has been suggested that the mite is inhaled into the respiratory passage or, alternatively, that it is ingested and reaches the lungs via the lymphatic system and the blood stream.

CONTROL. As the true pathogenicity of the parasites is in doubt, no real measures have been introduced, though cross-infection has been said to be preventable by dusting animals with DDT.

NON-SPECIFIC

After arrival in this country a number of young monkeys refuse to eat any appreciable quantity of food, even when a large variety of foodstuffs is offered. The animals behaving in this way are not necessarily suffering from any recognizable disease, though the refusal to eat frequently terminates in death due to inanition. Monkeys need more personal attention than most other laboratory species, and a special effort is often necessary on their arrival in the animal house to provide the extra care and interest needed. Following the critical period of acclimatization and adaptation to food and environment, losses from disease and inanition are usually negligible, providing the system of management is good. It is becoming common practice to treat batches of monkeys freshly arrived in this country with a mixture of antibiotics such as aureomycin and terramycin, fed in either the drinking-water or the diet. It is thought that these prophylactic measures aid in the control of specific disease and so-called non-specific complaints.

DOGS

The diagnosis and treatment of the diseases of dogs and cats require the services of a qualified veterinarian. It is, however, advisable that the animal technician should possess a knowledge of the more common diseases which may be encountered during the care and management of these animals, and they are described for this reason.

Distemper/hard pad

It is now accepted that distemper and hard pad are disease manifestations caused by the same virus, vaccination against distemper giving an immunity which prevents the occurrence of both disease syndromes.

Vaccination

Composite vaccines are now available for the simultaneous immunization of dogs against distemper, contagious hepatitis, and leptospirosis; primary vaccination normally being carried out at twelve weeks of age, followed by a second injection two weeks later. It is inadvisable to vaccinate before twelve weeks, as the young animal may possess until then a level of maternally transferred passive immunity sufficiently high to interfere with the development of active immunity following vaccination. However, maternal antibodies may fall to a low level or be absent before the twelfth week, and young puppies should not be exposed to infection until they are vaccinated. If it is impossible to avoid exposure to infection, then distemper vaccine should be given at eight weeks and followed by further vaccination at twelve and fourteen weeks. It is also advisable to revaccinate all animals at yearly intervals against distemper and leptospirosis, as the neutralising antibodies may fall to a low level and be insufficient to prevent infection. The three diseases are described below, and when vaccination is mentioned it is assumed that the vaccine used will be one producing an immunity against all three.

Quarantine

Dogs obtained from outside sources should, irrespective of their age and condition, undergo a period of quarantine of at least fourteen days. This period should be used to free the animals of skin parasites and intestinal worms and, in the absence of well-authenticated records of satisfactory vaccinations, immunization should be carried out against distemper, contagious hepatitis, and leptospirosis.

VIRAL

Distemper

CAUSAL AGENT. A virus.

SYMPTOMS. Distemper usually occurs in animals between three and twelve months of age. There is an elevation of temperature 103–104°F (39·4–40°C) and refusal to eat. The eyes discharge and show a conjunctivitis and keratitis. A nasal discharge is usually present consisting of serous or mucopurulent material. Distressed breathing occurs as a result of bronchitis and bronchopneumonia, and there is often vomiting. Ulcers may be found on the tongue and inside the cheeks. Diarrhoea may be present, and involvement of the central nervous system leads to convulsive fits and paralysis.

DIAGNOSIS. By the clinical symptoms.

SOURCE OF INFECTION. The infected secretions from the nose and eyes of infected or carrier dogs.

CONTROL. By the method of vaccination described above. Should distemper occur in the absence of, or despite, vaccination, then an anti-serum is available, from commercial sources, which is used in treatment. There is, at present, no specific therapeutic agent active against animal viruses, though, should secondary bacterial infection occur, sulpha drugs and antibiotics are often used.

Hard pad

CAUSAL AGENT. A virus.

SYMPTOMS. The disease received its popular name from the swelling and thickening of the pads of the feet which cause a tapping sound when walking. Other symptoms are as described under distemper.

DIAGNOSIS. As described under distemper.

SOURCE OF INFECTION. As described under distemper.

CONTROL. As described under distemper.

Contagious hepatitis/infectious canine hepatitis

CAUSAL AGENT. A virus.

SYMPTOMS. Overt-disease occurs most frequently among dogs between three and twelve months old, though many dogs apparently suffer a sub-clinical infection which passes unnoticed. In obvious clinical cases the temperature is raised 103–104°F (39·4–40°C). There is refusal of food, though the animal may drink copiously. Conjunctivitis and an accompanying discharge from the eyes and the nose is common. The blood clotting time is increased, leading to the appearance of petechial haemorrhages of the skin and to severe bleeding should the dog injure itself.

DIAGNOSIS. In the acute fatal form of the disease the dog may show few symptoms before dying. Viral inclusions can then be found in histological sections of liver tissue collected at autopsy. In less-acute cases it may be difficult to differentiate the disease from distemper and, indeed, it is not uncommon for the two diseases to occur simultaneously in the same animal. The onset of canine hepatitis is, however, usually more rapid than distemper, and prolongation of the bleeding time of diagnostic value.

SOURCE OF INFECTION. Dogs recovered from the disease may continue to excrete the causal virus in the urine for many months. During active infection the virus is present in all excretions from the body.

CONTROL. By the method of vaccination described above.

Leptospirosis (Stuttgart disease, infectious jaundice)

CAUSAL AGENTS. *Leptospira canicola* and *Leptospira icterohaemorrhagiae*.

SYMPTOMS. The animal refuses food and may vomit, the temperature is raised 104°F (40°C), and there is often unwillingness to rise owing to muscular stiffness and pain. There is yellowing of the eyes and mucus membranes of the mouth, and the latter may show haemorrhagic patches. Sometimes the gums bleed, and there may be a bloodstained diarrhoea.

DIAGNOSIS. By clinical symptoms. The causal spirochaete is excreted in the urine and can be demonstrated using the dark-ground-illuminated microscope or in dried smears of urine stained by a silver impregnation method. Antibodies are present in the serum and can be demonstrated by serological tests.

SOURCE OF INFECTION. *L. canicola* is found in the urine of infected or carrier dogs. *L. icterohaemorrhagiae* is harboured by wild rats; up to 40 per cent of groups of these animals have been found to excrete *L. icterohaemorrhagiae* in the urine, and the organism infects dogs following contact with materials contaminated by the urine of carrier rats. Access to the body is probably through the membranes of the mouth, though Leptospira can penetrate unbroken skin.

CONTROL. By the method of vaccination already described. Wild rats should never be allowed to exist in or around dog kennels. It should also be recognized that *L. icterohaemorrhagiae* is a human pathogen, and an infected animal may, unless care is taken, infect handlers.

HELMINTHS

Ascariasis

CAUSAL AGENTS. *Toxocara canis* and *Toxascaris leonina*. Nematode round worms.

SYMPTOMS. The mature worms are commonly found excreted in the faeces, and as they may attain a length of 90–180 mm, they are easily seen. Many adult dogs harbour round worms without showing noticeable signs of illness, though heavy infestations, particularly in young puppies, lead to loss of weight and condition and even death. A puppy may harbour so many worms that they cause occlusion of the small intestine.

DIAGNOSIS. Adult worms are seen in the faeces, and eggs can be detected microscopically using a flotation concentration method.

SOURCE OF INFECTION. Other dogs excreting round worm ova.

CONTROL. Piperazine acid citrate is effective in eliminating both species. Treatment usually needs to be carried out twice. Puppies *in utero* can be infected with *T. canis* from their mother, and bitches should be treated during the gestation period and young puppies a few weeks after birth. Fresh entrants into the kennels can be treated during the period of quarantine.

Tapeworms

CAUSAL AGENTS. Tapeworms of the Taenia genus. *T. pisiformis*, *T. hydatigena*, *T. ovis*, *T. multiceps*, and *Echinococcus granulosus*. A single worm of the Dipylidium genus occurs, *D. caninum*; this is probably the most common tapeworm found in dogs.

SYMPTOMS. In light infestations the animal usually shows no noticeable signs of illness. Heavy infestations can lead to loss of weight and condition and even death. Fits can be caused by helminthic toxins. Segments passing the anus may cause extreme irritation.

DIAGNOSIS. The finding of segments (proglottids) in the faeces; ova can be detected by microscopic examination.

SOURCE OF INFECTION. Taenia species require an intermediate host in which the ova excreted by the host animal, in this case the dog, develop as a cystic stage (coenurus), e.g. the cysts of *T. pisiformis* (*C. pisiformis*) are found in the rabbit. The dog eating that part of the rabbit containing the cyst becomes infected. A single species may be found in several intermediate hosts, e.g. the cysts of *E. granulosus* are found in man and domesticated mammals, and may be the cause of serious illness.

CONTROL. The use of cooked and tinned foods has led to a decrease in the infestation of dogs with certain tapeworms. However, this does not apply, for example, to *D. caninum*, where the intermediate hosts are the dog flea and louse. Control is by the exclusion from the kennels of animals or materials which may harbour cysts. Two drugs commonly used to expel tapeworms from infested animals are arecoline hydrobromide and acetarsol.

Ear canker

There is little value in attempting to tabulate the causes of canine ear canker. The examination, diagnosis, and treatment require the services of a veterinarian. It is commonly thought that ear canker is caused by a mite, *Otodectes cynotis*, and while it is true that this mite may be involved, it is far from being the sole cause of canine ear canker, which is often a complex condition, requiring lengthy and sometimes even surgical treatment.

Skin diseases (mange/ringworm/pruritus/eczema/dermatitis)

As in ear canker, skin diseases of dogs may arise from a variety of causes, e.g. sarcoptic mange (a mite, *Sarcoptes scabei*), ringworm (*Microsporum canis*), allergies, nutritional deficiencies, etc. Though a veterinarian is needed to attend to the skin diseases of dogs, it is nevertheless good practice to thoroughly bath all fresh intakes with a proprietary shampoo, such as a preparation of selenium sulphide in a detergent (Seleen), which is effective in killing fleas and lice, and also has some value in the treatment of demodectic mange and the so-called non-specific dermatoses. Lice can also be killed on the animal by using benzene hexachloride or DDT, though fleas will infest bedding and buildings and return to reinfest the animal unless thoroughly eradicated.

Nephritis

This is common in old dogs, and may cause serious illness. It is common practice in certain establishments to routine test the urine of dogs for the presence of albumin, a positive test indicating the necessity for further investigation.

CATS

Quarantine

Freshly imported cats should, like dogs, undergo a period of quarantine of at least fourteen days, during which time treatment can be given for fleas, lice, and helminths, and vaccination carried out to immunize against panleucopaenia (see p. 197).

BACTERIAL

Kittens and young cats reared in catteries may develop septicaemic conditions due to various bacterial invaders; examples of these being *Streptococcus haemolyticus*, *Pasteurella septica*, and *Haemophilus influenzae*. The author has personal experience of epidemic cat septicaemia where the causative organism was *S. haemolyticus*, and this condition is described. It is similar to septicaemias caused by other micro-organisms.

Septicaemia

CAUSAL AGENT. *Streptococcus haemolyticus* (Lancefield Group G).

SYMPTOMS. Generalized bacterial septicaemia occurs in kittens during the first few weeks of life, though older cats may be infected. When the disease is acute and fulminating the kitten may show no noticeable symptoms before death. In more chronic cases there is raised temperature, loss of appetite, loss of condition, and dyspnoea if the lungs become involved.

DIAGNOSIS. During life diagnosis of isolated cases is difficult. During an epidemic detailed examination of dead animals, at autopsy, usually results in the isolation and identification of the causative organism. Post-mortem examination shows infection of the liver and spleen and peritonitis; the lungs may become infected. The causal organisms can be seen in gram-stained smears of infected tissues and isolated on a laboratory medium such as blood agar.

SOURCE AND ROUTE OF INFECTION. Very young kittens can be infected via the navel cord, the mother carrying the causal bacteria in her respiratory or genito-urinary tract and infecting her offspring during post-parturition cleansing. It is also possible that infection may arise from biting injuries or possibly by the aural route (see Ear canker), or by the respiratory passages.

Humans can be carriers of *S. haemolyticus* and *H. influenzae*, and could be responsible for introducing infection.

CONTROL. The use of parenterally administered broad-spectrum antibiotics. Should breeder cats be known to harbour potential pathogens in their genito-urinary tract, it is of value to install into the vagina a topical application of a suitable antibiotic. For example, intramammary penicillin, containing 100,000 units per tube, instilled for two to three days before parturition will prevent the occurrence of streptococcal 'navel ill' in the youngster.

Pneumonia

CAUSAL AGENT. *Bordetella bronchiseptica*. (This is the same bacterium as that associated with pneumonia in the dog, rabbit, and guinea pig.)

SYMPTOMS. Dyspnoea, loss of appetite, and loss of condition.

DIAGNOSIS. During life by the clinical symptoms. At autopsy the lungs show a broncho-pneumonia with varying degrees of consolidation. *B. bronchiseptica* can be recovered by inoculating infected lung on to blood-agar and bile-salt-agar plates.

SOURCE OF INFECTION. Cats or other animals harbouring *B. bronchiseptica* in the respiratory tract.

ROUTE OF INFECTION. Respiratory.

CONTROL. Overt cases respond to parenterally administered antibiotics and injections of specific anti-serum. Vaccination can be of value as a prophylactic measure; the breeders being vaccinated annually, thus passing on a useful degree of passive immunity to their offspring. Vaccination can lead to a substantial reduction in the incidence of overt cases of *B. bronchiseptica* pneumonia.

VIRAL

Feline panleucopaenia/infectious feline enteritis

CAUSAL AGENT. A virus.

SYMPTOMS. The onset is often sudden; the animal refusing to eat: it may vomit and pass fluid stools. Panleucopaenia is highly contagious and fatal; mortality rates in young cats can be as high as 60–90 per cent.

DIAGNOSIS. The disease receives its name from the occurrence of aplasia of the bone marrow, this being reflected in a marked diminution of the leucocytes circulating in the peripheral blood. Leucocyte counts often fall below 2,000 per cu mm, and cases have been recorded where circulating white cells were absent. The clinical symptoms allied to a low leucocyte count and high mortality rates are highly suggestive of panleucopaenia.

POST-MORTEM. There is aplasia of the sternum bone marrow, the lymph glands are oedematous, and there is an enteritis, typhlitis, and colitis: the ileum being most affected.

SOURCE OF INFECTION. Other cats whose excretions are infected with the virus.

CONTROL. By the use of protective vaccination. An inactivated vaccine prepared from the tissues of infected cats is available from commercial sources. Kittens are inoculated at six weeks of age, followed by a second inoculation two weeks later. Adult breeders should receive a booster vaccination each year.

Feline pneumonitis

CAUSAL AGENT. A virus or viruses.

SYMPTOMS. Pneumonitis is a highly contagious disease of cats which is seldom fatal. There is loss of appetite, rise in temperature, conjunctivitis with a muco-purulent discharge, and a nasal discharge, the animal frequently sneezing.

POST-MORTEM. The anterior and diaphragmatic lobes of the lungs are consolidated; the trachea and larynx are inflamed and may contain thick mucus.

DIAGNOSIS. By the clinical symptoms. Elementary viral bodies contained in mononuclear cells can be seen in Giemsa-stained smears of mucus from the trachea and larynx and those areas of lung showing pneumonitis.

SOURCE OF INFECTION. Cats, either carriers or overt cases, who sneeze and disseminate the virus over large areas.

ROUTE OF INFECTION. Probably respiratory.

CONTROL. The respiratory tract of the cat is now known to harbour a number of viruses, at least two of these cause the disease syndrome detailed above. A vaccine has been prepared, from one of these viruses, grown in the yolk sac of the embryonic chick, but this, so far, is only available from commercial sources in the United States. Though there is no specific therapeutic anti-viral agent, pneumonitis is frequently complicated by secondary bacterial infection, particularly of the eyes and nose, and these respond well to topical installations of antibiotics. Parenterally administered antibiotics are often of value when there is bacterial invasion of the lungs. If it is possible to persuade the sick cat to eat well recovery usually takes place more rapidly.

FUNGAL

Ringworm/favus

CAUSAL AGENTS. *Trichophyton felineum*, *Microsporom felineum*, *Microsporom canis*, and *Achorion Quinckeanum.*

SYMPTOMS. Skin lesions may be distributed over the whole of the body, common sites are the head, neck, face, and the fore-paws. The lesions may vary from small bald patches to large areas of scaling and crusting.

DIAGNOSIS. Hairs and skin scrapings from affected areas are mounted in 10 per cent potassium hydroxide and examined microscopically. Spores and/or hyphae of the causal fungus can be seen in or around hairs and scrapings. Cultures are made by inoculating suspect material on to plates of a selective medium such as Sabouraud's. A Wood's light is of value in the diagnosis of *microsporosis*. Infected hairs when illuminated by this type of ultra-violet lamp exhibit the phenomenon of fluorescing, appearing as green coloured. Diagnosis, using a Wood's light, can be confirmed by microscopical and cultural examination of fluorescing hairs.

SOURCE OF INFECTION. Usually other cats and sometimes dogs. *A. Quinckeanum* infection can be contracted from wild mice.

CONTROL. Griseofulvin, a systemic antimycotic, is now used to treat ringworm. The occurrence of a case of ringworm should result in the isolation of the animal and adequate sterilization of the vacated quarters and the eating and drinking utensils, etc.

Body mange

CAUSAL AGENTS. Two mites *Notoedres cati* and *Demodex folliculorum* var. *cati.*

SYMPTOMS. These mites burrow into the skin, lesions being commonly found on the face and around the ears.

DIAGNOSIS. The mites and developmental stages are demonstrable by the microscopic examination of hair and skin scrapings mounted in 10 per cent potassium hydroxide.

SOURCE OF INFECTION. Other cats.

CONTROL. These parasitic manges usually respond to treatment with acaricidal preparations such as benzyl benzoate emulsion and suspensions of benzene hexachloride. Care must be taken during treatment to prevent the animal licking these preparations, which may be toxic to cats when ingested. Demodectic mange may be difficult to cure, and treatment is often prolonged.

Ear canker

CAUSAL AGENT. A mite, *Otodectes cynotis*.

SYMPTOMS. The animal may shake its head and scratch its ears, which contain a dark-brown waxy mass. Occasionally there is an underlying bacterial infection shown by the presence of free pus under the wax.

DIAGNOSIS. The causal mites and other developmental stages can be seen by microscopic examination of the aural exudate mounted in 10 per cent potassium hydroxide.

SOURCE OF INFECTION. Other cats.

CONTROL. A good measure of control can be achieved even when large numbers of cats are affected by *O. cynotis*, providing the system of treatment is adequate. A number of preparations are effective against *O. cynotis*, examples being dibutyl phthalate, benzyl benzoate emulsion, and 5 per cent piperonyl butoxide. Mixed otodectic and bacterial infections are not uncommon, and certain proprietary preparations contain both acaricidal and bactericidal substances and a locally acting analgesic agent which reduces discomfort. Cats' ears should regularly be examined and treated. It is convenient to examine adult breeders before parturition and kittens when they are vaccinated against panleucopaenia. Such a routine will eradicate ear canker or reduce cases to a minimum.

HELMINTHS

Roundworm infestations

CAUSAL AGENTS. The nematode worms, *Toxascaris leonina* and *Toxocara mystax*.

SYMPTOMS. Cats may harbour these worms in the gut without showing noticeable signs of illness. Heavy infestations may, however, cause severe loss of weight and condition.

DIAGNOSIS. Adult worms are found in the faeces, and ova can be detected microscopically after concentration by a flotation method.

SOURCE OF INFECTION. Other cats.

ROUTE OF INFECTION. Oral.

CONTROL. Piperazine acid citrate is efficient administered orally; treatment may have to be repeated once or twice.

Tapeworm infestations

CAUSAL AGENTS. *Taenia taeniaeformis* (*T. crassicollis*), *Taenia pisiformis*, and *Dipylidium caninum*.

SYMPTOMS. As in roundworm infestations, the cat may harbour tapeworms without showing noticeable signs of illness. Heavy infestations can cause loss of weight and condition. There may be vomiting and diarrhoea.

DIAGNOSIS. Tapeworm segments (proglottids) are found in the faeces, and ova may be detected by microscopic examination of faecal suspension.

SOURCE AND CONTROL OF INFECTION. The cystic stage of *T. taeniaeformis* is found in the liver of the mouse and the rat, and the similar stage of *T. pisiformis* is found in the abdomen of the rabbit. Therefore, infection can be prevented by not feeding raw rabbit meat and excluding wild mice and rats. *Ctenocephalides felis*, the cat flea, is an intermediate host for *D. caninum*. The larvae of the flea eat the tapeworm eggs and when the larvae become adult fleas the tapeworm embryos become cysticercoids; the cat being infected by eating these fleas. Therefore infection of cats with *D. caninum* does not take place in the absence of fleas. Drugs commonly used for the elimination of mature tapeworms from the cat are arecoline hydrobromide, drocarbil, arecoline acetarsol, and tetrachlorethylene. Great care needs to be taken in the estimation of dosage of these drugs and capsules of tetrachlorethylene can cause asphyxiation unless swallowed without chewing.

Flea and louse infestations

CAUSAL AGENTS. *Ctenocephalides felis*, the cat flea, and *Felicola subrostrata*, the cat louse.

SYMPTOMS. Heavy infestations may lead to discomfort, scratching, and inflammation of the skin.

DIAGNOSIS. The finding and identification of the causal parasite.

SOURCE OF INFESTATION. Other infested cats.

CONTROL. Fleas and lice can be killed on the infested animals by the use of pyrethrin dusts. DDT and BHC should not be used on cats. The insecticide should be rubbed into the animal's coat and brushed out. Lice can be eradicated by two treatments at ten-day intervals. The control of fleas, however, is more difficult. Apart from a thorough treatment of the animal, it is also necessary to treat bedding, floor cracks, wooden boxes, etc. Wherever possible infested bedding should be destroyed.

Urinary calculi

A common occurrence in breeding tom cats is the presence, in either the neck of the bladder or the urethra, of large calculi causing urinary obstruction. The cat is restless and irritable, and is often seen attempting to urinate and only succeeding in passing a drop or two of urine which may be bloodstained. Palpation of the abdomen shows the bladder to be full.

Urinary calculi are composed of substances, such as magnesium ammonium phosphate, urates, calcium carbonate, cystine, etc., and when a calculus is detected it is usually so large as to require surgical removal, and recurrences are not uncommon.

Various remedies have been suggested for the prevention of the formation of calculi: lowering the mineral content of the diet, increasing the vitamin A intake, altering the urinary pH by feeding substances such as sodium acid phosphate, etc. None of these remedies are of proven value, but the occurrence of renal calculi has been prevented in some colonies by feeding diets containing sodium chloride at a level of 1 per cent of the dry matter content. This appears to ensure fluid intakes and renal excretion rates that are high enough to prevent the formation of crystalline aggregates in the genito-urinary tract.

POULTRY

BACTERIAL

Pullorum Disease (Bacillary White Diarrhoea)

CAUSAL AGENT. *Salmonella pullorum.*

SYMPTOMS. *Chickens.* Infection of the egg can take place in the ovary of a hen that is a carrier of the causal organism. The chicken hatched from such an egg is infectious to others either hatched in the same incubator or reared with the infected animals during the first few days of life. Incubators and other equipment can harbour the organisms and be responsible for other outbreaks of disease if there is failure to clean and disinfect efficiently between hatchings.

In chickens up to seven days old no symptoms may be noted and the animals merely found dead. When the disease follows a less-acute course the chickens may be lethargic, fail to eat, continually chirp and have a white diarrhoeal exudate which forms a sticky mass around the vent. Though the overall mortality during an outbreak may be as high as 50 per cent, a number of chickens may show no symptoms and survive. These birds can then become carriers, and when adults lay infected eggs.

Adults. These are more resistant to infection than chickens, the majority of infected adults are symptomless excretors. These birds, when detected by the serological methods described below and subsequently culled, are often found at autopsy to have an infection of the ovaries. The developing ova may be obviously diseased and contain bloody fluid or inspissated material. Evidence of infection may also be found in the peritoneum and in the heart, causing a pericarditis and a focal infection of the myocardium. Sick adult birds may have a yellow diarrhoea, ruffled feathers, a jaundiced head and a shrunken comb.

DIAGNOSIS. The causal organism can be recovered by culturing on selective laboratory media pieces of infected organs and intestinal contents.

CONTROL. Until a few years ago it was considered advisable to slaughter any batch of chickens when *S. pullorum* was detected. The drug, Furazolidone, is now extensively used both to control and eradicate infection. Nevertheless, the practise of good hygiene is still essential, and incubators and other equipment used must be cleaned and sterilized after each batch of chickens is hatched and reared.

Carriers can be eliminated from flocks by application of the rapid, whole-blood, stained-antigen test. The method of testing is as follows:

A drop of fresh blood, obtained from a wing vein, is mixed on a glass plate with a drop of fluid containing dead *S. pullorum* organisms suspended in a dilute solution of a violet dye. The blood of carrier birds contains antibodies against *S. pullorum*, and these react with the stained bacterial suspension, causing the formation of visible clumps. Any bird whose blood gives a positive reaction is culled. As an alternative method, the blood can be sent to the laboratory, the serum removed and tested by the tube method, using an unstained suspension of organisms. This method is also used to examine the blood of birds whose blood gives a doubtful reaction to the rapid method.

Fowl Typhoid

CAUSAL AGENT. *Salmonella gallinarum*

SYMPTOMS. This disease usually occurs in adult birds. The infected animal is lethargic and has a profuse yellow-coloured diarrhoea; the head parts are often pale and jaundiced. As many as 50 per cent of a flock may die following exposure to infection and the survivors become carriers.

POST-MORTEM. The most characteristic lesion is congestion of the liver, the surface of the organ having a greenish-bronze appearance best seen when the carcase has been open for a short time. The lungs have a brownish discolouration and the spleen is enlarged. Nodules may be found on the heart and in the intestinal tract.

DIAGNOSIS. *S. gallinarum* can be recovered by culturing, on selective laboratory media, infected material collected at autopsy and from intestinal contents. *S. gallinarum* has the same antigenic structure as *S. pullorum*, but they can be differentiated by biochemical tests.

SOURCE OF INFECTION. Carrier birds can spread infection amongst non-immune flocks and, during acute disease, handlers and contaminated equipment help to spread infection.

CONTROL. The blood of carriers contains antibodies that react to the *S. pullorum* rapid whole-blood, stained-antigen test (see above) and the culling of positive reactors to the test will eliminate carriers of either organism. There is also available a commercially produced vaccine. The injection of birds when they are nine to ten weeks old results in a high degree of immunity. The disease can also be controlled by the feeding of diets containing Furazolidone.

Salmonellosis

In addition to *S. pullorum* and *S. gallinarum*, a number of other Salmonella serotypes are associated with disease in chickens. Infection follows a similar pattern to that seen in White Bacillary Diarrhoea, and can be differentiated by isolating the causal organism in the laboratory and identifying its antigenic structure. Control measures are as described for *S. pullorum* and *S. gallinarum*. The rapid whole-blood test using the stained antigen of *S. pullorum* is, however, of no value in eliminating carriers of other Salmonella serotypes.

Fowl Cholera

CAUSAL AGENT. *Pasteurella aviseptica*

SYMPTOMS. Comparatively few epidemics with high mortality rates occur in this country, the course of disease in infected birds being mild and/or chronic. Mild cases may show only a nasal catarrh and have distressed breathing. In chronic cases abscesses occur in the subcutaneous tissues, in the joints and the lungs. Infection can also be localised in the wattle, which swells and has a content of caseated pus.

DIAGNOSIS. The causal organism is easily demonstrated by the examination of stained smears made from infected material. It grows well on routine laboratory media. Pure cultures cause death when injected into mice by parenteral routes, and this method is used to establish the pathogenicity of a suspect isolate.

SOURCE OF INFECTION. It is thought that the disease spreads by the respiratory route following contact with carrier birds or materials infected by these birds.

CONTROL. Antibiotics and chemotherapeutic agents have been used to control mild outbreaks. These agents appear to be of little value in the control of epidemics.

Staphylococcus aureus infection

Two types of infection are caused by this organism. The first type, 'Bumblefoot', is an infection of the foot resulting in gross local swelling, which may spread upwards into the tissues of the leg. If the bird is valuable and the infection localized the lesion can be incised, the infected material removed and local antibiotic treatment applied. In the second type of infection the joints are affected, particularly the hock and tendons. Free or caseated pus is found surrounding these joints. The animals should be culled.

Fungal Aspergillosis

CAUSAL AGENT. *Aspergillus fumigatus* (a mould)

SYMPTOMS. Aspergillosis is usually found in young chickens. Infected chicks have respiratory distress and diarrhoea, and often die within a day or two of showing signs of illness. In adult birds infection is more chronic and there may be widespread involvement in the internal organs before the bird is noticeably distressed.

POST-MORTEM. *Chickens.* Small yellow-white nodules are found in the lungs, the breast bone and in the air sacs of the abdomen.

Adults. Green-coloured nodular growths consisting of spreading mycelium and typically mould-like in appearance are found in the lungs, the abdominal air sacs, the trachea and the bones. Care must be exercised when examining these animals at autopsy, as *A. fumigatus* is infectious to humans, and pulmonary disease can be caused by inhalation of the spores.

DIAGNOSIS. The fungus can be demonstrated by the microscopical examination of infected material mounted in a 10 per cent solution of potassium hydroxide. It grows readily on mycological media inoculated with infected material.

SOURCE OF INFECTION. Infection results from the inhalation of *A. fumigatus* spores disseminated by other birds or substances, such as mouldy food.

CONTROL. There is no effective cure. The provision of good ventilation and foodstuffs free from mould growths will substantially reduce the opportunities for infection to occur.

Chronic Respiratory Disease

The above term is used to describe certain disease processes found in the respiratory tract of birds. Agents involved can be *Escherichia coli;* Pasteurella spp; *Haemophilus gallinarum*; *Mycoplasma gallisepticum*. Readers requiring detailed knowledge of this syndrome are referred to the *Handbook of Poultry Diseases* (BVA, 1965).

VIRAL

Newcastle Disease

CAUSAL AGENT. A virus.

SYMPTOMS. This disease occurs in two distinct forms—mild and acute/virulent.

In this country the mild form of the disease is most commonly seen and is endemic. Infected young birds show symptoms of bronchitis, diarrhoea and lack of appetite. Paralysis of the wings, legs, neck and head may occur, with inco-ordination of movement. Death is common in young growing birds, but mortality may be low in established laying flocks.

The acute and virulent form of the disease follows exposure of flocks to infection borne by the carcases of poultry imported into this country. Infected animals may die without showing any noticeable symptoms, or they may exhibit great thirst, refuse food and produce a green watery diarrhoea. Breathing may be distressed and mucus discharged from the eyes, the nose and the mouth.

POST-MORTEM. Mild cases may only have a mucoid exudate in the trachea and bronchii.

Many acute cases have no obvious lesions, others have haemorrhages of the proventriculus, the gizzard, the intestines, the base of the heart and sternum.

DIAGNOSIS. This depends on isolation in the laboratory of the causal virus from infected tissue.

Evidence of infection can also be indirectly demonstrated by the Haemagglutination Inhibition (H.I.) test. Suspension of chicken cells when mixed with Newcastle virus show visible clumping of the cells. The addition of serum from infected or recovering birds contains sufficient specific antibody to prevent the formation of clumps. By performing the test with increasing dilutions of the serum under test an estimate can be made of the amount of antibody present.

SOURCE OF INFECTION. Other infected birds or materials contaminated by them.

CONTROL. Until 1963, when Newcastle disease was confirmed as present in a flock or flocks, all contact birds had to be slaughtered. This regulation is still enforced in Scotland, but not in England or Wales. Since the cessation of the slaughter policy in these two countries, inactivated vaccines have been widely used to control the disease.

Newcastle disease is still notifiable. Any person suspecting the presence of the disease must report this to the nearest veterinary officer of the Ministry of Agriculture, Fisheries and Food and, if the disease is confirmed as present, notification is made to the police. These authorities are then responsible for putting into effect compulsory orders regulating the movement of livestock to and from the infected premises.

Fowl Plague

CAUSAL AGENT. A virus.

SYMPTOMS. Similar to those found in the acute form of Newcastle disease. Mortality may be as high as 100 per cent.

POST-MORTEM. There is peritonitis and petechial haemorrhages in the proventriculus, duodenum, epicardium and sternum.

DIAGNOSIS. By isolation in the laboratory of virus recovered from infected material and the demonstration of specific antibodies in the sera of sick birds.

SOURCE OF INFECTION. Thought to be the ingestion of foodstuff contaminated with the excreta and other discharges of infected birds.

CONTROL. Fowl plague is uncommon in Great Britain. The disease, should it occur, is notifiable, and the official procedure is the same as for Newcastle disease.

N.B. Copies of the Fowl Pest Order can be obtained from H. M. Stationery Office. This details the procedure to be followed where Newcastle disease or fowl plague is confirmed as present in poultry. Fowl pest is a collective term used to describe Newcastle disease and fowl plague.

Avian Leucosis—Lymphoid Leucosis; Myeloid Leucosis; Erythroleucosis

CAUSAL AGENTS. Each of the above 'cancer-like' conditions is caused by a virus. These three viruses, though possessing distinct and specific properties, are related to each other.

SYMPTOMS. Affected birds lose condition and become debilitated. They may die, but are often culled before death occurs.

POST-MORTEM. *Lymphoid leucosis.* The blood cells involved are the lymphocytes. The condition is often referred to as 'big liver disease'. The liver and the spleen may be moderately or grossly enlarged, and obvious surface tumours may be present. Nodular tumours may occur in all the major organs, the skin and muscular tissue.

Myeloid leucosis. The blood cells involved are the early forms of the circulating granulocytes. Two forms are found, diffuse and discrete. In the former the liver and spleen are enlarged, in the discrete form white chalky tumours are found in the sternum, ribs, skull and pelvis.

Erythroleucosis. The blood cells involved are the red cells. Discrete tumours are not found. The organs principally involved are the liver, spleen and kidneys. These are enlarged, soft and friable.

DIAGNOSIS. By post-mortem appearance and the histopathological examination of diseased tissues.

SOURCE OF INFECTION. Infection probably spreads by young chickens being in contact with infected or carrier adult birds. The lymphoid leucosis virus has been found in eggs.

CONTROL. No drugs are of value in controlling or eliminating infections. There is evidence that susceptibility to the leucosis viruses is linked to genetical structure and that selective breeding can produce birds resistant to infection. It is also of obvious value to hatch and rear chickens free from virus and prevent these subsequently coming into contact with carrier adults.

Marek's Disease (Fowl Paralysis)

CAUSAL AGENT. A virus.

SYMPTOMS. There is usually a lameness resulting from involvement of the nerves of the legs. The wings may be dropped, the neck twisted and the bird unable to eat or drink.

POST-MORTEM. Swellings are found in the main nerves of the leg, wings, chest and other organs. These changes may be accompanied by tumours in the major organs.

DIAGNOSIS. By the clinical symptoms and post-mortem finding.

SOURCE OF INFECTION. As in avian leucosis—see above.

CONTROL. Ditto.

Avian Encephalomyelitis (Epidemic tremors)

CAUSAL AGENT. Entero-virus.

SYMPTOMS. Clinical signs are most marked in chicks from birth to six weeks. Movements are inco-ordinate as a result of damage to the nervous system. There may be partial or complete paralysis of the legs. Tremors may be noticed, the muscles of the head and neck being chiefly affected. A number of chicks may be unable to leave the eggs, and mortality at hatching can be high.

When pullets are exposed to infection the only sign of this may be a fall in egg production. Chickens hatched from eggs laid by these birds will, however, be infected and have clinical disease.

DIAGNOSIS. By the symptoms described above and the histopathological examination of the brain and lumbro-sacral cord.

SOURCES OF INFECTION. Actively infected, but asymptomatic layers infect their eggs, and the chickens hatched develop the clinical signs of epidemic tremors. These eggs are usually laid just before the drop in production noted under 'Symptoms'. The adults, as a result of the infection, then develop an active immunity and pass to their eggs a passive immunity. This protects the chickens hatched from these eggs for about the first six weeks of life, when the passive immunity is lost. The chickens can then become infected, but at this age the disease is likely to be asymptomatic.

CONTROL. Dead vaccines are available, and these produce a lasting immunity. They can be used to inject laying stock before selecting eggs for hatching and to inject the chickens after they have lost their passive immunity at about six weeks of age.

PROTOZOAL

Coccidiosis

CAUSAL AGENTS. *Eimeria* spp. At least eight different types of coccidia are found in the intestinal tract of poultry. The more common types are *E. tenella*, *E. necatrix*, *E. acervulina*, *E. brunetti*, *E. maxima* and *E. mitis*.

SYMPTOMS. *E. tenella* is usually responsible for the dramatic outbreaks of coccidiosis seen in young chicks aged from one to seven weeks. The infected chicks are in poor condition, and their droppings are frequently bloodstained. This results from massive infection of the caecal wall causing local haemorrhage. Mortality may be as high as 50 per cent.

In older birds, where infection is caused by *Eimeria* spp. other than *E. tenella*, the disease in confined to the small intestine and is usually more chronic than is chicken coccidiosis. Symptoms vary from a simple loss of condition to severe emaciation accompanied by a bloodstained diarrhoea.

POST-MORTEM. In acute *E. tenella* infections the caeca are enlarged, haemorrhagic and often contain blood. *Eimeria* spp. infections in adults cause a thickening of the small intestine with haemorrhagic areas and the formation of a catarrhal exudate.

DIAGNOSIS. By the microscopical demonstration in the droppings, or in material collected at autopsy, of large numbers of oöcysts. The different types of coccidia are identified by size and morphology after sporulation of the oocysts in 2·5 per cent potassium dichromate or 5·0 per cent formalin.

SOURCE OF INFECTION. Infected birds excrete oöcysts in their droppings. Under suitable conditions of temperature and humidity these sporulate after a few days and can cause infection when ingested.

CONTROL. Primary control should have as its aim the prevention of conditions that allow the existence of large numbers of sporulated oöcysts. Substantial reduction can be achieved by the regular cleansing and sterilisation of equipment and apparatus. It is not always practicable in large units such as deep-litter houses to carry out all the desirable measures. However, even under the worst conditions equipment should be kept clean and the birds always supplied with dry bedding: oöcysts require moisture to sporulate.

Oöcysts are not killed by germicides, this requires contact with either flowing steam at 100°C or a 10 per cent solution in water of 0·88 ammonia. Using industrial atomizers and pressure washers, the ammonia solution can be used for the treatment of rooms, buildings and bulky equipment. The ammonia solution is a skin irritant, and the inhalation of its vapour is dangerous. Personnel using it should wear protective water-proof clothing and respirators fitted with filters that absorb ammonia vapour. Suitable respirators are available from commercial sources.

A number of drugs, generally referred to collectively as coccidiostats, are used for both the control and treatment of coccidiosis. It is not possible to give a complete list of these, as new products are regularly marketed. Examples are sulphaquinoxaline, sulphadimidine and the nitrofurans. The coccidiostats are added to either the drinking water or the food.

HELMINTHS

CAUSAL AGENT. Nematode—*Ascaridia galli.* This is the common ascarid worm of poultry.

SYMPTOMS. Poor growth, poor egg production and occasionally death by inanition.

POST-MORTEM. Worms are found in the intestine. There may be a haemorrhagic enteritis caused by the larvae hatched from ingested eggs invading the mucosa of the gut wall.

DIAGNOSIS. By identification of the ova or worms in the droppings or at autopsy. *A. galli* grows to about 10 cm in length.

SOURCE OF INFECTION. Other birds whose excreta contain the ova of *A. galli.* Transmission is direct, there is no intermediate host. The ova can survive outside the host for long periods.

CONTROL. *A. galli* infestations respond well to treatment with piperazine compounds, such as piperazine citrate. This is obtainable as a preparation suitable for adding to drinking water or to wet mashes ('Pipricide'—Burroughs Wellcome).

CAUSAL AGENT. Helminths of the genus Capillaria. These are small slender worms found in the crop, oesophagus and small intestine of birds. Males measure 10–25 mm in length and females 14–80 mm.

SYMPTOMS. In heavy infections there may be difficulty in breathing caused by oedematous swelling of the crop and oesophagus, accompanied by sloughing of the oesophageal mucosa. Death may occur due to inanition.

Capillaria spp. invading the small intestine cause similar changes in the intestinal wall. This can lead to diarrhoea, emaciation and death.

POST-MORTEM. Swelling and sloughing of affected tissues.

DIAGNOSIS. The finding of *Capillaria* spp. in invaded tissues and their contents. Excreted eggs can be demonstrated by microscopical examination of the droppings.

SOURCE OF INFECTION. This is dependent on the species of *Capillaria* involved. Transmission can occur either directly from bird to bird or indirectly through earthworms.

In direct transmission the worm eggs are excreted in the droppings. Infected larvae develop in the eggs over four to five weeks, and these are then infective when ingested by other birds.

Indirect transmission depends on the ingestion of excreted ova by earthworms. A hen eating the earthworms then becomes infected.

CONTROL. Regular cleaning and sterilization will greatly reduce the opportunities for the occurrence of infection.

Treatments directed to the elimination of worms in the infected bird are not always successful. The worms lie close to the mucosa and are covered by a thick exudate. Some success has been reported using Methyridine ('Promintic' I.C.I.) and Haloxan ('Loxon' Cooper, McDougall and Robertson).

CAUSAL AGENT. *Hetarakis gallinae.* A small worm found in the caeca of poultry. The male measures 7–13 mm in length and the female 10–15 mm.

SYMPTOMS. The adults of *H. gallinae* do not cause active disease in the host bird. Growing larvae occasionally cause inflammation of the caecal wall.

POST-MORTEM. Worms and larvae are found in the caeca, caecal gland and the caecal mucus.

DIAGNOSIS. The finding of worms and eggs in the droppings or in the affected sites at autopsy.

SOURCE OF INFECTION. Transmission is direct: infected birds excrete ova in the faeces. Under favourable conditions larvae develop within about five days and are infective when ingested by other birds.

CONTROL. *H. gallinae* does not cause any noteworthy disease in birds. Its eggs, however, are a vector for the causal agent of a serious and lethal disease of turkeys and to a lesser degree of poultry. The causal agent of this disease, known as Histomoniasis (Blackhead), is a protozoan parasite, *Histomonas meleagridis*, which invades the liver and the caeca of the infected bird, causing severe lesions and functional impairment.

Birds infected with the protozoa, and the worm, excrete ova of the latter which, when artificially fed to turkey poults, cause Histomoniasis. Though direct transmission of the protozoa can occur, this parasite does not survive well outside the body, and infection is more likely to occur through the agency of the ova of *H. gallinae*. It is therefore important to control *H. gallinae* infections.

Cleanliness and good hygiene and, where possible, regular and effective sterilization substantially reduce the possibility that excreted ova will survive. The drug phenothiazine when fed in the food at a level ensuring the intake of 0·5–1·0 gramme per bird considerably reduces the numbers of *H. gallinae* carried by poultry and turkey flocks.

CAUSAL AGENT. *Syngamus trachea.* This is a small roundworm. The female measures 1–6 cm in length and the male 0·5–4 cm. Adult females and males are permanently attached in copulation.

SYMPTOMS. In heavy infestations birds find difficulty in breathing and may die. Worms are sometimes expelled during severe coughing.

POST-MORTEM. Red-coloured paired worms are found in the trachea.

DIAGNOSIS. At autopsy, by macroscopic evidence of infection. The identity of the worms can be confirmed by microscopical examination.

SOURCE OF INFECTION. Infection can pass directly between birds. The infected animal coughs up and swallows the worm ova. These are then excreted in the faeces. It takes about nine days after excretion for infective larvae to develop in the ova. When these are swallowed and enter into the intestinal tract the larvae pass through the wall of the intestine, enter the blood circulation and are deposited in the lungs. After further development the males and females pass into the bronchii, position themselves in copulation, then migrate to the tracheal wall, where they attach themselves sucking blood from the surface vessels.

The indirect method of transmission results from the ingestion of excreted ova containing infective larvae by earthworms, slugs, snails and insects. No further larval development takes place in these animals; they are merely transporters of infective ova. Ova can remain viable in the transporter animal for months and even years, protected from exposure to adverse climatic conditions, such as drying, etc. Hens are infected after eating the transporter hosts.

CONTROL. Small numbers of birds can be individually treated by exposing them, in closed boxes, to a powdered cloud of barium antimonyl tartrate. This powder, when inhaled, effectively kills the worms. The method is not practicable for the treatment of large numbers of birds.

It is obviously desirable that poultry should not eat earthworms containing infective ova. Where earth-floored houses are used, the floor should be covered with clinker to a depth of at least 3 in. If the birds have access to grassland the same piece of pasture should not be used for long periods. Some control of earthworms can be achieved by treating the land with copper sulphate.

EXTERNAL PARASITES

Lice. All lice infecting poultry are members of the sub-order *Mallophaga* (biting lice).

They live by eating the epithelial debris of feather sheaths, skin, etc. At least forty different species parasitize birds, and their identification requires some expertise. An important species in this country is *Eomenocanthus stramineus*. Adults and their eggs are found in large numbers on the skin, particularly round the anus. Badly infested individual birds can be treated by dusting with gamma-benzene hexachloride or DDT. This is not practicable in large flocks, when other measures have to be adopted (see below).

Mites

Three species of poultry mite are of importance in this country:

Dermanyssus gallinae: red mite
Liponyssus sylviarum: northern mite
Cnemidocoptes mutans: scaly leg mite

D. gallinae. The red mite, so-called because of its appearance when engorged with the blood sucked from the host animal. Adults and nymphs feed only on the bird at night; during the day they return to crevices and cracks in perches and other woodwork, nesting and laying eggs in these sites. The adults can live for many weeks without feeding and can infest successive batches of birds passing through a poultry house. Heavy infestations can cause severe and sometimes fatal loss of blood in the host bird.

L. sylviarum. The Northern mite is morphologically similar to *D. gallinae.* However, its entire life cycle takes place on the host bird. Heavy infestations can cause severe loss of blood and the formation of scabs which may become secondarily infected. Birds affected in this way lose weight and condition and give low egg yields. The mites frequently leave the host animal during handling, infesting the arms of the handler and sometimes causing a pruritis. *L. sylviarum* has also been found to transmit fowl pox.

Cnemidocoptes mutans. The mite responsible for causing 'scaly leg' in poultry. This species burrows into the skin layers of the lower leg, the feet and the toes, causing the formation of large yellow crusts. Limb movement can be affected, arthritis develop and the toes part from the feet.

CONTROL. Poultry lice and the three mites described above can be controlled in large units by the use of atomisers which disperse aerosols into the building. A solution containing 1·6–2·0 per cent piperonyl butoxide, 0·4–0·8 per cent pyrethrins and 0·3–0·6 per cent gamma-BHC is effective when atomised in this way. This solution can also be used to spray heavily infested birds individually.

C. mutans is more difficult to kill when the mites have burrowed deep into the skin layers and formed tunnels. The affected limbs may have to be treated by scrubbing with an emulsion containing 0·1 per cent gamma-BHC in water.

The use of aerosols has largely superseded the more laborious methods of painting perches with 40 per cent nicotine sulphate and the individual dusting of birds with DDT or gamma-benzene hexachloride.

NUTRITIONAL

It is unusual at the present time to find poultry suffering from severe nutritional deficiencies. Compounders are now aware of the necessity of producing well-balanced nutritionally adequate diets. However, a diet may be inadvertently wrongly compounded or birds may be prevented from eating adequate quantities as a result of overcrowding or fighting. A brief outline is given of the symptoms occurring in the more common deficiencies.

Vitamin A

Chickens. They grow slowly, feather poorly, appear sleepy and may have a nasal discharge with caseous material formed under the eyelids.

Adults. There is a marked loss of weight and a fall in egg production. The nose and eyes have a watery discharge. The fluid from the eyes becomes white, the eyelids seal and casts form over the eyes.

Vitamin D

A deficiency in chickens usually causes softening of the bones. The breast bone is bent, the beak and claws are soft and the ribs turn inwards. There may be failure to walk properly. Normal bone formation is dependent on adequate and balanced intakes of Vitamin D, calcium, phosphorus and magnesium. An imbalance of any of these may lead to orthopaedic abnormalities.

Vitamin E

Deficiency of this vitamin is now uncommon, as it is standard practice to add to diets stabilized forms of Vitamin E. In addition, antioxidants which protect the vitamin are frequently included in dietary formulae.

The disease attributed to Vitamin E deficiency is known as encephalomalacia ('crazy chick disease'). Young chickens show symptoms of ataxia, move in circles, fall backwards and lose the use of the legs and wings. Death is not uncommon.

Vitamin B2 (Riboflavin)

A condition known as 'curled-toe paralysis' may occur in chickens three to four weeks of age. The toes turn inwards, the affected chicken finds it difficult to walk.

Pantothenic acid

The deficiency may be seen in chickens from one to four weeks old. Growth is erratic and feathering is poor, scabs appear on the eyelids and around the mouth. The soles of the feet and toes become dry and crack.

This disease syndrome has been noted in chicks reared on diets containing, on analysis, apparently adequate quantities of pantothenic acid. The reasons for this are not clear, but may be due to imbalance rather than simple deficiency.

Manganese

The deficiency condition is known as Perosis or slipped tendon. It occurs in young growing birds which are noted to be partially or completely lame. On examination the hock joint is found grossly swollen and the Achilles tendon displaced from the condyles. The adjacent ends of the metatarsus and tibia are twisted and the foot displaced outwards.

Though the above syndrome can occur as a simple manganese deficiency, it is often associated with concurrent deficiencies of other minerals and vitamins of the B complex.

FERRETS

TUBERCULOSIS

CAUSAL AGENTS. *Mycobacterium tuberculosis* human type; *Mycobacterium tuberculosis* bovine type; *Mycobacterium tuberculosis* avian type.

SYMPTOMS. Rapid loss of weight and condition. The mesenteric glands are usually enlarged, and can be felt on palpation.

POST-MORTEM. Infection is found in the mesenteric lymph glands and the gastro-intestinal tract.

DIAGNOSIS. By the demonstration of alcohol/acid-fast bacilli in infected material. The organisms can be cultured and the type determined by the inoculation of infected type material on selective laboratory media and by animal inoculation.

ROUTE AND SOURCE OF INFECTION. *Oral.* By the ingestion of substances, chiefly milk, containing viable tubercle bacilli. Infection can also occur by the ingestion of materials contaminated by the droppings of birds which are excreting the avian type of tubercle bacillus.

CONTROL. The almost complete eradication of tuberculosis, both in humans and in cattle, in this country has greatly reduced the risk of infection to ferrets. Should infection occur, early cases can be detected by the presence of a palpable mesenteric lymph gland and/or by a positive reaction obtained when tuberculin is injected subcutaneously. All detected cases should be destroyed and associated equipment and quarters effectively sterilized. It should be noted that many germicides do not kill tubercle bacilli: 5 per cent Lysol is effective.

ABSCESSES

CAUSAL AGENT. *Staphylococcus pyogenes.*

Abscesses resulting from bites inflicted during fighting are not uncommon in male ferrets. The head and jaws are the areas usually attacked. If the infection remains localized, then topical and parenteral treatment with antibiotics is usually effective. However, should an associated osteomyelitis of the adjacent bone structures develop, then treatment may be protracted and largely ineffective. When this happens the animal should be destroyed.

DISTEMPER

CAUSAL AGENT. Canine distemper virus.

SYMPTOMS. The onset of disease is characterized by the appearance of purulent discharges from the nose and eyes. Nervous disorders appear within a few days, these being characterized by tremors and convulsions which terminate in death.

DIAGNOSIS. Clinical symptoms described above.

SOURCE OF INFECTION. Infected ferrets and dogs.

CONTROL. The canine distemper virus is usually introduced into ferret colonies by dogs excreting the virus, by materials contaminated by these dogs or by persons handling these dogs. Following the introduction of infection, the disease can spread from ferret to ferret. In these animals the disease is lethal and there is no worthwhile treatment.

Infected animals and their contacts should be destroyed and all equipment and quarters sterilized.

The management of ferret colonies should be such as to exclude the possibility of infection being introduced.

INFLUENZA

CAUSAL AGENT. Human influenza virus types.

SYMPTOMS. Similar to those found in human infections. There is a nasal discharge of mucoserous material, a mucopurulent conjunctivitis and an aural discharge of purulent material.

DIAGNOSIS. Clinical symptoms described above. Significantly high antibody levels against the responsible type of influenza virus can be demonstrated in the sera of sick or convalescent animals.

SOURCE OF INFECTION. Humans carrying the virus.

CONTROL. Humans suffering from an illness which presents the symptoms of influenza should be prevented from having access to ferrets.

There is no worthwhile curative treatment.

HAMSTERS

'WET TAIL'

CAUSAL AGENT. Two different bacterial species appear to have been implicated in two reported outbreaks, viz.: *Escherichia coli* and *Proteus vulgaris*, and in another, an unidentified organism.

SYMPTOMS. The name adopted for this disease arises from the soiling of the tail during an attack of acute diarrhoea.

POST-MORTEM. Inflammatory and ulcerative lesions may be found in the ileum and colon.

DIAGNOSIS. By the clinical symptoms. When so-called 'wet tail' occurs faecal material and pieces of the major organs collected at autopsy should be cultured in an attempt to establish if any particular bacterial type predominates in the gut flora or is recoverable from the body organs. A search should also be made for organisms of the Salmonella genus.

SOURCE OF INFECTION. This is unknown. Outbreaks have been associated with vitamin deficiency and stress conditions resulting from transportation or experimental procedures. It may be that stressing encourages, for some ill-defined reason, a large increase in the numbers of intestinal bacteria and that this leads to an acute reaction resulting in acute diarrhoea and in the development of inflammatory lesions in the gut.

CONTROL. Obvious stresses can be minimized by the feeding of nutritionally adequate diets, avoiding violent climatic changes, preventing overcrowding during transportation, etc.

Should cases of diarrhoea occur in a hamster colony, some degree of control can be achieved by adding to the drinking water or food a palatable antibiotic. Neomycin sulphate has been used at a level of 10 mgm per animal per day. The effectiveness of such treatments are dependent on the antibiotic sensitivity of the incriminated organism. This requires testing in the laboratory so that a suitable antibiotic can be selected.

SALMONELLOSIS

CAUSAL ORGANISMS. *Salmonella enteritidis.* (It is possible that other types of Salmonella can cause disease in hamsters.)

SYMPTOMS. Animals are seen to be obviously sick, and die soon afterwards.

POST-MORTEM. There may be areas of focal hepatic necrosis. The lungs are haemorrhagic and greyish in colour, but are not hepatized. To demonstrate clearly the changes in lung structure, it is necessary to cut histological sections. These show the presence of phlebothrombotic lesions, i.e. septic thrombi are found occluding the veins and venules.

DIAGNOSIS. By the isolation and identification of the causal organism from the infected organs.

ROUTE OF INFECTION. Probably oral by the ingestion of contaminated materials.

SOURCE OF INFECTION. The *Salmonella* spp. are ubiquitously distributed in feeding-stuffs.

CONTROL. See Salmonellosis in mice.

FIGHTING INJURIES

Hamsters sometimes attack each other with considerable ferocity. In particular, females can inflict severe bite wounds on males. These may go unnoticed and become grossly infected. It is of doubtful value to attempt treatment in such animals; they should be destroyed.

CHAPTER TWELVE

Pests of the Animal House

PART ONE

ANIMAL PESTS

There are many pests which can infect animals and animal houses. Pests feed on the blood (lice, fleas) or skin debris (mites) of living animals, or on undigested food present in excreta (house flies), or on foodstuffs intended for consumption by animals (wild rodents). Pests breed on living animals (lice, mites), in undisturbed dirt and bedding (fleas, house flies), or in cracks in walls or furniture (cockroaches, bedbugs).

The best precaution against infestations is a well-designed, constructed, and equipped animal house which is regularly and thoroughly cleaned. In a well-designed animal house the interior surfaces are hard, smooth, impervious, and washable; there are no sharply angled joins or corners where dirt can accumulate; service pipes and ducts are easily accessible for cleaning; woodwork is kept to a minimum; windows fit tightly. The objects of such a design are to minimize both the opportunity for access by pests and the provision of breeding places for them, and to facilitate cleaning of the building. The fabric of the building must be watched for signs of deterioration or damage, and such defects must be reported. Ideal breeding places for some pests are offered by cracks in plaster work (notably between walls and window- or door-frames), where service pipes enter walls, behind light fittings, flaking paint or loose or broken tiles, or in rotting wood. Wild rodents can enter a building by gnawing through door-posts, through the gap between the bottom of a door and a worn doorstep, through broken air bricks, or secreted in sacks or bales of food and bedding. Once established in a building, wild rodents will live and breed in hollow walls and floors and in service ducts. Buildings having solid walls and floors and unenclosed service pipes offer little shelter for these pests. Flies and mosquitoes enter through any fissure. Flies are attracted by the smell of rotting vegetation and excreta, so all refuse should be incinerated as quickly as possible. If flies are particularly troublesome and their breeding-place is unknown or unassailable the entrance to the animal house should be fitted with double screen doors and the windows kept shut or be screened. House flies do not attack laboratory animals directly, but they are vectors of disease, and their presence is therefore undesirable. Food and bedding should be inspected before delivery is accepted, and if the materials are infested the whole consignment should be returned to the supplier.

Before discussing in detail the major parasites of the animal house, it is as well to consider some basic principles for the prevention and control of infestations.

1. Most of the external parasites of animals are unable to leave the host for long periods. The first essential, therefore, is to start out with animals which are free from infestation. If it is necessary at any time to introduce new animals to the colony they should be isolated and several careful examinations made before they are moved into the main unit.

2. When an infestation is noticed the parasite should be carefully examined and identified, and this should be done before any treatments are undertaken. Some parasites are difficult to completely eradicate, even with frequent and thorough insecticide treatments, and it is often advisable to discard the animals. In some cases, however, the infestations may be of little importance; some fur mites, for example, are extremely difficult to control, and thorough examination of most small-animal units demonstrate their presence. They can be present in quite large numbers without adversely affecting the health of the animal, and excessive insect treatment is likely to do more harm than the parasite.

3. It is necessary after an identification of the parasite to know something of its life history. A little insecticide applied at the most vulnerable part of the parasite's life cycle can often be more effective than frequent doses at times when it is more difficult to control.

4. When the identification has been made and it has been decided to use an insecticide it is extremely important to choose an insecticide which will in no way affect the health of the animal. It should be realized that the animal technician is often unaware of the effect that some routine insecticide treatments may have on the research work which will later be carried out on the animals. Routine treatments with DDT and other hydrocarbon insecticides can leave quite high levels of insecticide in the fat and other organs of the body, so it is most important wherever possible to use a nonpersistent insecticide. It is also necessary to consider the health of staff using the animal units; continuous vaporizers using DDT and BHC are not to be recommended.

5. Food and bedding used for the animals should always be inspected thoroughly, and it is particularly important to change the animals bedding when an insecticide treatment is carried out. The reasons for this are discussed below. All waste materials should be removed and burnt immediately, as they provide a place for flies to breed.

6. Insecticides should not be left in the animal house to be applied at random, they should be used carefully and then locked away until they are required.

PARASITES

Pests which feed by attacking, without killing, other animals are called *parasites*. The attacked animal is called the *host*. Parasites which live on, or immediately beneath the surface of, the skin of animals are called *ectoparasites*.

Ectoparasites are harmful because they: (i) debilitate the host by sucking its blood; (ii) cause much irritation by biting or burrowing; (iii) carry diseases which can be transmitted to the host; and (iv) cause chronic diseases of the skin.

Insect pests include *flies, moths, beetles, bugs, fleas, and lice*. The bodies of insects are divided into three regions, the *head*, the *thorax*—from which arise *three pairs of legs*, and the *abdomen*; some have one or two pairs of wings. All insects are bi-sexual, and the females lay eggs. From the egg hatches either a miniature adult (nymph) or a grub (larva), which has to pass through a *pupa* stage from which the adult form emerges. In insects growth to the sexually mature form is achieved by *moulting* from stage to stage. The number of moults undergone varies from species to species.

The *Arachnida* are represented by the *mites* and *ticks*, neither of which is typical of this group. The bodies of mites and ticks are not so clearly divided into regions as are the bodies of insects. They have *four pairs of legs*, and do not have wings. They are bi-sexual and lay eggs from which *larvae* hatch. The larvae resemble the adults, but are smaller and have only *three pairs of legs*. Growth is achieved by moulting.

FLEAS

Fleas are blood suckers and can transmit disease through their bites. They are dark in colour and are about 2 mm long. Fleas are all parasitic on warm-

FIG. 1. Poultry flea (*Ceratophyllus gallinae*).
40 × actual size.

blooded animals in their adult stage. They are easily distinguished from other insect parasites, because their body is compressed laterally with backward-bending bristles.

Because of their shape they can travel very rapidly through the fur or feathers of the host, and they are also able to move by jumping.

Female fleas lay eggs anywhere, but often on the fur of the host, from where they fall, or are shaken, to the ground. Larvae hatch from the eggs and feed on decaying organic matter, particularly the excreta of adult fleas. The larvae spin cocoons around themselves in which they develop into pupae before finally reaching the adult form. In warm conditions the complete life cycle may take from three to four weeks, and a flea may live for three to four months. In cool conditions the flea may remain quiescent in the pupa stage for many weeks, and the adult life span may be as long as a year.

Each variety of flea has its favourite host species on which it prefers to feed, but each can, and does, feed from other hosts. Thus the Great Plague of the Middle Ages was spread from infected rats to man by the bite of fleas.

Murine typhus and the protozoan infection, *Trypanosoma lewisi*, are spread from rat to rat by fleas. Several tapeworms (Cestodes) spend part of their life cycle in the flea. (The flea is then said to be an intermediate host.) Tapeworm eggs are eaten by flea larvae and the tapeworm begins to develop. If an infected flea is eaten by a mammal the immature tapeworm is liberated to develop to the adult state within the gut of the mammal. Dog tapeworms are transmitted by this method.

Eradication

Animals must be treated to kill the adult fleas living on them.

Adult fleas, larvae, and pupae will be present in the bedding, which must be destroyed, and in the crevices of cages and trays, which must be sterilized. Walls, floors, fittings, and equipment must also be cleaned and treated with insecticide.

LICE

Lice are about 2 mm long, and vary in colour from creamy grey (unfed) to dark red (newly engorged with blood). Lice never leave the host animal and tend to be specific in their choice of host. The eggs are less than 1 mm long and are stuck to the hair of the host, on which they are just visible as a silvery scurf. Nymphs hatch from the eggs after five–eight days and undergo three or five moults to reach maturity. The whole life cycle is accomplished in two–three weeks.

Lice may be divided into two main groups—the blood-sucking lice, *Anoplura* (which occur only on mammals), and the *Mallophaga*, which feed on scurf, epidermis, and feather.

Lice are known to transmit murine typhus from rat to rat.

They may occur on all the common laboratory animals. Lice cause intense irritation to the host animal, which loses condition rapidly and may even die as the result of a heavy infestation. Lice spread from host to host by direct contact, and the infestation spreads quickly among overcrowded animals.

Eradication

The host animal must be treated, but clean cages and bedding should also be given as a precautionary measure. A second insecticide treatment should always be given 7–10 days after the first to kill the newly hatched larvae which will have hatched from eggs present at the time of the first treatment.

FIG. 2. Biting louse of cattle.
100 × actual size.

BEDBUGS

Bedbugs are rare today, but are a potential danger, because a colony of considerable size might become established before its existence was suspected.

The bedbug is dark brown in colour, and is a flat oval in shape, about 3 mm × 4 mm. Bugs live and breed in cracks in walls or behind loose wallpaper or flaking paint. They emerge from their hiding-place only for a short time at night to feed. Bugs feed only on the blood of mammals. Eggs are laid in the hiding-places. Nymphs hatch from the eggs and undergo five moults to reach sexual maturity. The nymphs also feed by sucking the blood of mammals. The life cycle of the bug may be completed in any period from one month to one year, depending on the environmental temperature and the opportunities for feeding.

Bedbugs found in this country are not known to transmit disease, but they can cause severe anaemia in the host, and their bites are intensely irritating.

Bugs can become established only in dirty, neglected premises.

Eradication

Control measures should be focused on locating and treating the hiding- and breeding-places.

MITES

Mites may be classed in two main groups: (i) *blood-sucking mites*, those which live and breed in dirt and crevices and visit the host only to feed, and (ii) *mange mites*, which live and breed on, or in, the skin of the host.

Mites lay eggs from which larvae hatch. After two or three moults the adult form is reached. The whole cycle may be completed within about twelve days (Psoroptes) or about seventeen days (Sarcoptes).

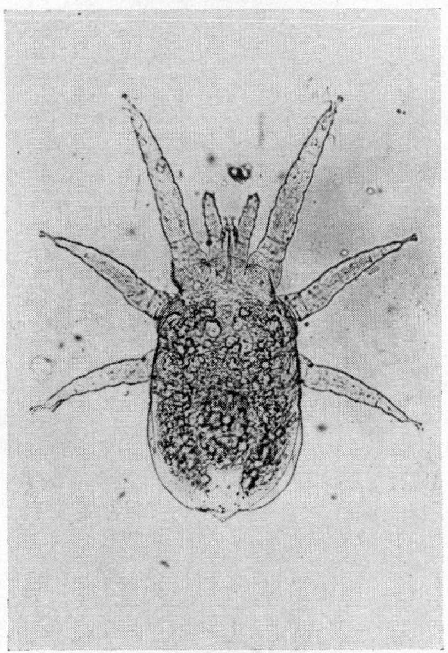

FIG. 3. Poultry Red Mite (larva).
250 × actual size.

Mites are barely visible to the naked eye. They tend to be host specific, and are most troublesome on poultry, mice, and rabbits, but they also occur on other species.

The blood-sucking mites (*Liponyssus bacoti*) prefers to feed on the rat, but will also feed on other rodents and on man. The bite is extraordinarily irritating. This mite is said to be a vector of tropical typhus.

The 'Red Mite' (*Dermanyssus gallinae*) is an important pest of domestic fowls. These mites feed only at night, and hide in the crevices of poultry houses during daylight.

Mange mites are of the family Sarcoptidae, and for mammals, may be subdivided into three groups—mites which live *on* the skin (Psoroptes), those which burrow *in* the skin (Sarcoptes and Notoedres), and those which burrow in the sebaceous glands and hair follicles (Demodex).

Mange mites living and breeding on skin debris cause flaking of the skin. Ear canker in rabbits is caused by such a mite (*Psoroptes communis cuniculi*).

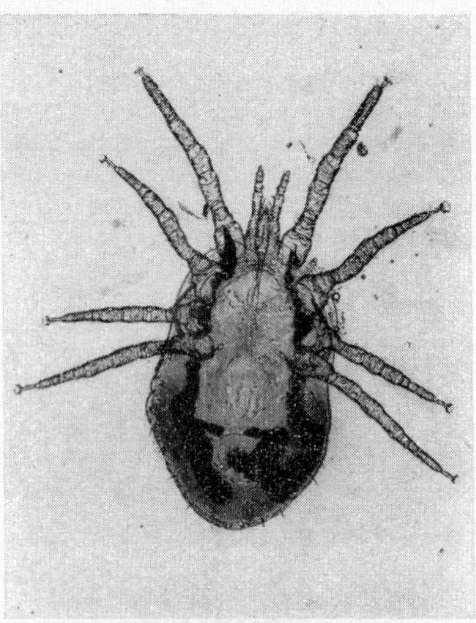

FIG. 4. Poultry Red Mite (adult).
60 × actual size.

Burrowing mites can be difficult to eradicate, because their eggs are laid, and hatch, and mature, and die within the burrows in the skin. Some mites have a preference for certain areas of the host body, such as the nose,

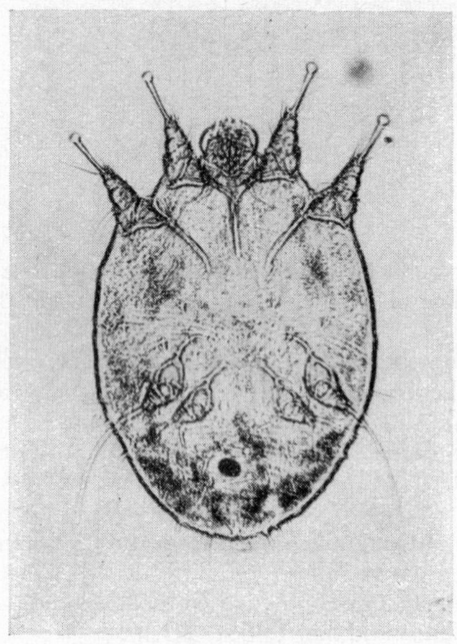

FIG. 5. Sarcoptic mange mite.
300 × actual size.

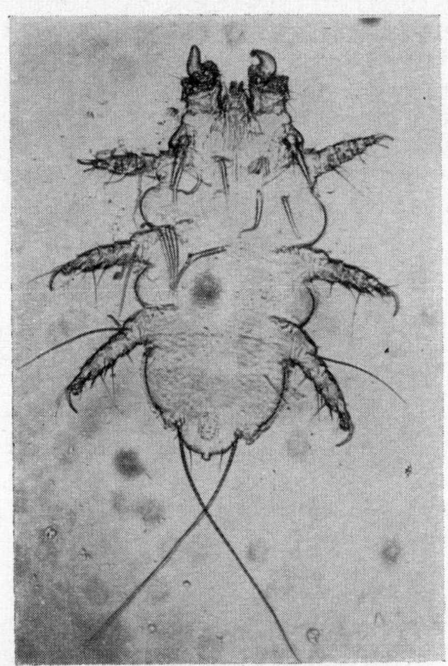

FIG. 6. Mouse fur mite (Mycoptes sp.).
250 × actual size.

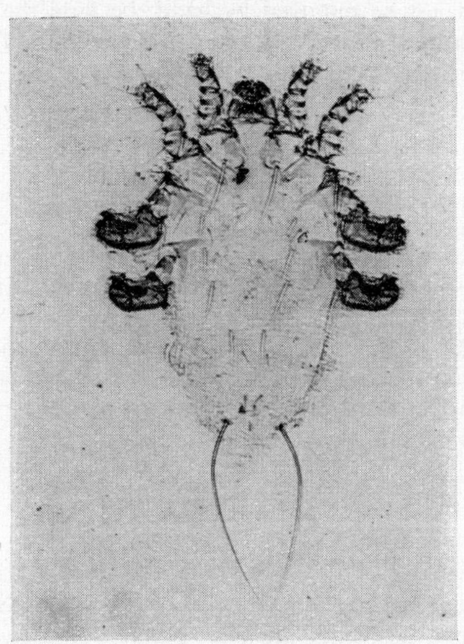

FIG. 7. Mouse fur mite (Myobia sp.).
250 × actual size.

ears, legs, or tail. Scabies (now almost non-existent) in rats is due to such a mite (Notoedres). Rat scabies is transmissible to man. *Cnemidocoptes mutans* causes 'scaly leg' in poultry, and *C. laevis* causes 'depluming itch', so called because irritation from mites living in the skin round the base of the feathers causes the birds to pull at, and break, the feathers.

Mites which enter the sebaceous glands and hair follicles (Demodex) are very difficult to eradicate. These mites may affect any of the common laboratory animals.

Eradication

The presence of scabs, 'warts', pustules, or flakes of dead skin may indicate an infestation with mites. These excretions must be removed before treatment, to permit good penetration of the medicant.

Animals suspected of mite infestation must be isolated and not returned to the general stock before it is certain that treatment has been completely successful.

Infestations are spread by direct contact. Overcrowding greatly facilitates the spread of infestations.

TICKS

Ticks are very rare on small animals, but are occasionally found on larger animals, such as dogs, sheep, and horses. They are large, 3 or 4 mm long. Ticks affix themselves to the host animal, and because they are so easily seen, a severe infestation could not develop unnoticed. Small numbers of ticks found on animals should not be removed by hand, because the mouth parts are frequently left embedded in the skin and cause irritation. The animal should be treated with an insecticidal wash or dust.

FLIES

Flies can act as the vector of any disease which can be picked up from excreta and transported to contaminate food or water.

The larvae of some flies burrow in the skin of mammals, including man, e.g. *Bot* and *Warble fly* larvae affect sheep, cattle, and horses, and the infestation can occur on man.

Blood-sucking flies visit their hosts only to feed. Horse and cattle flies breed in marsh land. Blood-sucking flies are not troublesome in laboratories in this country, but they are much to be feared in the tropics, where they are vectors of trypanosomiasis (sleeping sickness).

CONTROL OF PESTS

Insecticide treatments can be applied in several ways. As the majority of present-day synthetic pesticides are not soluble in water, they have to be specially formulated either as wettable powders, emulsions, miscible liquids, dusts, or aerosols.

Dusts

These are formulated by mixing the insecticide with an inert material, usually kaolin or talc.

Wettable powders

These are similar to dusts, but a wetting agent is added so that a suspension in water can be made.

Emulsions

The insecticide is dissolved in a solvent and is already in the form of an emulsion so that water can be added.

Miscible liquids

These formulations form an emulsion as soon as they are added to water. The insecticide is present in a solvent together with emulsifying agents.

Aerosols

The chemical is dispersed as extremely fine particles either by combustion or by passing through an orifice under gas pressure. The commercially produced aerosol pressure packs do not produce fine particle size (less than $\frac{1}{1000}$ mm) and are therefore not true aerosols.

It is important for the animal technician to appreciate the differences between these formulations. Insecticides formulated specially for animal treatments should be used whenever possible. These are often wettable powders or dusts because of the inert nature of the dispersion agents. Miscible liquids and emulsions can be used, but preparations manufactured for agricultural use should not be used, as the choice of solvent for these materials is frequently not satisfactory for application to animals, and they may cause severe skin irritation.

See also:

PAGE, K. W., 'The Ectoparasites of Laboratory and Domestic Animals', *Journal of Animal Technicians Association*, **3**, No. 2, 34 (1952).

CUSHNIE, G. H., 'The Life Cycle of Some Helminth Parasites of the Rat, Mouse and Rabbit', *ibid.*, **5**, No. 1, 22 (1956).

PART TWO

PESTS OF FOODSTUFFS

The storage of foodstuffs presents many problems, one of which is to keep it free from insect and other pests. The food of laboratory animals (which is rich in cereal products and is often fortified with proteins) is particularly susceptible to infestation by pests.

The common invaders of the food store can be grouped under the orders to which they belong, namely:

Lepidoptera	(moths)
Coleoptera	(beetles)
Orthoptera	(cockroaches)
Acarina	(mites)

LEPIDOPTERA

The adult moths are short-lived. The body and wings are covered with scales, and the mouth is adapted for sucking up juices. The larva of the moth is a grub or caterpillar. The mouth parts are adapted for biting, hence only the larvae damage foodstuffs. Some of the more common moths are:

Ephestia kuhniella (The mill moth or Mediterranean flour moth)

This moth usually infests meals and flour. The adult moth is about 13 mm. long, and is a pale grey colour with wavy markings on the fore-wings. The female lays 50–350 eggs, which hatch in seven–fourteen days, when the young caterpillars crawl about the foodstuffs trailing a silken thread which is produced from an opening near the mouth. In about ten weeks the larva is fully grown. It then leaves its source of food and prepares a cocoon, usually in the angles of walls and ceilings, or in any other protective crevice. The larva pupates in the cocoon and the adult emerges after about twenty days.

Foodstuff invaded by moths is often caked together by the silken web produced by the larvae.

Ephestia elutella (The cacao moth)

This moth is another common invader of grain products, such as wheat and wheat offals; it is also found in cacao beans, ground nuts, linseed, and other oilseeds.

The adult moth is about 7–9 mm long and is grey in colour, with a pair of lighter bands across the fore-wings. The eggs are laid loose on the surface of the grain or seeds. The larvae attack and penetrate the germ of grain. Foodstuffs which have been attacked by *Ephestia* smell sour and are contaminated by the droppings and webbing of the caterpillars.

COLEOPTERA

Beetles have two pairs of wings, the hind pair are used for flight and the front pair form a hard, protective cover. The front, protective wings (elytra) are joined together in beetles which have lost the power of flight.

The larvae vary in type from active mealworms to sluggish, fleshy grubs. Both larvae and adults have mouth parts for biting, enabling them to feed on and damage foodstuffs.

Calandra granaria (The grain weevil)

This is a polished, dark-brown or black insect about 2·5–5 mm long. The hind-wings are not developed, and it is unable to fly. The adult lives for seven to eight months. The female lays about 100 eggs, which are deposited in small holes bored into the grain. The larva is a small, white, legless grub which tunnels into the endosperm of the grain. After several moults it becomes a pupa within the grain, and the adult finally eats its way out of the grain. The whole life cycle takes from twenty-eight to forty days, according to the environmental temperature. This weevil not only attacks wheat, oats, barley, maize, rye, rice, and other seeds but it can also breed in flour.

Calandra oryzae (The rice weevil)

This has a similar life history to the grain weevil. It can be distinguished by its four red or yellow spots on the wing-cases, and by the presence of hind-wings under the elytra. It has been known to fly in Britain in warm, sunny weather. The rice weevil will attack the same range of grains as will the grain weevil.

Stegobium paniceum (The biscuit weevil or bread beetle)

The adult beetle is from 1·75 to 3·75 mm long, is cylindrical in shape and a light-brown colour. The life cycle occupies two to seven months according to temperature. The larvae are active, and when fully developed form cocoons from a gelatinous secretion mixed with the surrounding food. The larva turns into a pupa and finally into an adult beetle which bites a hole in the cocoon and emerges. This beetle has been found in a variety of foodstuffs.

Dermestes lardarius (The bacon or larder beetle)

The beetle is dark brown in colour, 7–9 mm long, and has a fawn, six-spotted band on the elytra. The female lays up to 175 eggs, which hatch out into brown, hairy grubs in about twelve days. The fully grown grubs leave the food and tunnel into surrounding material, such as woodwork, where they pupate, to emerge as adult beetles after about ten days. The bacon beetle requires food of animal origin to complete its life cycle, which takes about fifty to sixty days. An adult beetle may live for twelve months. Damage is caused to foodstuff by the tunnelling of the larvae, and by contamination with skin casts and faeces.

ORTHOPTERA

The cockroach, popularly called the black-beetle, though it is neither black nor a beetle, is a pest found in bakeries, kitchens, and food stores. The two species commonly found in Britain are the Common or Oriental cockroach and the German cockroach. Both are natives of Africa; the common cockroach reached England about the time of Queen Elizabeth I and the German cockroach some time later.

The females lay large numbers of eggs in cases; these cases are carried around by the females. The young are called nymphs; they resemble the adult cockroach, but are wingless. Wings grow in the final moult, except in the case of the female, *Blatta orientalis*. Cockroaches develop slowly and have a long life. They require moisture and heat for satisfactory development. They infest heated buildings, living behind hot-water and steam pipes and in other suitable crevices, only emerging at night in search of food. They contaminate foodstuffs with their faeces and may carry infection. Cockroaches are the intermediate host of a nematode parasite of rats, *Gigantorhymmochus*.

The three common species are:

Blatta orientalis (Oriental cockroach)

A large dark-brown insect, the so-called black beetle, with a life cycle of some 300 days. The female is wingless.

Blattella germanica (German cockroach)

A small, yellowish-brown insect known as the steam fly is the most troublesome of the cockroaches. It develops rapidly, completing its life cycle in about fifty days.

Periplaneta americana (American cockroach)

A large, dark-brown insect with a life cycle of approximately 200 days.

ACARINA

These are minute animals related to spiders, but without the well-marked division of the body seen in the latter. Probably the most troublesome mite found in foodstuffs is *Tyroglyphus farinae* (the flour mite). It infests stored wheat, flour and offals, and other cereals and their products. A greyish dust is the first sign of infestation, but careful examination will reveal mites. Food is damaged in several ways: the germ is eaten away, grain is eaten or bored, and the food is fouled by excreta, cast larvae skins, and dead mites. The adult female mite lays twenty to thirty eggs scattered about the foodstuff. After four days the six-legged larvae emerge from the eggs, and after a few days feeding they become inert and moult to become eight-legged nymphs; this stage lasts from six to eight days. After further moults the adult is produced.

The life cycle takes from seventeen to twenty-eight days, but is influenced by both temperature and humidity. Mites breed much more rapidly in damp conditions, and cannot thrive on material containing less than about 12 per cent moisture. However, this can be a local condition, and the bulk of the sample may have a much lower moisture content than that of the portion where the mites are living. The optimum temperature for mites to breed lies between 64° and 77°F (18° and 24°C), but mites are very resistant to cold and cease to feed at only a few degrees above freezing point. Below this temperature they hibernate.

PREVENTION AND CONTROL

Most of the pests mentioned require both warmth and moisture to enable them to complete their life cycles satisfactorily. Further, they require some form of cover in which to hide if the food is removed or disturbed. It follows that a food store should be cool, dry, and well ventilated, and should be free from crevices or corners where food and dirt can accumulate and remain undisturbed.

The room and all food containers should be kept clean, and spilled foodstuff should not be allowed to remain lying on the floor.

Food containers should be emptied and sterilized before refilling.

Unnecessarily large stocks should not be carried, and all foodstuffs should be used in strict rotation of delivery.

All empty food sacks should be returned or burnt as soon as possible, and surplus foodstuffs should be burned. Both these items offer suitable and secure places for breeding. All fresh deliveries of foods as well as existing stocks should be inspected for signs of infestation.

To prevent insect infestation of food stores is almost impossible, and some methods of control must be undertaken.

Both heat and fumigation can be used to destroy insect invaders. Exposure to temperatures of 140°F (60°C) will kill many of the common pests, but it may be necessary to expose a mass of food to temperatures of 212°F (100°C) for 1 or 1½ hours to get complete heat penetration. At such temperatures some of the constituents of the food may be so altered, or even destroyed, as to render it useless as a feeding material.

Dusting powders are useful in preventing infestation, and in cases of slight infestation. Powders containing gamma-BHC are very effective.

Sprays of various types are useful, and insecticides containing pyrethrum are particularly suitable against moths. Continuous-flow aerosols prevent the spread of infestations.

In cases of heavy infestation it may be necessary to fumigate the whole food store. For this purpose hydrogen cyanide and ethylene oxide are used commercially.

The former is very poisonous to man, and should only be used by an experienced specialist operator.

Ethylene oxide has the advantages of having a marked odour and of being less toxic to man than hydrogen cyanide; but has the disadvantage of being highly inflammable, and a mixture of the vapour with air is explosive. The gas is used with an excess of carbon dioxide to reduce the explosive risk, but again it should be used only by an expert.

Small quantities of material may be fumigated by placing it in a suitable bin and pouring in a mixture of ethylene dichloride and carbon tetrachloride, sealing it up, and leaving it for 48 hours. Such food cannot afterwards be fed to animals.

Smoke canisters of various sizes containing gamma-BHC are available. They are suitable for use in lofty buildings.

It should be remembered that prevention is better than cure. A careful watch should be kept on all stored foodstuffs, and at the first sign of infestation immediate action should be taken to eliminate the pest. It is often easier, cheaper, and more effective to destroy a batch of infested food than to try to disinfest and use it.

CHAPTER THIRTEEN

Humane Killing

Untrained personnel should, in no circumstances, be permitted to kill or administer barbiturates to animals. The animal technician must have complete confidence in his ability to kill cleanly, swiftly, and humanely, and, therefore, must be thoroughly conversant with the correct methods of handling animals and must have the requisite knowledge of the equipment and materials at his disposal.

The need for killing animals may be classified in four groups:

(i) to alleviate unnecessary or prolonged suffering because of disease or accident;
(ii) because an animal has become redundant;
(iii) as part of an experiment;
(iv) slaughtering for food.

In the third case the decision to kill is the experimenter's, although it is the duty of the animal technician to keep the experimenter informed of any change in the animals' condition. Slaughtering for food is carried out by slaughtermen at licensed premises, in accordance with the laws relating to food hygiene and humane killing.

In groups (i) and (ii) the onus of taking the decision to kill falls upon the animal technician.

Euthanasia

This is the term for painless killing. If an animal passes quickly and quietly into the unconscious state and death ensues before consciousness is regained, then it may be assumed that the animal has been correctly and humanely killed.

The following basic principles should be adhered to:

(i) Handle the animal carefully and gently, taking care not to frighten or antagonize it unnecessarily.
(ii) Remove the animal from the animal room.
(iii) Do not kill in the presence of another live animal.

Physical methods

Such physical methods as stunning are distasteful to the operator, but these methods, if applied efficiently, are often less distressing to the animal than some more complicated method. After stunning the animal usually bleeds from the nose and/or mouth. It is important that blood is cleaned from all surfaces, especially the hands, before another animal is touched, so that it may not be distressed by the smell of blood. Physical killing may also be carried out by dislocation of the neck, thus breaking the continuity of the spinal cord in the cervical region.

Inhalation of poison gases

Strict precautions must be observed when killing by means of poison gas, because these gases present a health hazard to other animals and to personnel. Lethal chambers should therefore be used. These should be situated in a room apart, completely isolated from the animal rooms. The room itself must have ample ventilation and contain no naked lights. Carbon monoxide, in the form of coal gas, is probably the most commonly used lethal agent, and its inhalation causes little or no distress to animals or human beings. It should be noted that the minimum of distress is caused to the animals if the lethal chamber is filled with coal gas before the animals are placed in it. Because of its characteristic smell, any escape of coal gas is easily detected, though it should be remembered that carbon monoxide itself is odourless. Where coal gas is not readily available carbon monoxide may be obtained from the exhaust gases of a four-stroke petrol or diesel engine. The amount of carbon monoxide gas in exhaust fumes from a normally running engine is negligible, but if the air/fuel ratio is altered (i.e. if the choke is used) the carbon monoxide content of the exhaust fumes can be raised to about 14 per cent. Exhaust gases are hot and dirty, so they must be cooled and filtered by passing the fumes through a large volume of cold water and a metal gauze and a cloth filter. This method is in common use at poultry stations for killing large numbers of unwanted chicks.

Nitrogen may be used, but for general purposes it has no great advantage over coal gas, and is much more expensive. It should be administered at the rate of $\frac{1}{2}$–1 ft^3/minute, and the air vent should be left open until the animal's respiration ceases.

Coal gas, nitrogen, ether, and chloroform may all be administered in the same type of lethal chamber.

Inhalation of volatile anaesthetics

Ether and chloroform are the two volatile anaesthetics most commonly used. Ether is highly inflammable, and a mixture of ether and air may be exploded by a naked flame. The use of ether should therefore be confined to cases of necessity. These vapours should be used in a lethal chamber. A large, lidded glass jar or a glass dessicator is satisfactory as a chamber for killing small animals. A chloroform-soaked pad is placed at the bottom of the jar and a fine-mesh wire grid placed over and above it. Chloroform is a skin irritant, and therefore on no account should the liquid anaesthetic come into contact with the animal's skin.

A metal (or even wooden) chamber is used for larger animals, such as guinea pigs, rabbits, or cats. This chamber should be fitted with a large *glass* (*N.B.* Perspex is soluble in chloroform) observation panel and must have an inlet point through which a chloroform-soaked pad can be inserted. It should also be fitted with an adjustable air vent. A wire-mesh grid must be fitted on the floor of a metal chamber, as the animal may become panic-stricken if it cannot retain its foothold. The animal is placed in the chamber and the lid closed but the air vent left fully open, and the chloroform-soaked pad is then inserted. In a few moments the animal sinks into unconsciousness and falls on its side; the air vent is then closed. Note the difference between the gradual administration of chloroform and the instant administration of undiluted coal gas. The animal should be left in the chamber until death is certain.

A mixture of equal parts of chloroform and carbon tetrachloride may be used instead of chloroform alone.

The room temperature should be about 70°F (21°C) to obtain the quickest results.

Chloroform must never be used in a mouse room (see p. 116).

Carbon dioxide as a lethal agent

The Universities Federation of Animal Welfare (UFAW) has been investigating the use of carbon dioxide for anaesthesia and for euthanasia. Carbon dioxide is safer in use than either coal gas or ether, as it is non-inflamable and, being heavier than air, it sinks to form a layer at floor level. Carbon dioxide is a colourless, odourless, non-irritant gas and is non-toxic in low concentrations.

A mixture of 80 per cent carbon dioxide and 20 per cent oxygen is regarded as being suitable for use as an anaesthetic. For euthanasia, a concentration of at least 60 per cent carbon dioxide in air is recommended. (Such a mixture would contain not more than 8 per cent oxygen). Commercial-grade carbon dioxide, supplied in cylinders, is cheaper than pure carbon dioxide, and is suitable for euthanasia.

Carbon dioxide may be applied in a lethal chamber, and it may also be applied to animals confined inside a polythene bag. Any plastic bag could be used, but transparent polythene is recommended so that the operator may observe the state of the animal within the bag. Very small, quiet animals, e.g. the mouse, may be put loose directly in the bag. Large groups of animals or single larger animals may be left in their cage and the bag be put round the cage. Intact bags can be efficiently filled with gas, but bags having small tears may still be used, as any escaping gas is non-toxic at low concentrations.

The gas should be applied from above the animal. The carbon dioxide will sink to the floor, and the animal will gradually become enveloped in a layer of the gas. The animal sinks gently and peacefully into unconsciousness, and death follows quickly. The gas should be administered slowly until the animal becomes unconscious, after which the rate of gas flow may be increased until the operator is satisfied that the container is full of carbon dioxide. Ten minutes after the animal appears to be dead it may be removed into open air, but if this is done the animal must be kept under observation for 20–30 minutes in case recovery occurs. It is safer and more convenient to leave the animal exposed to carbon dioxide for 20–30 minutes after breathing ceases.

The application of carbon dioxide is probably the most humane way of killing animals. They are unaware of the presence of the gas and lapse into the unconscious state easily, quickly, and quietly. The simplicity and safety of the method also commend it.

Rats, mice, rabbits, chickens, and mink have been killed by the application of carbon dioxide. Commercially, swine are slaughtered by exposure to carbon dioxide. The gas is not recommended for killing dogs.

The oral administration of barbiturates

Barbiturates in capsule form may be given to cats and dogs, but only under qualified supervision. The method is fairly satisfactory, but the animals' reaction is largely dependent on the amount of the stomach contents. Thus, this method should be avoided if the animal has been fed recently. Sodium pentobarbitone is widely used for this purpose, and is packed specially in capsules containing 1½ grains of the drug.

Killing by means of injections

Animals which can be conveniently injected intravenously may be killed swiftly by means of the injection of a saturated solution of magnesium sulphate. The size of the dose depends on the size of the animal and may vary from 5 to 20 ml. Magnesium sulphate is much cheaper than most other killing agents, so it is common practice to sedate the animal with sodium pentobarbitone and then kill with an intravenous or intracardiac injection of saturated magnesium sulphate solution. No other route for the injection of magnesium sulphate solution is effective.

Intravenous or intracardiac injections of barbiturates are much swifter in effect than intrathoracic or intraperitoneal injections, but it will be realized that the former are the more skilful operations. The intraperitoneal injection of sodium pentobarbitone apparently causes little discomfort to the animal and, providing a large enough dose is injected, the animal will quickly become unconscious and death will ensue. The speed of the response of animals to intraperitoneal or intrathoracic injections of sodium pentobarbitone varies considerably, but the simplicity of these methods commends them.

The lethal dose for injected sodium pentobarbitone ('Nembutal') is 42·5 mg/kg body weight (or 1 ml/1·5 kg body weight of a solution containing 64 mg/ml).

Electrocution

In old methods of electrocution about 20 seconds elapsed before unconsciousness and death occurred. The use of an 'Electrothanator' results in unconsciousness and death occurring in 1½–2 seconds. The instrument is designed to pass an electric current strong enough to stun and kill, simultaneously, through the head and the heart. Properly constructed electro-medical equipment with adequate insulation for the operator must be used. The current required has to be stepped-up from mains voltage through a transformer.

Electrocution is suitable for dogs but not for cats, because cat fur is a poor conductor of electricity.

Electric stunning

The electric stunning of larger animals as an integral part of humane killing is an important feature of killing for food. At present this method is mainly used for stunning pigs. Sheep and calves may be dealt with in this manner, and research is being carried out to include its use for cattle, horses, and large mammals in general. After exhaustive tests it has been found that in about 10 per cent of the animals stunned in this manner there has been some slight lung haemorrhage. Otherwise all organs and muscles were undamaged. The animals usually remain unconscious for about 150 seconds, but researchers found that the ideal time to dispatch the animals is from 4–6 seconds after stunning. Prior to this time and after there is likely to be convulsive movement, which makes the task of killing more difficult. There are various instru-

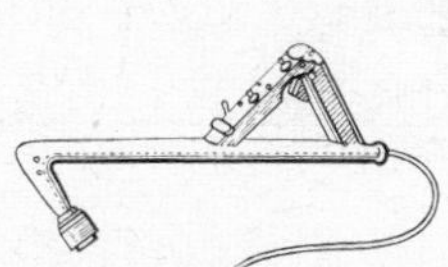

FIG. 1. Triangular shaped electric stunner.

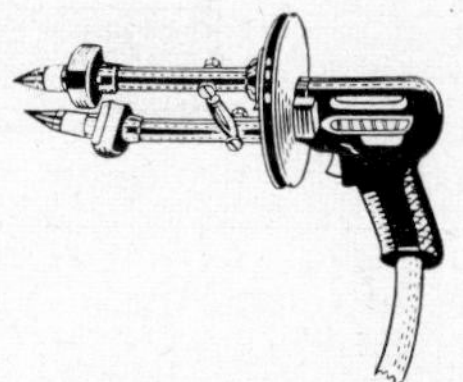

FIG. 2. Pistol-grip type of electric stunner.

ments on the market for electric stunning. One is triangular in shape, with one leg extending to the applicator head (Fig. 1), another is of the pistol-grip type with twin electrodes running parallel and separate (Fig. 2). These instruments nearly all utilize tattooing needles to ensure contact with the skin of the animal, and may be applied to any part of the head with equally effective results.

Compressed-air stunning

Stunning by compressed-air gun. This may be accomplished by using equipment which consists of a compressed-air system, furnishing a working pressure of approximately 180 lb per sq. in., and either a captive-bolt or a concussion-knob instrument (Figs. 3 and 4). In the captive-bolt instrument a sharpened, hollow bolt is driven through the skull into the brain by a piston activated by compressed air. The concussion-knob stunner consists of a solid head bolt activated in the same manner, which renders the animal unconscious purely by means of a high-velocity impact with the skull, causing concussion.

Captive-bolt pistol

Penetrating captive-bolt pistol. This instrument resembles a pistol in appearance and is fired by the detonation of a blank cartridge (Fig. 5). It is called the 'Cash X' pistol and is widely used. It weighs less than 6 lb, and the muzzle has a castellated face to reduce the possibility of movement when in the firing position. The cartridge is detonated by squeezing the trigger. The captive bolt is projected into the brain to a depth of 3–3½ inches, as it is with the compressed-air-gun captive bolt. Cartridges are ·22 calibre and are made in several strengths. It is advisable to use the correct strength cartridge designed for the species and size of animal to be killed, as the continued use of unnecessarily heavy charge cartridges causes needless stresses to the bolt and barrel of the pistol. There is also a cylindrical-type stunner similar in action to the pistol type, but the cartridge is detonated by means of a side lever (Fig. 6). This instrument may be fitted with a long handle having the detonating lever at its base, thus allowing the operator to work at a distance from the animal.

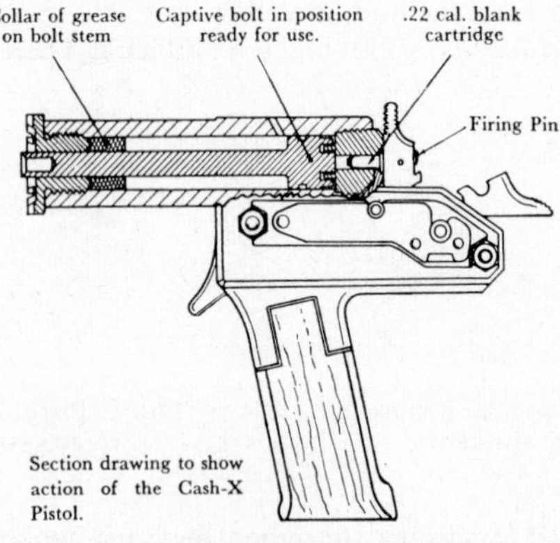

FIG. 5. Penetrating captive-bolt pistol.

The 'knocker' stunner

This is a modern interpretation of the older, hammer-method of stunning (Fig. 7). It is a pistol-like instrument having a flat stunning bolt (the impact mechanism) projecting slightly so that it touches the animal's head before the stunning bolt. The instrument is swung in the manner of a hammer, and on impact with the animal the firing mechanism drives the stunning bolt at high velocity against the beast's skull. This does not cause permanent unconsciousness, but will deprive the animal of all feeling for about 5 minutes, thus allowing sufficient time for the animal to be bled out. The force of the blow can be adjusted to suit the species and weight of animal to be killed. This method demands considerable skill and experience, and is not recommended for the novice.

Humane stunning cartridges

These are ·22 calibre rimfire cartridges which may be fired from a ·22 calibre rifle. These cartridges, which should never be used in any cartridge-powered or piston-type stunner, have a velocity of 2,000 feet per second, and they disintegrate on impact with hard surfaces. This means that the hazard from a stray bullet is somewhat reduced. They are also lead-free, which is an extremely important factor in the food industry.

It should be noted here that any of the guns, stunners, captive-bolt pistols, or the humane stunning cartridge, or, indeed, any instrument using cartridges as their means of power, must be covered by a firearms licence, and should be used only under qualified supervision.

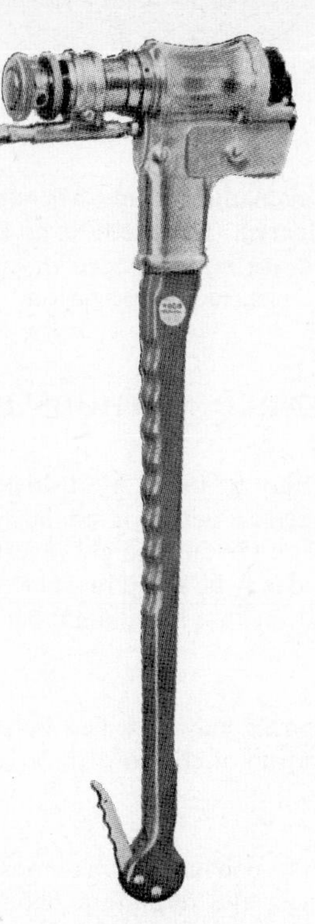

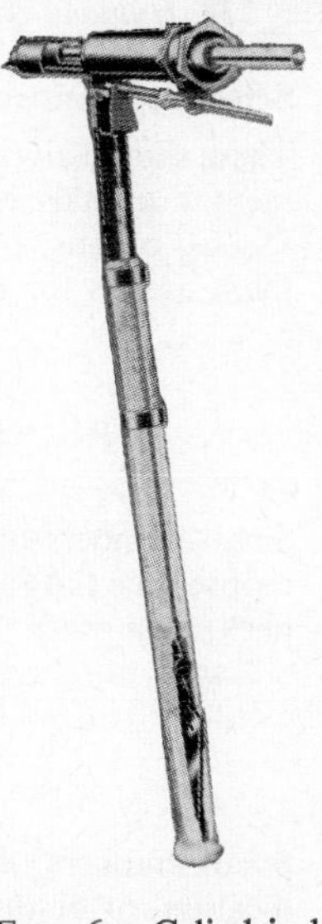

FIG. 6. Cylindrical type stunner with or without long handle.

FIG. 7. The 'Knocker' stunner.

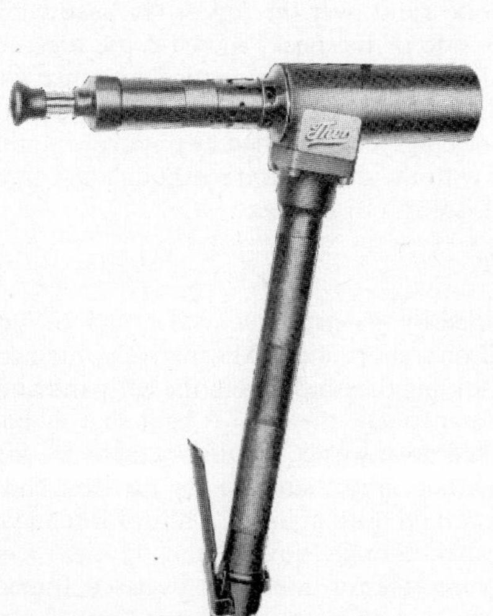

FIG. 3. Compressed air gun stunner.

FIG. 4. Compressed-air gun stunner. Quarter actual size.

Scientific requirements

These should always be taken into account when the question of killing experimental animals is considered. The method of killing should be such that tissues, organs, or blood are not damaged in a manner that would render them useless for further scientific investigation.

RECOMMENDED METHODS FOR KILLING

MICE

Small numbers of mice may be killed by dislocation of the neck. Place the mouse on a flat surface, press a pencil or similar object across the back of the neck, then grasp the base of the tail and give a jerk to dislocate the neck.

Larger numbers of mice may be killed in a glass jar or dessicator containing a chloroform-soaked pad, or in a lethal chamber with coal gas.

RATS

Small numbers of young rats may be killed by physical stunning or by dislocation, or by the inhalation of chloroform or coal gas in a lethal chamber.

HAMSTERS

Inhalation of chloroform or coal gas. Intravenous or intraperitoneal injection of sodium pentobarbitone, the minimum lethal dose being 13·5 mg per 100 gm body weight.

GUINEA PIGS

Kill by dislocation of the neck. Place the guinea pig on a flat surface facing the operator. Place one hand over the top of the head with the first and second fingers on either side of the neck. Increase the pressure on the fingers and swing the animal so that the body is vertical and then let the arm drop swiftly downwards to the side. The weight of the falling guinea pig's body will cause dislocation. Guinea pigs may be killed by physical stunning, but this method is crude compared with the dislocation method. They may also be killed by the inhalation of chloroform or coal gas.

RABBITS

These may be killed by a sharp blow at the back of the neck either with the edge of the hand or a short stick. Alternatively, the neck may be dislocated. This is done by holding the hind-legs in the left hand and the head in the right hand, then, simultaneously, the head is bent sharply backwards and the legs and body wrenched downwards. Inevitably there are some post-mortem convulsions. This method is not suitable for novices, and skill in it should be obtained by practice on dead animals. Chloroform and coal gas administered in a lethal chamber, though slower than physical methods, is quite satisfactory, and is preferable for injured or diseased animals.

Intravenous injection of a saturated solution of magnesium sulphate or sodium pentobarbitone may also be used.

MINK

These animals should be killed in a lethal chamber with coal gas or nitrogen.

CATS

Cats may be killed by administering barbiturates, in capsule form, by mouth; this method is especially satisfactory for kittens. The capsule or capsules may be passed over the back of the tongue with the aid of a finger, the blunt end of a pencil, or forceps. 'Nembutal' is supplied in capsules containing 1½ grains (100 mg) of sodium pentobarbitone. A lethal dose will vary from two capsules upward, so that a medium-size cat may need from four to seven capsules to kill it by this method. Alternatively, sodium pentobarbitone may be given by intrathoracic or intraperitoneal injection at the rate of 60 mg (in 1·0 ml) per 3 lb body weight. Probably the easiest method is to administer either chloroform or coal gas in a lethal chamber. Young kittens may be killed by stunning by striking the base of the skull on a hard surface, such as the edge of a sink or bench. Though this method sounds barbaric, it is, in fact, swift, and is certainly preferable to killing by drowning.

DOGS

Sodium pentobarbitone may be injected intraperitoneally, intravenously, or intrathoracically. Alternatively, a sedation dose of sodium pentobarbitone may be followed by the intravenous injection of a saturated solution of magnesium sulphate. The dog may be killed with a captive-bolt pistol, but this method is not recommended for persons unskilled in the use of these weapons. Coal gas may be administered in a lethal chamber. Chloroform is not recommended for use with dogs, and should only be resorted to in an emergency. A very large dose of chloroform is needed to kill a dog.

MONKEYS

Monkey cages are often fitted with a movable, vertical back grid, which can be pulled forward to trap the animal against the front of the cage. It is then quite a simple matter to inject sodium pentobarbitone intraperitoneally. After the injection the back grid is released to its normal position, and the monkey becomes unconscious in a few minutes. Providing the cage is fitted with a movable back grid, this is undoubtedly the simplest method, and, though it is comparatively slow, the monkey dies peacefully. The animal may be killed swiftly by an intravenous injection of sodium pentobarbitone, given either as an overdose or as a sedative followed by the intravenous or intracardiac injection of a saturated solution of magnesium sulphate. An intravenous injection entails handling the animals, and great care should be exercised (especially with newly imported stock) to avoid bites, which can transmit disease from monkey to man.

FOX

The fox may be killed by the oral administration of sodium pentobarbitone to produce sedation, and then chloroform or coal gas may be administered in a lethal chamber. The intravenous or intrathoracic injection of sodium pentobarbitone is a swift method of killing, providing the animal can be adequately restrained while the injection is given.

FERRET

The animal may be killed with chloroform in a lethal chamber, or sodium pentobarbitone may be injected intraperitoneally or intravenously.

HEDGEHOG

These animals are best killed by the intraperitoneal injection of sodium pentobarbitone.

To unroll a hedgehog hold it close to the back of the head and lift it clear of the table, and, by moving the wrist, rock the animal gently up and down; it will then unroll. When the hind-legs touch the table, place a pencil across them and hold it in position. The head may then be lowered until the animal is lying on its back. The hedgehog cannot roll itself up while its head is held and the pencil is kept across its hind-legs.

FOWL

Physical dislocation of the neck is the most common method of killing fowls. The legs are taken in the left hand and the head between the first two fingers of the right hand with the thumb under the beak. A sharp stretching movement, pulling the head backwards over the neck, will part the spinal cord in the cervical region. Older birds, or large birds, such as turkeys and geese, may first be stunned by means of a sharp blow from a short, heavy bar on to the base of the skull. They should then be hung, head downward, and the throat cut with a sharp knife.

Large fowls may be killed by exposure to carbon dioxide or by the intramuscular injection of sodium phenobarbitone (Luminal sodium), the dose being not less than 1·5 ml of a 15 per cent solution per kg body weight.

CHICKS

Chicks in small numbers may be killed by pressing the neck against the sharp edge of a table to part the vertebrae. Alternatively, they may be killed by the administration of carbon dioxide, chloroform, or coal gas in a lethal chamber.

LIZARDS, SNAKES, CROCODILES

Lizards, snakes, small caymen, or crocodiles are best killed by administering ether or chloroform in a lethal chamber.

FROGS AND TOADS

These should be placed in a dessicator or jar over a chloroform-soaked pad. They may be anaesthetized before killing by placing them in a beaker containing 1–2 per cent Urethane solution; the heads should be left above the level of the liquid, but the beaker must be covered, as the solution is volatile.

FISH

Small fish in aquaria may be anaesthetized by putting them in a 1–2 per cent Urethane solution, or they may be killed by being thrown forcibly on to a hard floor. Larger fish should be struck on the head with a stone or heavy stick. Eels should be decapitated after stunning.

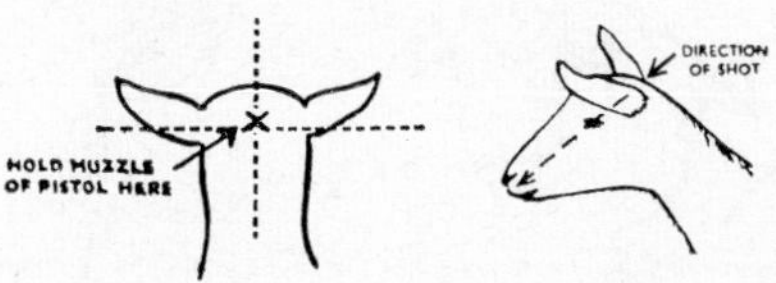

FIG. 8. The Kid: using captive-bolt pistol.

KIDS

These may be humanely killed by the administration of chloroform. This task may be done single-handed, but it is easier if two people are available—one to restrain the kid and one to administer the anaesthetic. The kid's head is restrained with one hand, and a wide-mouthed jar containing a chloroform-soaked pad is held loosely over the animal's face. There is little struggling if the anaesthetic is given slowly. When the animal becomes insensible the jar should be placed tightly against the face to exclude air.

A captive-bolt pistol may be used. The firing position is illustrated in Fig. 8. Sodium pentobarbitone may be administered intravenously or intrathoracically either as a lethal dose or as a sedative, followed by the injection of a saturated solution of magnesium sulphate.

CALVES

Electric stunning methods may be used. Alternatively, the captive-bolt pistol, compressed-air gun, or humane stunning cartridge methods are satisfactory, but, of course, the animal must be bled out immediately after stunning to kill it.

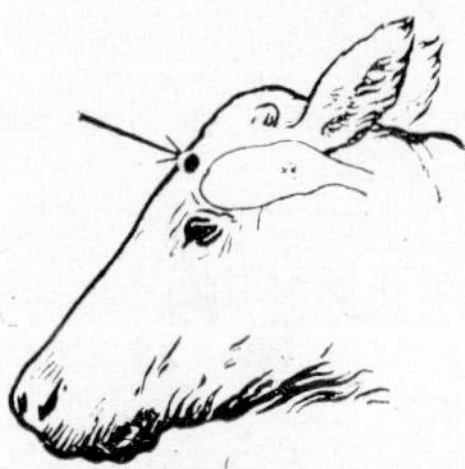

FIG. 9. Calves: using captive-bolt pistol.

CATTLE

The captive-bolt gun is used for killing cattle, but the firing position should be slightly higher than for calves. When shooting bulls or other thick-skulled beasts the muzzle of the gun should be placed $\frac{1}{2}$ in. to one side of the ridge that runs down the centre of the face.

FIG. 10. Cattle: using captive-bolt pistol.

PIGS

Animals small enough to be restrained adequately can be killed by an injection of sodium pentobarbitone solution into an ear vein.

Pigs may be killed by bleeding out immediately they have been stunned or anaesthetised. The use of an anaesthetic tunnel containing a mixture of air and carbon dioxide is becoming as common, in commerce, as the electric stunning, compressed-air gun, or captive-bolt stunners, now used extensively. Special care must be taken to position the captive-bolt pistol accurately when killing pug-nosed pigs, as the brain surface is rather small.

FIG. 11. The Pig: using captive-bolt pistol.

HORNLESS SHEEP

Captive-bolt guns or electric stunners may be used for these animals.

FIG. 12. Hornless Sheep: using captive-bolt pistol.

HORNED SHEEP OR RAMS

A captive-bolt gun may be used, but the muzzle must be placed behind the long ridge that runs between the horns.

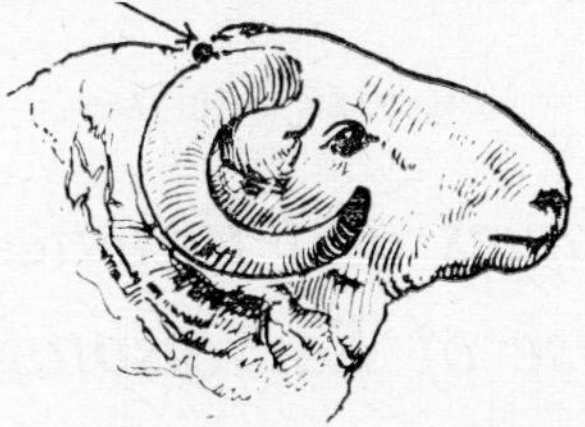

Fig. 13. Horned Sheep or Rams: using captive-bolt pistol.

GOATS

Electric stunning or captive-bolt pistols may be used as for sheep, but the muzzle of the gun should be directed towards the mouth of the animal instead of the gullet.

Fig. 14. Goats: using captive-bolt pistol.

HORSES AND DONKEYS

A captive-bolt stunner may be used, the muzzle being placed immediately below the roots of the forelock and firing roughly parallel to the line of the neck. The operator should stand to one side of the animal, as it will almost certainly fall forward on shooting.

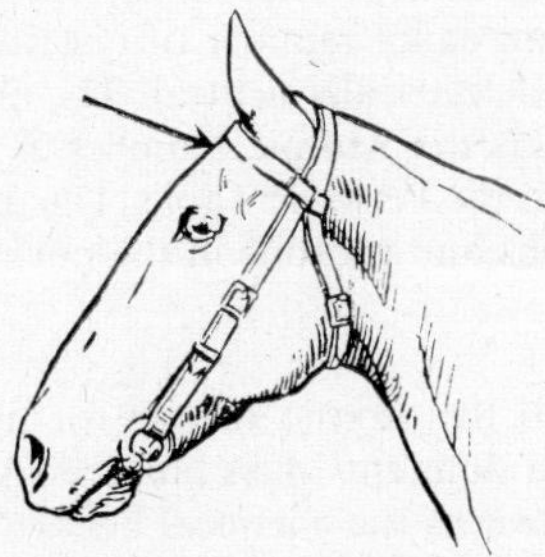

Fig. 15. Horses and Donkeys: using captive-bolt pistol.

CHAPTER FOURTEEN

Techniques and Practice in the Use of Radioisotopes

INTRODUCTION

Although the emphasis in this chapter will be on the practical aspects of handling radioactive animals, it is first necessary to give a simplified explanation of the physics of the atomic nucleus.

THE ATOM

Three fundamental particles make up the structure of the atom:

(*a*) The proton, which has a mass of 1·0076 atomic mass units (a.m.u.) and is positively charged.

(*b*) The neutron, with a mass of 1·0089 a.m.u., but no charge.

(*c*) The electron, with a mass of 0·00055 a.m.u. and a charge equal in magnitude to that of the proton, but negative in sign.

The atomic mass unit is defined as $\frac{1}{16}$ of the mass of the oxygen-16 atom, and is equivalent to $1{\cdot}6603 \times 10^{-24}$ gm.

The model suggested by Bohr and Rutherford provides a satisfactory picture of atomic structure. This supposes a small central nucleus of protons and neutrons, surrounded by electrons travelling in fixed orbits, the orbits being grouped together in successive shells. Since the protons and neutrons are much heavier than the electrons, the nucleus represents by far the major part of the mass of the atom. The protons in the nucleus are positively charged, and this is balanced by an equal number of negatively charged orbital electrons, so that the atom is electrically neutral. The number of protons present in the nucleus is known as the Atomic Number Z, and this also defines the position of the element in the Periodic Table. The mass number A is the sum of the number of neutrons and protons in the nucleus.

Isotopes

This is the term given to the atomic species (or nuclides) having the same Atomic Number Z, but a different Mass Number A (i.e. the number of neutrons differs). Since Z governs the chemical behaviour of the atom, different isotopes will behave identically from the chemical point of view.

Radioactivity

The variation in the relative number of neutrons to protons in the nucleus gives rise to 'stable' or 'radioactive' atoms. Certain combinations of protons and neutrons are stable, but as the excess or deficiency of neutrons increases, so does the probability of the atoms being unstable or radioactive.

The unstable nucleus undergoes a rearrangement of its particles and emits its excess energy in one of several forms. If the nucleus has an excess of neutrons the emission of an electron (or β-particle) from the nucleus accompanies the change of a neutron into a proton. This process is β-decay, and it is often the case that further energy in the form of electro-magnetic radiation (γ-ray) is emitted from the nucleus. The isotope decaying in this manner is one often to be encountered in normal isotope work.

In certain nuclides of high atomic weight, e.g. radium, disintegration occurs with the emission of a heavy particle having a positive charge twice that of the electron and a mass of about 4 a.m.u. Such α-particle emitters are not in wide use, and will only be referred to briefly.

Nuclides with a deficiency of neutrons can decay by the emission from the nucleus of a positively charged electron (positron), which rapidly interacts with the electrons of surrounding matter, suffering annihilation with the subsequent emission of 2 γ-rays. Alternatively, the nuclide can capture an orbital electron (electron capture) and subsequently an X-ray, identical in effect to a low energy γ-ray, is emitted.

Thus to summarize the types of decay, we have:

α-emission—heavy particles emitted. Not in common use.

β-emission. By far the commonest type of decay, and often accompanied by γ-emission.

γ-ray emission. Electro-magnetic radiation.

Positron emission. Usually detected by the annihilation radiation γ-rays.

Electron capture. Detected by the X-rays emitted.

Individual radioactive isotopes are characterized by the energy of the emitted radiation, and the rate at which disintegration occurs.

Energies

These are measured in terms of the electron-volt (eV), defined as the energy acquired by an electron in falling through a potential of one volt. The multiple terms keV (thousand eV) and MeV (million eV) are often employed. The energies encountered with common isotopes can range from that of a very weak β-emitter of 18 keV up to γ-emitters with an energy of 2·8 MeV. Since the latter is difficult to handle, and the former requires special equipment for measurement, it is more usual to use isotopes with energies of 100 keV up to about 1·7 MeV. However, the current availability of very sensitive equipment has begun a trend towards more use of the lower-energy isotopes.

Half-life

While it is impossible to predict when any particular atom of a given isotope will disintegrate, the average rate of disintegration of a large number of such atoms is fixed and invariable. It is convenient to introduce the term half-life, which is the time taken for the number of radioactive atoms present to fall to one-half of the original value. The passage of a second half-life reduces the number to $\frac{1}{4}$ of the original value, and in 7 half-lives we have less than $\frac{1}{100}$ of the original radioactivity present ($2^7 = 128$). The unit of disintegration rate is the Curie, defined as the quantity of any radioactive material disintegrating at the rate of $3{\cdot}7 \times 10^{10}$ disintegrations per second (d.p.s.). Sub-units of more practical interest are the millicurie (mCi) having $3{\cdot}7 \times 10^7$ d.p.s. and the microcurie (μCi) with $3{\cdot}7 \times 10^4$ d.p.s. (or $2{\cdot}22 \times 10^6$ dis. per minute). Animal experiments generally involve quantities ranging from a few up to perhaps several hundred μCi.

Half-lives exist in a range from less than a second up to thousands of years. The very short ones are of no practical use, but some long-lived isotopes are of immense biological importance, a particular case being Carbon-14, of 5,600 years half-life.

A few of the commonly used isotopes, with their main characteristics are listed below:

SYMBOL	HALF-LIFE	β-ENERGY	γ-ENERGY
H-3	12·3 years	18 keV	—
C-14	5,600 years	158 keV	—
S-35	87 days	167 keV	—
P-32	14·3 days	1·7 MeV	—
I-131	8 days	610 keV	640, 364 keV
I-132	2·3 hours	1·53, 1·16 MeV	960, 760, 670 keV
Fe-55	2·7 years	—	6 keV
Fe-59	45·0 days	460 keV, 270 keV	1·3, 1·1 MeV
Co-60	5·3 years	310 keV	1·3, 1·2 MeV
Na-24	14·8 hours	1·39 MeV	2·8, 1·4 MeV

Applications

The name isotope means ‘same place’, implying the same chemical properties. Thus a radioactive isotope will follow the same chemical route as the stable nuclide, but the radiation it emits makes it an easily detectable label. Usually small tracer doses are used to follow the metabolism of a substance or the distribution in tissue of a drug or other compound. The label's behaviour in the body can be followed without the necessity of killing the animal. The excretion or retention of a substance can equally easily be followed, such experiments requiring tracer amounts of isotope only.

In studies on the effect of internal radiation on animals much larger amounts of radioactivity are injected, but this type of work is unusual, and will not be considered here.

While radioisotopes are undoubtedly powerful tools of research, they can present handling hazards which need to be balanced against these advantages. It is easy to exaggerate the dangers, and even easier to forget them.

BIOLOGICAL EFFECTS OF RADIATION

When ionizing radiations are absorbed in living tissue, either from an external source or from an internally deposited isotope, profound changes can occur in individual cells. Although the absolute amount of energy involved is very small, its effect is concentrated in a few molecules only. Cell death or damage may result, and gene mutations may also occur. The results of this damage may manifest itself in the individual, or the effect may be a genetic one, inherited in future generations.

PROPERTIES OF RADIATIONS

In order to appreciate the relative hazards of different isotopes, consideration must be given to the behaviour of the radiations emitted.

α-PARTICLES, having a relatively large mass and being doubly charged, have an extremely short path length in tissue. Along that path, however, considerable ionization and damage can occur. Although a sheet of paper will completely stop α-particles, so that they cannot be an external radiation hazard from an injected animal, they can be extremely dangerous if once deposited and retained in the body.

β-PARTICLES have a well-defined range in matter, their small mass causing them to follow a tortuous path, analogous to a ping-pong ball ejected into a set of billiard balls. The range of 1·7-MeV β-particles in air is about 650 cm, or in tissue about 7 mm. Their ionizing power is less than that of α-particles by a factor of about 300.

Their larger path in tissue means that an animal injected with mCi amounts of a high-energy isotope could deliver a measurable β-dose during handling. This, however, would be an extreme case, and it is difficult to imagine a tracer experiment using a pure β-emitter where the handling time of an animal would be sufficiently long for any significant total dose to be delivered to the hands.

γ-RADIATION, AND X-RAYS. These are the most penetrating and least ionizing of the radiations. The ionization produced depends on the original γ-ray energy, but is of the order of 100 times less per centimetre of path than a β-particle. There is no definable range, their absorption by matter causing an exponential fall in intensity. Where the intensity is reduced to one-half by a given thickness of material (half-thickness), the addition of a second similar thickness reduces it to one-quarter. A 100-keV photon will have a half-thickness in tissue of about 4·0 cm. In this case the injection of mCi amounts of a high-energy isotope into several small animals could produce an external level of radiation which might require additional shielding if personnel are to remain close to such animals for long periods. This, again, would be an unusual situation in normal tracer experiments.

Dose units

The units of dose measurement are still subject to discussion, though from a practical point of view the several units in existence may be considered as identical.

THE ROENTGEN (r) is defined in terms of the ionization produced in 1 cc of dry air at normal temperature and pressure (N.T.P.), and by definition is applicable to X-rays and γ-rays. It is a measure of exposure dose.

THE RAD is defined as the absorption of 100 ergs in each gramme of tissue, following exposure to ionizing radiation. It is the release of energy in tissue which eventually causes radiobiological effects, but this effect also depends on the nature and energy of the incident radiation. α-particles produce a dense distribution of ions along a short track, and to allow for such differences between particles a factor known as the Relative Biological Effectiveness (R.B.E.) is used to modify the rad. For α-particles the R.B.E. is 10, so that the dose in rads $\times$ R.B.E. gives the Roentgen Equivalent Man (rem), i.e. rad $\times$ R.B.E. $=$ rem. For β-rays, and moderate energy γ-rays, R.B.E. $= 1$, so that the rad and rem are equivalent for these cases.

Small differences also occur between the values of the roentgen and the rad, since materials of different chemical composition receiving the same exposure dose will experience different absorbed doses. Although academically these differences are important, from a protection point of view they may be ignored.

Control of dose

The dose actually received by an individual from a given source depends on three factors:

TIME. i.e. the duration of the exposure.

DISTANCE. Where air absorption is unimportant, i.e. with γ-rays, the dose rate at a distance d from a point source is inversely proportional to the square of the distance:

Dose rate $= \frac{K}{d^2}$, so that doubling the distance will reduce the dose rate by a factor of four.

SHIELDING. β-particles are completely stopped by light materials, the thickness required depending on the β-ray energy. Below are a few values of absorber required for complete shielding.

SUBSTANCE	ENERGY MEV		
	0·5	1·0	2·0
Perspex	2·0 mm	4·0 mm	7·0 mm
Glass	1·0 mm	2·0 mm	4·0 mm
Wood	4·0 mm	7·0 mm	14·0 mm

A γ-ray shield needs to be of a material which is both dense and of high atomic number. Lead is probably the commonest, but by no means the cheapest, and in tracer experiments where an occasional shield is required steel can often be an adequate substitute, although greater thicknesses are needed. The half-thicknesses for lead for a few isotopes are given below:

ISOTOPE	HALF-VALUE LAYER CM LEAD
Bromine-82	1·0
Chromium-51	0·2
Cobalt-60	1·2
Copper-64	0·4
Gold-198	0·3
Iodine-131	0·3
Iron-59	1·1
Potassium-42	1·2
Sodium-24	1·5

HEALTH HAZARDS

The health hazards to be encountered by personnel may be divided into two types:

(*a*) External radiation, principally from γ-ray sources.

(*b*) Contamination of the body, either externally or by ingestion or inhalation.

As stated previously, the tracer amounts normally used in animal experiments are unlikely to present a radiation hazard.

The greater hazard is undoubtedly from contamination. External contamination from handling of excreta, cages, or animals can easily be avoided by the use of rubber gloves, or plastic disposable ones. Operation gowns should be donned before any radioactive work is commenced, and removed after monitoring when work is complete. Rubber gloves should be washed and monitored before setting aside, and care should be taken not to contaminate the inside when removing them. Heavily contaminated gowns, such as might arise following an accidental spill, or urination, should be set aside for decay if the half-life of the isotope is not too long. Rinsing with water will often remove the major portion of activity before sending the gown to the laundry. It is to be emphasized that spills are usually the result of poor technique, and as such are avoidable. Finally, rubber boots worn in the laboratory are easily washed down, and reduce the risk of spreading contamination.

Monitoring

Frequent monitoring with a suitably sensitive detector will prevent the build-up of unsuspected contamination. It should be carried out after any procedure where there is the slightest possibility of any contamination occurring, since besides the potential danger to personnel, an experiment can also easily be ruined.

Internal contamination by breathing radioactive gases or vapours is more difficult to control, but less likely to occur. Although radioactive carbon-14 and tritium-3 can be expired by injected animals, the concentrations reached, when mixing with the laboratory air has occurred, are normally so low as to be negligible. Isotopes excreted in urine are usually chemically bound and unlikely to create a breathing hazard, but it must be the responsibility of the worker concerned to point out any special precautions necessary if such products are likely to be volatile.

It is to be remembered that much higher levels of radiation obtain during the dispensing of an isotope, and also during its actual injection into an animal, than during the subsequent feeding and handling. Separate facilities are required for dispensing, and shielded syringes may be necessary for injections, but these aspects are the responsibility of the scientific worker planning the experiment. Once an animal is injected, the hazards are reduced to those stated above.

Film badges

Both the total body radiation and the levels of isotopes in the air which will deliver small permitted maximum doses are laid down in the Report of the International Committee for Radiation Protection. These levels are maxima for a 40-hour week, and in a well-run laboratory it should never be necessary to approach these figures.

In order to check radiation levels film badges should be worn by persons in contact with γ-emitters and high-energy β-emitters. They also serve as a useful indicator of splashing or spraying which occurs but is not detected by normal monitoring. Badges are supplied by the Radiological Protection Service, Sutton, Surrey, who also process them and report the result.

RADIOACTIVE ANIMAL LABORATORY

Wherever possible, a separate laboratory should be set aside for the use of radioactive animals. The concentration of such work in one area minimizes the spread of contamination and generally makes control easier. Preferably it should not be too close to counting-equipment laboratories.

Basically, such a laboratory needs to be easy to clean and needs to be kept clean. For this purpose, it should have smooth, non-absorbent surfaces to benches, and walls should be painted with a high-gloss paint. Cracks in any surface whatever, which are difficult to clean, must be avoided. The junction of floor and walls should be suitably faired, and a gulley in the floor provided, since at least one daily wash of the floor is necessary. The floor surfaces must also be smooth, and painting with an acid-proof paint has been found to be satisfactory.

A changing lobby at the entrance, with a hand-basin for washing, complete with arm-operated taps, should be provided. There should be a supply of paper towels, and sufficient space for a trolley to carry monitoring equipment.

Sinks and drains

For cleaning purposes a deep sink with arm-operated taps is desirable. A rubber hose attached to the water inlet and reaching to the sink floor will considerably reduce splashing. The sink U-bend should have an easily openable trap, and experience has shown that normal lead plumbing does not lead to a build-up of contamination in the trap if small amounts of activity are always flushed away. It is preferable to use only one sink for active waste, and if more than one is necessary in the laboratory the 'active' one should be so marked.

Cage-washing tank

An open water tank (100 gallons) with a high-level outlet pipe is useful for washing contaminated cages. These may be left to soak in continuously running water, the large volume of which is also an effective shield against external radiation. Cages must not be re-used until completely clear of activity, as a small residual contamination can completely vitiate a subsequent experiment.

Furniture

This is best kept to a minimum. A cupboard for the storage of syringes, chemicals, etc., is needed, and an operating table and lamp are usually necessary. The table is most suitably made in stainless steel throughout, and the cupboard is also best made of metal. Any surface to be used for handling animals or radioactivity should be freshly covered with two layers of blotting-paper or large squares of Whatman filter-paper. This simple precaution saves a considerable amount of time should any spillage occur. When sources for injection are being handled the operator should use a second containment vessel on the table in case of breakage.

Daily routine

The day-to-day routine of a radioactive animal laboratory differs only little from an ordinary animal room. It must be kept spotlessly clean, being washed down daily. The change of clothing must be used, and smoking, eating, drinking, and the application of cosmetics strictly forbidden.

Metabolism cages are often required, or at least cages with open-mesh bases, so that faeces and urine fall into suitable collection pans or bottles. It must be the responsibility of the worker to warn technical staff if highly active excreta is to be anticipated in any experiment.

Coprophagia is always a possibility, even in metabolism cages. Its importance in any experiment obviously depends on whether any of the metabolites occur in faeces. If necessary rabbits may be collared, but smaller animals are more difficult, and may have to undergo whole body restriction.

If total recovery of the isotope is required it may be necessary to trap expired carbon dioxide or water vapour.

Finally, hay or sawdust is to be avoided in radioactive animal cages.

Waste disposal

There is no entirely satisfactory method of disposing of radioactive wastes. The matter has been discussed in the Stationery Office publication *Control of Radioactive Wastes* (HMSO Cmnd 884, 1959), The Radioactive Substances Act (8 & 9 Eliz. 2 Ch. 34, 1960) is now in force and requires that all premises where radioactive materials are kept or used must be registered with the

Ministry of Housing and Local Government. The Act also calls for an authorization from the Ministry for the disposal of any radioactive wastes, and the two necessary Certificates must be displayed on the premises, where they may be read by all persons who may be affected.

Liquid wastes

Certificates under the Act must be obtained before any radioactive wastes are discharged. The quantities involved must be discussed with the Authority, but the following approach is a useful one. A level of 10^{-6} μCi/ml is below the maximum permissible figure for drinking-water for all β-emitters except strontium-90. If the amount of water used daily in the building is such that the average concentration in effluent does not exceed 10^{-4} μCi/ml there will be no significant radiation problem, since no one regularly drinks sewage, and also since further dilution occurs. The concentration quoted amounts to about 5 mCi per 10,000 gallons, and this is usually far beyond the figure required.

Since transient figures will be higher, it is best to ensure that at the sink the concentration does not exceed 0·1 μCi/ml. An inactive 'carrier' solution, i.e. containing the inert form of the substance disposed of, should be added to every solution, and followed by copious amounts of water.

Solid wastes

Waste paper, glassware, etc., should be temporarily placed in lined steel buckets, clearly marked 'radioactive waste'. The activity levels of solid wastes are usually lower, and combustible materials should be carried directly to the incinerator and burnt. Some storage facilities may be necessary for glassware contaminated with the shorter-lived isotopes.

In most animal experiments the level is sufficiently low for combustion of the body to be safe, although a storage method has been described by Boursnell and Gleeson White.

All incineration methods require that gaseous products are carried above surrounding buildings, and that a heavy ash is produced. Considerable care is necessary in handling active ash. The organization of a safe procedure during and after combustion is the responsibility of the Radiation Protection Officer. Waste-disposal machines capable of macerating animals as large as dogs and which deliver directly to the sewers are now available. Although expensive, they avoid the potential risks of handling radioactive ash, and are probably the best means of radioactive animal disposal.

BIBLIOGRAPHY

Recommendations of the International Commission on Radiological Protection. Pergamon Press (1959).

KINSMAN, S., *Radiological Health Handbook.* U.S. Department of Health, Education, and Welfare.

BOURSNELL, J. C. and GLEESON-WHITE, M. H., 'Temporary Preservation of Carcases of Small Laboratory Animals Containing Radioactive Isotopes', *Nature*, **179**, 54 (1957).

DUNSTER, H. J., 'Protection of Personnel Working with Radioactive Materials and the Disposal of Radioactive Waste', *Medicine Illustrated*, **8**, No. 11, 1 (1954).

SHERWOOD, R. J. and DUNSTER, H. J., *A Short Course in Radiological Protection.* AERE Report HP/L. 23. HMSO.

BOURSNELL, J. C. *Safety Techniques for Radioactive Tracers.* Cambridge University Press (1958).

BARNES, D. E. and TAYLOR, D., *Radiation Hazards and Protection.* G. Newnes Ltd. (1958).

FAIRES, R. A. and PARKS, B. H., *Radioisotope Handling Techniques.* G. Newnes Ltd. (1960).

VEALL, N. and VETTER, H., *Radioisotope Techniques in Clinical Research and Diagnosis.* Butterworth and Co. (1958).

Hazards of the Animal House. Lab. Animals Centre. Collected Papers, Vol. 10.

COMAR, C. L., *Radioisotopes in Biology and Agriculture.* McGraw-Hill, New York (1955).

CHAPTER FIFTEEN

Techniques for Infected Animals

Infections in the animal house may arise from diseases occurring in the normal animal population or from agents used in diagnosis or in research projects, which have been introduced experimentally in the animals. Of all the possible infections encountered in the animal house some may be dangerous to man and animals, while others may be dangerous only to the animal population. The precautions which have to be taken are therefore not only for the safety of the animal technicians but also to reduce the risk of infection spreading among the animal population. The latter precautions are essential for the maintenance of a strong and healthy animal population. Because of the very wide range of infectious agents which may arise in handling different kinds of animals or may arise from the type of diagnostic or experimental work, no attempt will be made in this chapter to deal with the subject systematically. Instead, the basic principles which can be applied to any of the infections likely to be encountered in a small or large animal house will be outlined. In addition, a general description will be given of special isolation-units which can be used for housing large numbers of infected animals, where the risk to the animal technicians or other animals is greater. For the successful running of a special isolation-unit I will relate our own experiences during the last ten years at the National Institute for Medical Research, London, where we have safely handled many thousands of animals experimentally infected with tuberculosis.

HOW INFECTIONS ARE TRANSMITTED

An infection is a disease which is caused by a germ or infectious agent which may be a fungus, a bacterium, or a virus. In the animal house these infectious diseases can spread or be transmitted from animal to animal, animal to man, or even man to animal. The method of spread depends on the type of disease. It may be direct contact, and is then often referred to as a contagious disease, as occurs in many skin infections, for example ringworm. On the other hand, the disease may affect only the gut or intestine, as in food poisoning, diarrhoea, or dysentery, and then the germs enter the body in already infected food or water at the time of eating or drinking. Or again, if the hands are soiled there is a serious risk of introducing the germs when smoking a cigarette or handling food. A third route of infection is that of inhaling the germs into the lungs. This method of spread occurs when small droplets of moisture or dust containing the germs enter the lungs during breathing, and is most likely to occur when infected dust from the floor or animal cages is disturbed. An example of this type of infection is tuberculosis of the lung.

From these examples of how infection can be spread it is seen that preventive measures for the animal-house personnel depend on common-sense measures of hygiene. Similarly, the spread of infection among the animals can also be prevented by the same simple principles of hygiene which must be applied by the animal technician. The animal technician therefore can not only safeguard his own health but can play an essential part in maintaining the health of the animals in his care.

GENERAL PRECAUTIONS FOR ANIMAL TECHNICIANS

The animal technicians should be given simple instructions in bacteriological principles with particular reference to the type of infections handled in their own animal house. It is essential that the animal house be provided with adequate washing facilities, including soap, scrubbing brushes, and disposable towelling. A number of skin conditions which arise from the animals can be prevented by thoroughly washing the hands after handling these animals. Rubber boots and gowns should be worn in the room housing infected animals. These gowns should not be taken out of the infected rooms. There should be no smoking, eating, or drinking in the infected part of the animal house. No unauthorized persons or casual visitors, such as friends or relations, should be allowed in rooms housing infected animals. All injuries which occur in the animal house, particularly bites from animals or scratches from cages, should be reported to the senior animal technician or, if available, to the medical staff. Any cuts on the hands or other exposed parts of the body should be adequately covered during the time of work in the animal house. Where special infections are handled protective vaccination should be given to the animal personnel. For example, in units handling poliomyelitis or tuberculosis the specific vaccines should be administered to the personnel before employment in the unit.

GENERAL PRECAUTIONS FOR THE ANIMALS

1. Ventilation

That part of the animal house in which infected animals are maintained should be especially well ventilated.

2. Cleaning of animal rooms

It should be accepted that all dust, animal food, and bedding on the floors is infected, and therefore all such dirt should be removed regularly. All floors should be kept clean by hosing with water and if necessary by swabbing with disinfectants, such as 1 per cent Tego in hot water or 5 per cent Lysol (see chapter on Sterilization and Disinfection for further details), but never by dry-sweeping. Similarly, wet-dusting should be used for cleaning racks, bench tops, window-sills, and all other furnishings and equipment. Dry-sweeping will automatically produce in the air particles of dust which may be loaded with germs and which could be a hazard to both the technicians and the animals. Wet-sweeping with or without disinfectant will prevent the formation of potentially dangerous dust.

More complete disinfection should be carried out in the rooms when an experiment is completed and before starting another experiment. During this procedure the walls and ceilings should be disinfected, most conveniently undertaken using Tego applied from a spray-machine (see chapter on Sterilization and Disinfection).

In all operations where disinfectants are being used in rooms housing animals it should be remembered that some disinfectants can damage or be toxic to the animals, and therefore care should be taken to avoid splashing the animals with the disinfectants.

3. Cleaning and sterilization of cages

It is important to keep the cages clean by regularly changing the soiled bedding and removing stale green or root vegetables. Bedding and food heavily soiled with excreta will favour the spread of disease among the animals. Furthermore, all excreta from experimentally infected animals are likely to be contaminated. It is important that the litter from these cages should not be disturbed, and therefore at the time of cleaning the animals should be carefully placed in a clean cage. The dirty cages, trays, and bedding should be sterilized by steam *before cleaning and disposal of the litter*. If the autoclave is in another building the infected cages should be transported to the autoclave in suitable metal bins with properly fitting lids. If no steam-sterilizing apparatus is available the infected cages should be soaked in disinfectant before cleaning and disposing of litter. Unless the infected bedding is first treated by steam-sterilization or by antiseptics there is a serious risk to the animal personnel from contaminated dust particles.

4. Feeding and watering animals

In order to avoid the possible contamination of food by infected excreta, the standard diet should be provided in hoppers. Where it is necessary to supplement with fresh vegetables these should be changed regularly. Water should be provided freshly from bottles. It is also important that each cage should have its own bottle in order to avoid the carrying over of infection from one cage to another.

5. Reporting of sick or dead animals

Depending, of course, on the type of investigations, any unexpected sickness or deaths among the animals should be reported to the Officer in Charge. A prompt notification of sickness or death may avoid a serious outbreak of infection among the animals.

6. Eradication of wild animals

Mice and rats are the most likely wild animals to be encountered in the animal house. These must be eradicated, because there is a serious risk that they may bring disease to the experimental animals or spread infection from cage to cage. If such wild animals are present in the animal house there is the added risk that they may contaminate the main stocks of food.

ISOLATION TECHNIQUES

The same basic principles dealt with in the previous sections apply equally to the more specialized animal houses dealing with larger numbers of infected animals. Nevertheless, the greater risks involved in dealing with larger-scale work, particularly when handling several different experimental infections in the same animal house, require additional safety precautions and isolation techniques. The precautions to be taken are again not only for the safety of the animal technicians but also for the well-being of the animals and the success and reliability of the experimental work. Where several different experimental infections are being studied there is a greater risk of cross-infection among the animals, which if it was allowed to occur would lead to completely erroneous results. Unless strict isolation techniques are enforced cross-infection is likely because many of the species of animals are highly susceptible to the particular experimental infections, and furthermore, there may be in the same animal house similar species of normal animals which could also become infected.

General isolation precautions

A separate room should be provided for each experimental infection. Rubber boots and gowns should be worn in the infected areas. A large tray containing Lysol or some other suitable disinfectant should be placed in front of the door in the infected room so that the technician can wash his boots before leaving. The doors to each room should be self-closing, using a strong spring device, and should preferably be of the 'push-open type' (from the inside), thus avoiding the use of a handle on the infected side of the door. Bins should be provided for the dirty gowns, which should not be taken outside the infected area. Furthermore, where the same animal technician is responsible for the care of rooms housing different infections, a clean gown should be used in each infected room. The same gown should never be worn when handling animals with different infections. Cages and water bottles must never be taken from one infected room to another room housing a different infection. Similarly, any apparatus or materials that are likely to be contaminated with one infection must never be used in a room with a different infection.

Type of cage

Although no type of cage will eliminate completely the escape of infected material to the outside, the risk can be considerably reduced by housing mice in metal boxes and rabbits or guinea pigs in cages with the back and side and bottom 3 in. of the front made of sheet metal. The larger animals are kept most conveniently in cages with a wire-mesh base standing on or above a detachable tray. In this type of cage only the tray is changed, and the animals can remain in the same cage throughout the experiment. In order to reduce the risk of infection from dried excreta the trays can be kept moist with dilute disinfectant. By using this simple device the dirty trays can be changed without disturbing or handling the animal and therefore avoiding the spread of infection by contact or by disturbing the dusty bedding.

Cage sterilization

In the more specialized infected units all infected cages, trays, and bedding should be sterilized by steam before cleaning and disposal of the litter.

Post-mortem examination of animals

Post-mortem examinations should be carried out in sterilizable trays, and the fur of the animals should be kept damp with diluted disinfectant. After examination of the animal all the instruments should be sterilized by heat, and the carcase should be placed in a waterproof bag and incinerated.

Syringes and needles

In order to avoid the risk of a needle becoming detached from the syringe during animal inoculation of infected material a Luer–Lock type of syringe should be used. If it is necessary to remove air from the syringe which has been filled with infected material the air should be extruded into a tightly packed wedge of cotton wool.

In the foregoing sections the general principles concerning the spread of infection and the general and special safety precautions which should be used when handling experimental infections have been outlined. The methods that have been suggested are all those that should and can be applied in the ordinary type of animal houses where this more specialized type of experimental work is undertaken. Where larger-scale work is undertaken or where new animal houses have been built for this type of work, special isolation units have been designed. In the last section is a general description of the type of isolation units used at the National Institute for Medical Research, Mill Hill, and particularly the one used by the Tuberculosis Research Unit.

SPECIALLY DESIGNED ISOLATION-UNITS

The rooms housing the animals infected with tuberculosis are situated in a self-contained unit in the main animal house. The unit is provided with its own double-ended autoclave and its own changing room for the animal technicians. The rooms housing the infected animals have their own ventilation system providing fresh air to each room and extracting the stale air. Because the extracted air may contain living virulent tubercle bacilli, it is sterilized by dry filtration, followed by ultra-violet irradiation before being discharged to the outside. The rate of extraction from the isolation unit exceeds the rate of inflow of fresh air in order that the whole isolation unit is placed at a slightly negative pressure to the rest of the animal house. This ensures a general flow of air towards the isolation unit, so that when the main door of the unit is opened air will not pass from the isolation unit back into the normal animal house.

Safety screen for the inoculation of small animals

Mice infected intravenously with suspensions of virulent tubercle bacilli are used on a large scale in many of the experiments. In order to protect the operators against the hazard of splash with infected material during the inoculations, a simple screen has been designed. This apparatus is shown (Fig. 1), together with the wire-mesh cage for holding the mouse and the Luer–Lock type of syringe used for injecting the bacilli into the tail vein of the mice.

FIG. 1. Safety screen for protection against splash during the inoculation of small animals.

Method of housing the animals

When large numbers of animals, such as rabbits, guinea pigs, rats, and mice, are heavily infected with virulent tubercle bacilli the danger of their contaminating the environment is very considerable. This risk has been reduced by housing the animals in special cabinets which are connected to the extract system. The type of cabinet is shown (Fig. 2). It is a metal-framed cabinet with three shelves for animal cages, and each shelf is provided with a pair of sliding glass windows. The upper part of the cabinet is connected to the extract duct, and the lower part houses a trough of Lysol. Fresh air enters the room through the grille shown in the top right-hand corner of the plate. When all the windows are closed there is a sufficient flow of air between the panes of glass for the animals. As soon as one of the glass windows is opened the resistance in the cabinet is reduced and air enters freely through the open window. The ventilation system has been balanced so that when one of the windows to the cabinet is opened air enters at the rate of approximately 100 ft per minute, which is sufficient to carry away from the operator all small droplet-nuclei and even particles of dust or sawdust. This type of extracted cabinet therefore provides complete protection for the animal technicians in the room both against droplet-nuclei from the animals during the resting

FIG. 2. Ventilated cabinet for housing infected animals. Sliding glass windows provided for each shelf.

phase and at the time of feeding or changing the animal cages. All infected bedding or excreta that may escape from the cages will fall into the pool of Lysol at the bottom of the cabinet.

Post-mortem examination of the animals

All the experimental animals are injected with tubercle bacilli in the isolation unit to avoid moving infected animals from the laboratories to the animal house. When necessary *post-mortem* examinations are carried out in the isolation unit; no infected animals ever leave the unit. *Post-mortem* examinations are carried out in a special hood which is connected to the extract system.

The type of *post-mortem* hood is illustrated (Figs. 3 and 4). The cabinet, with a glass front, is connected to the extract duct, shown entering the top of the cabinet. A 6-inch-wide opening below the edge of the glass screen admits the operator's arms for carrying out the *post-mortems*. The air velocity through this opening is not less than 300 ft per minute. Two cork-covered turn-tables are provided for pinning out the dead animals (mice, rats, or guinea pigs).

FIG. 3. Ventilated cabinet for the *post-mortem* examination of infected animals showing mice pinned to cork-covered turn-tables.

Each animal can be brought in front of the operator by adjusting the turn-table. When the examination of the animals is completed the turn-tables with the attached animals can be immersed into a bath of Lysol by turning the handle shown on the right-hand side of the cabinet (Fig. 4). The animals are left immersed in Lysol for 2 hours before being removed. The cabinet is provided with ordinary strip-lighting and also ultra-violet light (Fig. 4), the latter can be switched on for a predetermined time by the time-switch shown in the switch box at the right-hand side of the *post-mortem* cabinet (Figs. 3 and 4).

FIG. 4. *Post-mortem* cabinet with screen raised showing turn-tables lowered into tank of Lysol.

Cage sterilization

All infected cages, trays, and bedding are sterilized in the autoclave provided for the unit. The autoclave is a double-ended type with one end opening on the infected side and the other end on the clean side of the unit. The cages are finally cleaned and their contents disposed of only after autoclaving. Every effort is made to pass most of the items from the isolation unit out through the autoclave, including animal carcases, notes, and syringes and tubes containing the suspension of tubercle bacilli used to inoculate the animals. No living infected animals are allowed out of the isolation unit, and when the experiments are completed the animals are killed, autoclaved, and then after leaving the unit the carcases are incinerated.

SPECIAL PRECAUTIONS FOR PERSONNEL

There remain two additional precautions that should be taken for animal technicians and personnel working in animal houses where there is a risk of contracting tuberculosis. This applies to personnel dealing with experimental or diagnostic animals and also personnel handling large numbers of monkeys. The latter particularly applies to special units involved in the production or the testing of poliomyelitis vaccine. First, all such personnel should be given the opportunity to have periodic X-ray examination of their chests. The frequency of these examinations will depend on the degree of risk, but should be every six or twelve months. Periodic X-ray examination is, of course, a benefit to the staff whatever their risk, and there is the possibility that an animal technician might have active disease and, in fact, be a hazard to the experimental animals! This hazard is of considerable importance in animal houses dealing with large numbers of monkeys.

In addition, the resistance of the worker against contracting tuberculosis can be increased by vaccination with BCG. BCG vaccination can give very considerable protection to those who are most susceptible, namely those who are tuberculin negative. It is therefore important that all personnel, before being employed in the animal houses, should be tuberculin tested and if they are negative they should be vaccinated. In fact, tuberculin negative workers who refuse to be vaccinated should not be employed in the animal house where there is a risk of contracting tuberculosis from the experimental animals. The decrease in the incidence of tuberculosis in the population of this country is associated with a steadily increasing proportion of tuberculin negative reactors in our younger population. It is therefore more than likely that younger new recruits for the animal house will be tuberculin negative, and therefore they will be at greater risk than their predecessors, who were more likely to be tuberculin positive.

Although in the foregoing section the description of a specially designed isolation-unit for housing animals infected with tuberculosis is given, the general principles apply equally for other infections.

CHAPTER SIXTEEN

Techniques for Injections and the Meaning of Terms

INJECTIONS

Animal experiments in this country, many of which involve simple injections, may only be performed under a licence granted by the Home Secretary (The Cruelty to Animals Act, 1876). In addition, certain certificates have to be obtained according to the nature of the experiment and the species of animal used.

Although the animal technician himself may not be required to perform an experiment, it is vital that he should be able to assist when necessary. A knowledge of the techniques used and the care of instruments is essential. In many experiments it is necessary to inject a substance into an animal, and therefore it is necessary for the animal technician to know the routes normally used for a particular species of animal.

There are many routes by which substances may be introduced into the animal body; the route depends on the nature of the material and the mode of action, particularly that affecting its transfer to the experimental site.

Enteral injections

The substance is introduced directly into the alimentary canal, either into the mouth (orally or *per os*) or into the stomach using a catheter.

Parenteral injections

The introduction of substances by routes other than the alimentary canal. These may be listed as follows:

1. SUBCUTANEOUS. A fold of skin is lifted and the point of the needle is introduced into the base of the fold so that it enters the subcutaneous tissue (i.e. immediately beneath the skin).

2. INTRAVENOUS. This method is used when material has to be introduced directly into the circulatory system.

3. INTRAPERITONEAL. The needle is passed through the subcutaneous tissue, through the abdominal wall, and into the peritoneal cavity.

4. INTRACUTANEOUS OR INTRADERMAL INJECTION. Hair is removed using a depilatory or clippers. A short hypodermic needle is used, and care must be taken to see that the needle passes only into the dermis, as near the surface as possible, and not into the subcutaneous tissue, by holding the syringe nearly parallel to this surface.

5. INTRAMUSCULAR. Substance injected directly into muscular tissue.

6. INTRANASAL. Substance injected directly into the nose.

7. INTRATHECAL. Substance injected directly into the spinal canal. Care must be taken to avoid damaging the spinal cord.

8. INTRAOCULAR. Usually substance administered as eye drops; in special cases injected into the cornea.

9. INTRATRACHEAL. Substance, in small quantities, is usually injected into the trachea by means of a fine catheter which is inserted through the nose.

10. INTRACEREBRAL. Injected directly into the brain through the skull.

11. PERCUTANEOUS INJECTIONS. The substance is placed on the surface of the skin for a given time and protected from the animal's reach.

The common laboratory animals will now be considered.

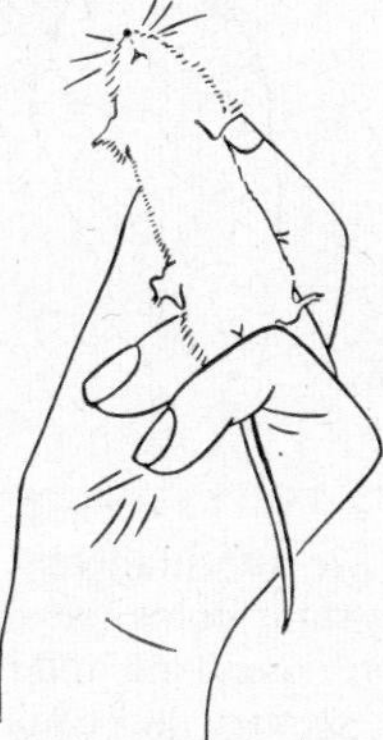

FIG. 1

(*a*) Mice

The most common routes of injection are subcutaneous, intraperitoneal, intravenous, and intracerebral. For intraperitoneal injections the loose skin at the nape of the neck is held between the thumb and index finger and the tail is held between the third and fourth finger (Fig. 1). Subcutaneous injections are made into the loose skin of the back.

Intravenous injections are normally made into the lateral veins of the tail as near to the tip as practicable. Before attempting this injection the whole mouse should be *warmed* to dilate the tail veins.

For intracerebral injections the mouse is first anaesthetized and the substance introduced to the crown of the head slightly to one side of the centre.

(*b*) **Rats**

Common routes of injection in the rat are intraperitoneal, subcutaneous, and intravenous. Intracutaneous injections are again made into the loose skin of the back.

For intraperitoneal injections the rat is immobilized by grasping gently, but firmly around the shoulder region, the tail is held with the other hand and the animal kept taut (Fig. 2). Intravenous injections are made into the tail vein or the saphenous vein in the hind-limb. For percutaneous injection the substance is protected by either placing the animal in a special cage, or by the use of collar, sleeve, etc.

FIG. 2

(*c*) **Guinea pig**

Common routes of injection are subcutaneous, intraperitoneal, intracutaneous and intravenous. An assistant holds the animal during the operation, holding the shoulders firmly with one hand and restraining the lower part of the body with the other hand, so that the required portion is exposed. For the intravenous route, the saphenous vein in the hind-leg is used, and for both intracutaneous and intravenous injections the skin is shaved.

For oral dosing, the guinea pig is best held wrapped in a cloth. A gag and flexible polythene cannula, with a well-rounded smooth tip, are used.

(*d*) **Rabbits**

A subcutaneous injection is normally made into the loose tissue of the flank or at the back of the neck. The marginal vein of the ear is the most convenient site for intravenous injections, for which the ear is shaved and the vein is dilated. Injections are made towards the head of the rabbit. For intracerebral injections the animal is first anaesthetized for an incision in the skull using a trephine. After injection the skin is sutured.

Although these are the most common ways of handling animals for injection, it may be that the animal technician can adjust these methods if he finds a method which affords greater comfort to himself and the animal.

Injections can be made difficult if hypodermic needles are not in good condition. Blunt or hooked needles cause damage to the tissue into which they are forced, pain to the animal and strain on the temper of the operator. The needle point should always be examined (even if the needle is a new one); for this purpose a small hand-lens or binocular microscope should be used. For rapid examination it may be found useful to draw the needle across the back of the hand, before sterilization. The MRC Memorandum No. 41 *The Sterilization, Use and Care of Syringes* give practical advice on the sharpening of needles. Consideration of the shapes of needle points is discussed in a paper by Franz and Tozell in *Anaesthesiology*, **17**, No. 5, 726 (1956).

The MRC Memorandum, already mentioned, also gives details of the different ways by which syringes may be sterilized. Many people now use sterile disposable plastic syringes which may be used once and then thrown away.

CHAPTER SEVENTEEN

Preparation for Surgical Procedures and Post-operative Care

A surgical procedure with subsequent recovery may be performed on an experimental animal only by a person who holds a Home Office licence and the appropriate certificates.

A technician may administer the anaesthetic or provide manual assistance as directed by the licence holder during the operation. Condition 5 attached to the Home Office licence requires that all such operative procedures shall be conducted under strict antiseptic or aseptic conditions and with adequate anaesthesia.

The following details must be considered for all operations:

(i) preparation and premedication of the animal;
(ii) the antiseptic or aseptic precautions which will be needed;
(iii) choice of anaesthetic;
(iv) general care immediately after operation;
(v) any special treatment.

Preparation and premedication of the animal

All animals used for operations with recovery must be perfectly healthy. They should be housed in the animal house for some time prior to operation so that they become accustomed to their conditions and surroundings. Each animal should be cared for and handled by the same person before and after the operation. The fur must be clean and the skin free from ectoparasites or any skin disease. The technician should inquire from the licence holder if any pretreatment with antibiotics or other drugs is required, but these must never be administered without the licence holder's knowledge.

Removal of hair (depilation)

This is essential for all species. For some procedures it may be sufficient to clip the hair very carefully around the operation site, but if the skin must be absolutely free from hair, then the remaining hair can be removed either by shaving or with a depilatory paste. Since hair regrows very quickly, complete removal of hair can be carried out only on the day of the operation, probably with less trouble to the animal after anaesthesia has been induced.

Proprietary preparations, such as 'Sleek' (Elizabeth Arden), are recommended for depilation; they are expensive but gentle in action. A cheap, but harsh, depilatory paste with a most unpleasant smell may be made to the following formula:

Barium sulphide 35 g, flour 35 g, talc 35 g, powdered soap 5 g. The powder keeps indefinitely, and a sufficient quantity may be mixed into a paste with warm water as required. It should be applied carefully and washed off *even more carefully*. The skin can easily become sore, and healing will be impaired. For the comfort of the animals the more expensive proprietary preparations are much to be preferred. After clipping or depilating, the skin should be cleaned with an antiseptic solution such as Cetavlon or Hibitane.

PREPARATION OF SYRINGES, INSTRUMENTS, ETC.

Antisepsis is the prevention of sepsis by the destruction of micro-organisms and infective matter. To achieve such a condition various methods can be used, and although not all will produce complete asepsis, the precautions taken must prevent the wound becoming infected, with the subsequent risk of general infection. All animal houses which are intended to cope with a large number of operative procedures should have a specially designed and equipped operating theatre.

Sterilization of syringes

Any anaesthetic or any other drug administered with a hypodermic syringe during an operation may introduce infection if sterile solutions and sterile syringes are not used. A wide range of disposable plastic syringes and disposable needles (sterilized by irradiation) are now available sufficiently cheaply to be used in animal houses. If these are not available syringes and needles should be packed in sealed metal or glass tubes and sterilized in a hot-air oven at 160°C for 1 hour.

Sterilization of instruments, etc.

Instruments can be packed into sealed metal tubes or wrapped in paper and packed in cardboard boxes. Towels, gowns, gloves, and swabs can be made into packets containing sufficient materials for each procedure and then either wrapped in two layers of paper or packed into cardboard boxes. The sealed packets or boxes, marked very plainly to show the contents, must then be autoclaved. Scalpel handles can be autoclaved, with other instruments, and sterile disposable scalpel blades can be purchased cheaply. Nylon, thread, silk, or catgut with attached (atraumatic) needles can be bought in sterile tubes or packets. Sufficient sterile equipment must always be available to cope with any emergency which may occur either during or after an operation.

CHOICE OF ANAESTHETIC

The anaesthetic to be used depends on the animal and the experimental procedure required. The anaesthetic may be administered by inhalation or by injection, or may require the use of both methods. The anaesthetic used must provide adequate anaesthesia for the duration of the operation, but since all anaesthetics are toxic, recovery should occur as soon as possible. The quantity of anaesthetic given by injection is judged according to a definite dosage scale based on the body-weight of the animal. While this scale gives some indication of the probable effects of a particular dose, every animal reacts differently to anaesthetics, and the utmost care is needed if this method is used.

The classical stages of anaesthesia are not always easily recognizable in laboratory animals. The early stages—up to unconsciousness and through the excitable period (during which struggling occurs)—should be passed through as rapidly as possible to reach the stage of full relaxation, which is accompanied by regular respiration. Anaesthesia is maintained at this level. Overdosing with anaesthetic at this stage leads to shallow, irregular respiration, which may cease altogether, and death will ensue.

INHALATION ANAESTHETICS

ETHER and other inhalation anaesthetics cause excessive salivation and bronchial secretion in some species, and there is always a danger that these secretions will pass to the lungs and cause pneumonia or interfere with respiration. If these difficulties are expected it is possible to prevent or minimize them by the injection of *atropine* at least a quarter of an hour before the administration of the anaesthetic. *Morphine* in conjunction with atropine is sometimes given, prior to operation, in order to produce some degree of narcosis before the general anaesthesia is started. Morphine must never be given to cats.

Ether is a good anaesthetic for most small animals, and is easy to use.

Ether should not be given too slowly, but a high concentration should be produced in the early stage. This can be reduced and anaesthesia can be maintained with a much lower concentration. If the anaesthesia does become too deep the concentration can easily be reduced by removing the mask, and after a few breaths the animal will have become much more lightly anaesthetized. Care must be taken when ether is being used. It is very volatile, highly inflammable, and ether–oxygen mixtures are explosive. The anaesthetist should take care not to inhale dangerous amounts of ether himself. Ether decomposes rapidly on exposure to bright light; it should be kept in small dark bottles, and all apparatus used to administer ether must be cleaned after use.

Small animals (rats and mice) may be placed under a bell jar (or large funnel) and the ether admitted to the jar either by dropping it on to a pad of cotton-wool placed under the jar or by blowing air first through ether contained in a Woulff's bottle and then into the jar. The jar or funnel must be of glass so that the animal may be observed. When unconscious the animal can be restrained on a board and anaesthesia can be maintained with a small mask or funnel over the nose. Administration of ether to rabbits and guinea pigs is more difficult, as they get a laryngeal spasm very easily and resist the anaesthetic. It is not advisable therefore to put them into a box, as they will struggle a great deal and may injure themselves. They should be gently restrained by an assistant, and ether should be administered with a mask, increasing the concentration as rapidly as possible, but always allowing some air intake.

Cats, dogs, and monkeys may be put into a box, and anaesthetic can be pumped in by blowing air through ether in a Woulff's bottle. Anaesthesia can be maintained by using a mask, but for operations of any length in dogs it is better to pass a Magill's cuffed endotracheal tube. When the cuff is inflated the tube is held in position. The endotracheal tube can be attached to a Woulff's bottle or preferably a vaporizer, which will enable a known concentration of ether to be given.

ETHYL CHLORIDE is a volatile anaesthetic sold in tubes which are graduated. It is highly toxic to the heart. It is used chiefly in cats and kittens to produce unconsciousness rapidly. About 5 ml and not more than 10 ml should be sprayed on to the mask. When anaesthesia has been induced ether should be used, since ethyl chloride cannot be used to maintain anaesthesia.

TRILENE is a volatile anaesthetic which is inflammable and has no advantage over ether in animal experiments.

CHLOROFORM is a volatile anaesthetic which is not explosive. It produces deep anaesthesia, but is dangerous because of its toxicity. In comparable concentrations it is ten times more toxic to the heart than ether, and it also produces liver damage. Mice are extremely susceptible to liver damage caused by chloroform, and it is wise to ban the use of chloroform for any purpose in an animal house containing mice. Chloroform is used in the veterinary field in preference to ether where large amounts have to be used and the risk of explosion would be present.

HALOTHANE (FLUOTHANE) is a new volatile, non-explosive anaesthetic. Induction is smooth and fairly rapid, and it is rapidly excreted. It is suitable for all animals and provides surgical anaesthesia of long duration, but the drug itself and the equipment required are costly. The administration of halothane is a skilled technique which should never be attempted by a technician without expert tuition and supervision.

CYCLOPROPANE is a gas which must be administered together with oxygen. Since the amounts must be controlled, a flow meter is necessary. The mixture is explosive. It is expensive, and should be given in a closed circuit, i.e. a certain amount of a known mixture is put into a bag, which can be attached to the animal. Since this is again a skilled technique, there is no indication for the general use of cyclopropane in animal surgery.

NITROUS OXIDE is a gas which is of low anaesthetic potency. When used to produce unconsciousness it is given without oxygen, and part of its effect is due to anoxia. Anaesthesia cannot be maintained with it. It has been used for monkeys to provide temporary unconsciousness, for which it is safer and less irritant than ethyl chloride.

ANAESTHETICS GIVEN BY INJECTION

THIOPENTONE SODIUM (PENTOTHAL). This is an irritant, and can be given only by intravenous injection, and must on no account be allowed to leak perivascularly during injection. It lasts only a short time, and it was thought that this was because it was rapidly detoxicated and excreted. It has been shown, however, that thiopentone is rapidly absorbed from the blood into the fat, and although anaesthesia will have passed off, a great deal of the original dose is still present in the fatty tissue.

PENTOBARBITONE SODIUM (NEMBUTAL). This can be given intravenously or intraperitoneally. It will produce a greater depth of anaesthesia than pentothal, and lasts much longer. When giving an intraperitoneal injection be sure that the needle is within the peritoneal cavity, particularly if the animal is fat, and take care to avoid the bladder and the liver.

URETHANE and CHLORALOSE should not be used for recovery experiments; their excretion is very slow, and the animal may develop respiratory complications before recovery from the anaesthesia.

Since the dose of injection anaesthetics is based on body weight, it must be emphasized again that there is great individual variation between animals of the same species. Once an injected anaesthetic has been given, it cannot be taken out, which may create difficulty with the intraperitoneal route. When a barbiturate is injected intravenously it must be given slowly, the rate and depth of the respiration being the anaesthetist's best guide.

STIMULANTS. After an excess of barbiturate, respiration may be stimulated by the intravenous or intraperitoneal injection of Bemegride (10 mg/kg) or Leptazol (5 mg/kg). If respiration ceases artificial respiration must be applied, preferably by a rocking motion. Raise and lower the back legs so that the viscera fall against and away from the diaphragm at regular intervals.

POST-OPERATIVE CARE

At completion of the operation the wound will have been sutured, but for most operations no dressings should be applied. Nobecutane or some other form of collodion dressing should be sprayed thinly over the site of the operation. If abdominal operations have been performed on cats or dogs a coat of some material can be made which will help support the weight of the abdominal contents, but this should not be put on until the animal recovers consciousness. All operations produce a degree of shock, so that the animal must be kept warm, and it may be considered necessary to give glucose–saline intravenously or intraperitoneally. While the animal is unconscious it should be placed on corrugated cardboard in a cage either on a grid or on a bare tray. No bedding should be put in a cage with an unconscious animal, which may asphyxiate itself or tangle itself up as it recovers consciousness. Following inhalation anaesthetics where recovery is rapid the animal should not be left until it has recovered consciousness and can walk normally. Following barbiturate anaesthesia where recovery is slow it must be remembered that animals of the same species will vary considerably in their rates of recovery. Rats and mice with a natural tendency to cannibalism will tend to nibble at or smother their slower-recovering cage-mates if housed in groups. Water must be available in the cages always, but the experimental procedure will determine how soon food should be provided.

If the technician who is to care for the animal has not been present during the operation it is essential that he should be given a complete description of the procedure carried out. The technician must be given details of any post-operative symptoms or complications which may be expected to occur and instructions as to any measures which may have to be taken in an emergency.

GENERAL SUMMARY OF REQUIREMENTS

MOUSE AND RAT

Withhold food for 1 hour prior to commencing anaesthesia, but allow water.

Anaesthetics: Ether. Pentobarbitone intraperitoneally.

Recovery usually rapid; keep warm; offer normal diet unless nature of operation suggests a simpler diet.

Watch for respiratory infections. No special treatment required unless necessary as a result of the operation, e.g. cortical extract to adrenalectomized rats.

GUINEA PIG

Withhold food but not water for not more than 4 hours prior to commencing anaesthesia. Atropine 2·0–2·5 mg/kg subcutaneously 15 minutes before anaesthetic.

Anaesthetics: Ether. Pentobarbitone intraperitoneally.

Recovery from barbiturate may be slow; essential to keep warm.

Watch for respiratory infections. Encourage to eat by offering green food.

RABBIT

Withhold food on day of operation, but allow water.

Anaesthetics: Ether. Pentobarbitone sodium intravenously.

Ether anaesthesia is not an easy procedure, and pentobarbitone sodium dosage very variable, but is usually to be preferred.

Recovery from barbiturate fairly quick to sitting position, but may take several days to behave and eat normally.

HAMSTER AND FERRET

Withhold food on day of operation. Make sure hamster has no food in cheek pouches.

Anaesthetics: Ether. Pentobarbitone sodium intravenously.

Recovery rapid.

CAT

Accustom to being caged for some days before operation. Groom fur with a pyrethrum-based insecticide if necessary. Fast overnight, but allow water. Atropine 0·5 mg/kg subcutaneously 15 minutes before anaesthesia. Morphine must never be given to cats, since it induces maniacal tendencies.

Anaesthetics: Ethyl chloride followed by ether. Pentobarbitone sodium intravenously or intraperitoneally.

Only a very quiet animal can be injected intravenously, whereas most cats can be injected intraperitoneally without the animal's knowledge. Cats must not be left after barbiturate anaesthetics have been given until anaesthesia is complete, since they all go through an unco-ordinated stage when they may injure themselves.

Recovery variable in time. Offer milk as well as water to drink; a cat can be encouraged to eat with food it can smell. Fish is preferable to meat, and sardines or liver will tempt most cats.

DOG

Accustom to be kennelled for some days before operation. Accustom to handling and being placed in the required position for anaesthetic administration. Bathe with an antiseptic solution. Withhold food overnight, but allow water. Exercise to encourage dog to pass faeces and empty bladder immediately before operation.

Anaesthetics: Ether. Pentobarbitone sodium intravenously. Ether suitable only for small dogs; pentobarbitone sodium can usually be administered with little stress to the animal.

Recovery slow from barbiturates. Keep wounds clean, especially free from faecal contamination. The dog will need to be allowed to take gentle exercise to encourage bowel and bladder emptying. Offer solid food on first post-operative day unless nature of procedure requires liquid diet. The dog more than any other animal requires sympathetic care to encourage quick recovery.

Detailed accounts of Anaesthesia for laboratory animals may be found in:

CROFT, PHYLLIS G., *An Introduction to the Anaesthesia of Laboratory Animals.* UFAW, 2nd ed. (1964).

The UFAW Handbook on the Care and Management of Laboratory Animals. 2nd ed. (1957).

WRIGHT, J. B. and HALL, L. W., *Veterinary Anaesthesia and Analgesia.* 5th ed. (1961).

CHAPTER EIGHTEEN

The Handling, Care, and Management of Farm Animals

Domestic animals belong to several species, each of which embraces a number of different breeds. These species, and to a lesser degree the breeds that make them up, differ in body conformation, in physiology, and in temperament. It follows that techniques of handling and management must be suited to the particular kind of animal involved. Efficient methods with one species may be quite unsuitable for another. Again there are important differences between the sexes. The male is usually stronger and more difficult to handle than the female, and is often rendered more amenable by castration. In the female striking differences in behaviour are often associated with oestrus, pregnancy, and nursing. Finally, it is important to remember that each animal is an individual, by virtue of both its inherent make-up and its experience, and should be treated as such.

To a large extent the successful control of large and powerful animals depends upon training them to react in particular ways to man-made routines and instructions. Usually a well-trained animal is easily controlled. The extent to which training can be effective is well illustrated by circus animals, for even those that are normally wild and highly dangerous can be made to perform complicated acts. As a result of training, patterns of behaviour may become so well conditioned that lay observers often attribute them to the influence of an intelligent mind, and believe that some animals possess a near human capacity to reason and to act in the interests of their handlers. This is not so, and it is well to remember that all animals are fundamentally selfish and self-centred, and most probably have no conception of right and wrong. The provision of adequate food and shelter by man leads them to modify their behaviour and plays a large part in training, and a good knowledge of these requirements is essential. Some knowledge of animal behaviour and a good deal of common sense are also necessary. The handler should always give warning of his approach. He should always be kind, firm, patient, and assured. He should never make sudden changes of routine, should never show fear, and should never take risks. He should have confidence in his own ability to handle and control the animal. Animals remember instances of cruelty for a very long time, and many difficulties in management result from this. Some animals are quick to sense fear in their handlers and are quick to take advantage of it. Some that are normally quiet and manageable may suddenly become unmanageable and dangerous.

A further important aspect of handling large animals concerns the use of various methods of physical restraint which must be resorted to when training alone does not suffice. These devices are designed to secure animals in such a way that manipulation of various kinds can be carried out without injury to either the animal or the operator. Foolhardiness and the taking of risks when dealing with the larger and stronger animals nearly always becomes corrected with time and experience.

THE HORSE

The horse is used chiefly for traction and for riding. There are big differences in temperament between the various breeds, and thoroughbreds are much more likely to be nervous and difficult than are heavy draught horses. The mature stallion can be difficult and dangerous to handle, and males not required for breeding are usually castrated at about eighteen months to two years of age. Earlier castration results in a loss of physical strength. The gelding is not only more docile and amenable than the stallion but also behaves in a less active and 'intelligent' way. For this reason circus horses are never castrated.

Training

Training should start at an early age, and should be characterized by kindness. Pleasure should be associated with early lessons and obedience inculcated by firmness. Much of the horse's disposition is established during his early training, and only good training produces the prompt and willing responses to commands which endow the horse with so much of its value. Cruelty often results in a sullen, spiritless submission, and lack of firmness often gives rise to temper and obstinacy.

Handling

When approaching a horse the handler should always speak before touching the animal. He should then pat the animal's neck and attempt to reassure it. Strange horses should be approached with special care, but should never be allowed to suspect that the handler is afraid.

Before examining a horse a halter or head collar should be fitted and held by an assistant. For simple operations like blood taking, the assistant stands on the opposite side to the operator and holds the horse's head, one hand on the nasal bone and with the other hand, a good grip on the ear. No additional restraint should be necessary. If the fore-quarters of a horse are to be examined the horse should be turned so that his hind-quarters are in a corner, or at least against a wall. In this position the horse is unable to back away and the examiner is left with maximum freedom of movement.

If the hind-quarters are to be examined it is wise to have an assistant pick up the front leg on the side which the examiner will stand. The front foot should be raised so that the leg is well flexed and does not provide the horse with support. In this position the horse will find it difficult to kick with the hind-leg that is on the same side as the raised front leg.

When painful operations, such as the dressing of wounds, are to be carried out it may be necessary to apply a counter-irritant that will divert the horse's attention. This is best done by applying a twitch loop to the sensitive area of the upper lip. The twitch loop consists of a loop of $\frac{3}{8}$-in. cord (total length about 16 in.) attached to the end of a stout wooden pole about 3 ft long. The twitch is applied by slipping the loop over the upper lip of the horse, care being taken to leave the nostrils free, then twisting up the lip by turning the pole until slight pain is inflicted. The operator should stand close to the horse's shoulder, in which position he cannot be kicked by the front feet, and should hold the pole pointing backwards. The horse should be held by an assistant by means of a halter. (The horse should never be tied up when a twitch is applied.) The twitch should only be applied immediately before restraint is required, and should be used only when absolutely necessary. There are several other types of twitch, most of them meant to bring pressure on the sensitive area of the lip, but the twitch loop is more generally used, and is simple to apply.

For a number of manipulations a horse must be cast and its feet secured by ropes and hobbles. This is carried out on a suitable surface, such as a grass meadow, a bed of peat moss, or straw, and in such a way that the horse falls on the required side and so that the available light can be used to the best advantage. It is important that adequate man-power should be available and that there should be as little noise and fuss as possible. It is common practice to administer tranquillizing drugs beforehand. These minimize struggling, and hobbles are applied easily and enable the horse to be cast more safely and by fewer men. There are several methods of casting, and the method chosen should be the one suitable to the job in hand. This, of course, depends on the type of horse and the particular operation. A method which has been used for many years and with good effect is the 'London Method'. This method is suitable for horses weighing up to about 1,100–1,200 lb. Two ropes (about $\frac{5}{8}$ in. × 25 ft) are required, each with a spliced noose at one end. Three hobbles made of strong woven cotton webbing with leather piping at the loop end and a strong brass 'D' at the other end are used. To cast the horse on to its left side, a fixed loop is placed over the neck. The rope is then passed round the chest, where a half-hitch is made. The hobbles are put on both fore and the near hind pasterns. A loop is made in the other rope and put round the right fore cannon bone of the leg which will lie uppermost when the horse is down. The other end is then threaded through the brass loop of the near hind-leg, then through the brass loop of the near fore-leg, than through the brass loop of the off fore-leg and held on the right side. The head should be held by a man and, when all is ready, the horse is brought down by a steady pull on both ropes. When down, the rope used about the chest is then looped round the right hind pastern and passed through the neck loop so that the right hind leg can be pulled well forward and secured by a half-hitch. The three hobbled legs are made secure by taking a couple of half-hitches round the right fore pastern with the rope used in the hobbles.

Vices

The chief vices of the horse are temper and obstinacy. Stable vices are many, such as kicking the heel post, crib-biting, wind sucking, licking walls, tearing clothing, weaving, and sleeping standing. Once these vices are established, they are difficult or near impossible to remedy. The stable vices result mainly from lack of work, boredom, and lack of human attention, and they will spread rapidly from one horse to others in the same stable.

Food

Horses have small stomachs and should be fed frequently; three or four feeds a day are usual. The staple indoor diet consists of oats, bran, chaff, and hay. The total amount of food and the proportion of its constituents vary with the type of horse and the nature of its work. For heavy slow work the proportion of grain is less and the proportion of roughage is greater. Horses weighing between 1,100 and 1,200 lb require about 26 lb of food daily. A van horse working at a slow rate would get about 10 lb of oats, 1 lb of bran, and 15 lb of hay. A hunter doing fast work would get about 15 lb of oats, 1 lb of bran, and 10 lb of hay.

Water

If water is not continuously available it should be offered before feeding. If this is not possible, then at least 1 hour should elapse after feeding before water is offered. Some horses drink 8–9 gallons a day and more in hot weather.

Bedding

Good bedding is necessary for horses, for providing them with warmth and comfort. It encourages them to lie down and rest their legs. Wheat straw is the best bedding material, because it is longer than other straws, and because its toughness and bitter taste discourage most horses from eating it.

Grooming

When not out at grass horses must be groomed regularly. This removes dirt, loose hair and scurf, stimulates the circulation of the skin, and discourages parasites.

Reproduction

The breeding season normally extends from early February to the end of June. During this time oestrus recurs every 21 days, unless pregnancy intervenes, and on each occasion lasts 5–6 days. The gestation period is 340 days. The foal is usually suckled for about 6 months. It reaches maturity at about 4½–5 years.

CATTLE

In this country cattle are kept for two main purposes: beef production and milk production. There are again differences in behaviour between breeds. Mature bulls are often difficult and dangerous, and the utmost care should always be taken in handling them. Males not required for breeding are usually castrated at about four to six months of age. Both sexes are often de-horned within a few days of birth. The details of management vary greatly with different systems of farming.

Training

Cows do not have the capacity to respond to training that horses have. They do, however, easily become conditioned to regular routines, and patterns of behaviour established in this way play an important part in management. At milking time, for example, dairy cows at pasture usually gather at the field gate, and when this is opened they proceed, unguided, to the cow byre and into their own stalls. Training does not normally extend beyond the establishment of simple patterns of behaviour except in countries where cattle are still used for traction. Temperament, especially in bulls, may be greatly influenced by early treatment.

FIG. 1

FIG. 2

Handling

Cows are usually driven rather than led. Bulls are usually led by a pole attached to a ring through the nose. The ring is usually inserted at about nine to twelve months of age. Dangerous bulls are led by two handlers. One takes the pole and the other takes the rope that is fastened round the horns and then passed through the ring.

Cows should be approached quietly. When the head is to be examined a hand is placed on the animal's back and rubbed or scratched along the back until the head is reached. A horn or ear is then grasped with one hand and the nostrils with the other. The nostrils are held by placing the thumb in one nostril and the first two fingers in the other (Figs. 1 and 2). If the examination is prolonged, or if the animal becomes restive, a 'bull-dog' holder is attached to the nostrils. This relieves strain on the fingers and thumb, and the swivel handle allows the animal to twist its head without twisting the operator's wrist.

Cattle do not usually allow their feet to be picked up by hand. To lift a front foot it is best to attach a rope above the fetlock and pass the free end of the rope over a sack (for protection on the animal's back) to an assistant on the other side. As the operator leans against the shoulder of the leg to be raised the assistant pulls on the rope. To lift a hind-leg it is best to pass a pole in front of the hock to be lifted and behind the other one. The operator and his assistant each take an end of the pole and lift it backwards and upwards at the same time, leaning towards each other.

When taking blood from a cow a loop of cord is passed around the lower part of the neck and drawn tight enough to obstruct and raise the jugular vein. An assistant holds the cow by the right horn and the nostrils and pulls the cow's head to the right and backwards. This leaves plenty of room for the operator to take blood from the left jugular vein.

Casting cattle as a means of restraint

One of the best methods of casting cattle is by the side-line method of Reuff. A rope about 40 ft long is tied to the horns, or to a head collar, and is passed backwards along one side of the animal. A half-hitch is made around the neck, a second half-hitch around the chest, and a third around the abdomen. The remainder of the rope is trailed behind the animal, and all slack rope is taken up. The cow is brought down by two men pulling with a strong and steady pull on the rope. Hobbles are then quickly placed around the cannon bones and all four legs are pulled together and secured by two half-hitches.

Feeding

Cattle are fed in a variety of different ways, depending on the need. A cow needs about 3 acres of land to support her; $1\frac{1}{4}$ acres for grazing and $1\frac{3}{4}$ acres for hay, oats, and roots. Dairy cattle are usually fed having regard to the amount of milk they produce and the foodstuff available. The common practice is to feed 4 lb of concentrates to each gallon of milk produced, with about 16–20 lb of hay per day. Intensive feeding results in cows giving three times more milk than is needed to feed a calf, and lactation ailments are common. Drinking-water must always be available.

Reproduction

The cow comes into oestrus throughout the year, but more regularly in summer. Heat lasts from 6 to 30 hours and recurs every three weeks. The gestation period is 280 days. The cow licks the calf as soon as it is born. A strong calf will be up on its feet in 20–30 minutes. Some dairy farmers wean the calf at once, some about 3–4 days after the calf has had the first milk (colostrum). The calf depends entirely on milk for the first 3 weeks of its life. Subsequently the calf should have access to small amounts of good hay and clean water. If it is to be weaned the calf must be taught to drink from a pail. It is usual to start it drinking by dipping a finger in the milk and letting the calf suck the finger. The hand is then lowered into the milk, the calf still sucking the finger and at the same time taking a little milk. Gradually the finger is withdrawn as the calf learns to drink without help. Calfhood is usually reckoned to be up to about 6 months of age. Maturity is reached at about 18 months, at which time heifers are usually served so as to calve at about $2\frac{1}{2}$ years of age.

SHEEP

Sheep are chiefly used to supply meat and wool. There are many breeds, and the choice of breed depends mainly upon the type of country on which the sheep are to be kept. There are again many differences in temperament between breeds. Mountain sheep, for example, tend to be much more active and excitable than lowland sheep. Males not required for breeding are usually castrated at about two to three weeks of age, the operation usually being carried out at the same time as 'docking'. 'Docking' consists of removing about two-thirds of the tail. This is done to prevent the accumulation of dirt and faeces and to reduce the risk of blow-fly strike. 'Docking' may be done by cutting the tail with a sharp knife. Bleeding is usually negligible, but if it persists the wound may be dressed with Friars Balsam. Sheep are dipped usually twice a year, in August and November, to eradicate parasites. Some farmers dip their sheep once before shearing and a second time when the fleece has grown long enough to retain in it a certain amount of dip. By law all sheep in some areas must be dipped twice a year in the presence of a police officer.

Handling

Sheep tend to flock together, and for this reason can be driven easily in flocks by shepherds and trained dogs. They quickly learn to recognize their handlers, and will follow people well known to them. Sheep should never be caught by holding on to the wool, but by gently sliding one hand under the base of the neck and lifting the chin high. Horned sheep may be caught by the horns. To restrain a sheep the handler stands astride the animal and lifts its head high by the chin or horns (Fig. 3). To turn a sheep 'up' the handler reaches well

FIG. 3 FIG. 4

under the sheep's belly with his right hand and lifts the chin with his left hand. The sheep is then lifted clear of the ground, fore-quarters first, and is lowered into a sitting position with its back against the handler's knees. The handler then holds the fore-legs (Fig. 4).

Bedding

Wheat straw is the best bedding for sheep, but sawdust and peat moss can also be used. Concrete floors are suitable for pens, and can be hosed regularly. If it is necessary that the sheep eat nothing but the ration given, then slatted floors should be used with sawdust beneath. The slats should be about $\frac{3}{4}$ in. apart.

Feeding

A normal ration consists of rough fodder, such as hay, and concentrates. Adult sheep will eat up to about 4 lb per day. Drinking-water must always be available.

Reproduction

The normal age for the first mating of female sheep is about eighteen months, so that the first lamb is dropped at nearly two years of age. The gestation period is 147 days. The oestrus period varies from breed to breed and with local climatic and food conditions. Some sheep come into oestrus as early as July, while others may not do so until October.

Mountain sheep rarely have more than one lamb per year, but lowland flocks may be expected to give 50 per cent twins. The lambs are usually weaned at about three to four months of age and are then transferred to good grazing.

PIGS

Pigs are kept chiefly to supply bacon and pork. There are again many breeds and systems of management. Males not required for breeding are usually castrated at about six weeks of age, or two weeks before weaning.

Mature boars may be extremely dangerous on account of their powerful tusks, and should always be handled with great care. Nursing sows will attack anyone interfering with the litter, and should always be secured before the piglets are handled.

Handling

Pigs can usually be driven, but when this is not effective they can be backed into any desired position by putting a large bucket over the head. Small pigs may be caught by the hind-legs above the hock, or by their ears. They should not be held by the tail, for in young pigs the skin of the tail tears easily. Large pigs are best restrained by driving them into narrow crates, or by pinning them in a corner with a hurdle or table top. Restraint can also be obtained by applying a twitch loop around the upper jaw.

Bedding

The best type of bedding for pigs is short straw or sawdust. This kind of bedding will not become entangled about the legs of the piglets. With suitable bedding pigs will keep themselves as clean as do other animals.

Feeding

At weaning time the young pigs should weigh about 25–30 lb. They will eat about $1\frac{1}{2}$–2 lb of meal per day, and they should gain weight at the rate of 1 lb per day. At about 16 weeks of age the rate of growth increases to 10–12 lb per week up to bacon weight, which is about 200–210 lb and the intake of meal per day will rise to about 6–7 lb. The meal can be fed dry or as a mash, with water *ad libitum*.

Reproduction

Sows will breed throughout the year, but oestrus is more prolonged in the spring and autumn. Oestrus recurs every 21 days, and lasts for 3–4 days. The gestation period is 115 days, and the litter is usually weaned at about 8 weeks. The average litter size is about 8, but numbers of up to 20 have been known. Maturity is reached at about 12 months of age. Piglets should be given iron (usually by injection) during the first few days of life to prevent anaemia.

GOATS

Since the turn of the century, the goat has been used increasingly as an experimental animal in Physiology, Pharmacology, Endocrinology, Biochemistry, Dairy Research, and latterly experimental orthopaedic surgery. For orthopaedic purposes the goat is preferred to the sheep and dog (Barnett, 1958).

Handling

The goat learns quickly, providing it is handled gently and firmly. Careless handling and the use of force will encourage stubbornness in the animal. Horns are an advantage when handling goats, as they afford a useful handgrip. It is satisfactory to run horned and non-horned goats together for exercise, provided there is enough room for the non-horned goats to take evasive action. It is advisable to train newcomers to the herd to walk with a collar and lead, remembering that the goat is a very inquisitive animal (as well as a nervous one) and should be allowed to satisfy its natural curiosity.

Housing

Goats are best kept in individual pens that are warm, dry, draught-proof, and strong. Beard and Duncombe (1951) recommend a size of 8 ft × 5 ft, with 4 ft 6 in. × 4 ft 6 in. as the minimum. These pens should be adjacent to an extensive exercise yard, which is essential in winter. In summer strip grazing in grass paddocks is desirable.

Bedding

Goats need plenty of bedding; straw is recommended, but peat moss and sawdust may also be used.

Feeding

Food may be divided into maintenance and productive rations. For maintenance goats require 0·6 lb starch equivalent and 0·15 lb protein daily, while for production 0·3 lb starch equivalent and 0·07 lb protein are needed per pint of milk. A good maintenance ration per day is supplied as follows:

5 lb good meadow hay
3 lb marrow stem kale
1 lb of any of the following meal mixtures—

	MIXTURE 1	MIXTURE 2	MIXTURE 3
Linseed cake	4	1	1
Flaked maize	3	—	1
Bran	2	—	2
Crushed oats	1	1	1

Parts by weight (Beard and Duncombe, 1951)

For milk production 6 oz of the meal should be fed per pint of milk produced.

Reproduction

Sexual maturity in the female is reached at 5–6 months of age, but the first mating should be delayed until the animal is 15–18 months old. The goat has an oestrus cycle of 21 days and the heat period lasts 2–3 days. The gestation period is 5 months, and the breeding season is from September to February. Recurrence of oestrus after the young are born is next season.

REFERENCES

BARNETT, *Journal of Animal Technicians Association*, **9**, No. 2 (1958).
BEARD and DUNCOMBE, *Journal of Animal Technicians Association*, **2**, No. 1 (1951).

The substance of this chapter was originally a paper read before the 15th Laboratory Animals Centre Congress, and was printed in J. anim. Tech. Assoc. *vol 13, No 1 (1962).*

CHAPTER NINETEEN

Care and Management of Amphibians, Reptiles, and Fish

AMPHIBIANS *

There are some 1,500 species known to exist in the world today, with a few new species being discovered each year, and it is hardly an exaggeration to say that they all require different treatment if the aim is to establish a breeding colony. Really adequate knowledge exists only in respect of the few species which have already been bred in captivity in large numbers. Such general information should only be regarded as a basis for further detailed research with selected species, and mention will be made of a number of problems which deserve further investigation.

Modern amphibians are divided into three orders:

(i) Apoda or Gymnophiona

These are the caecilians—legless, wormlike, burrowing amphibians, found only in certain tropical areas.

(ii) Caudata or Urodela

This order comprises the tailed amphibians, including the salamanders and newts as well as a number of permanently aquatic forms.

(iii) Salienta or Anura

The tailless amphibians, i.e., the frogs and toads.

CAECILIANS

Very little indeed is known about these primitive, secretive creatures, of which some fifty species have been discovered in the tropics of Africa, South-East Asia, and the New World. They all have the shape of a worm, are limbless and blind, and spend almost their entire lives underground. In most cases they lay large-yoked eggs in an underground burrow and the eggs are guarded by the female until they hatch. Knowledge of their requirements in captivity is almost non-existent, but presumably they would need to be kept in moist, soft soil or leaf-mould at a reasonably high temperature. Their food, as far as is known, consists of worms and grubs. Besides being difficult to obtain and probably unsuitable in general as laboratory animals, their requirements as such would

* J. W. Steward, 11 Churchill Rd., St Albans, Herts.

need to be determined in far more detail than is already known, and further speculation at this stage might well be misleading.

TAILED AMPHIBIANS

For practical purposes regarding their requirements in captivity, these may be divided into three broad groups: the salamanders, the newts, and the permanently aquatic forms.

Salamanders

The salamanders proper spend most or all of their adult lives on land, and some New World species even live to a greater or lesser extent in trees. Many, but not all, resort to water for a limited breeding season each year. They require a certain amount of humidity at all times, and most of them are secretive, at least in the daytime. These requirements are best met by some such container as a normal glass aquarium with a few inches of damp soil or preferably leaf-mould on the bottom. One or two flat stones or pieces of wood under which the creatures can hide is advisable, and a plant or two in a pot could be added.

It is important to keep the soil at the right humidity, for if it is allowed to dry out the salamanders may well perish. At the same time, a very damp or wet soil can be dangerous, particularly if it becomes sour. Little or no research has been carried out to ascertain the pH values most suitable for salamanders, but it is well known that in a wild state some can live in tolerably acid soils, while others are limited to limestone regions. With many species it may well be a factor of some importance to provide optimum pH conditions, and a study of the natural habitat should give some indication of these. What is certainly important is to avoid conditions which promote the growth of fungus, as certain forms are dangerous to salamanders. They are, in fact, susceptible to a particular kind of fungoid infection which can quickly kill and which appears to be highly contagious, overcrowding will lead to a higher incidence of this infection. Periodic changing of the soil is to be recommended to avoid this and other infections, as amphibians are prone to various skin troubles.

Some salamanders lay their eggs on land, the Alpine Salamander, *Salamandra atra*, even gives birth to living young in the adult form. Such species do not require water containers, as they are often poor swimmers and can easily drown. Most species, however, enter water to lay their eggs, which hatch as tadpoles. The well-known European Salamander, *Salamandra salamandra*, deposits its tadpoles alive in water at a fairly advanced stage of development. For the breeding of these species a water container is necessary; only a few inches of water is necessary, and the container should be arranged so that the salamanders can climb out of the water if they so wish. Small plastic bowls are suitable if a few stones or crocks are placed in the water to overcome the difficulty of the smooth sides. The water should be kept clean, and pond water is generally better than tap water, though this varies according to the district. The eggs or tadpoles should be removed immediately, and with most species it is advisable to keep the tadpoles in separate containers (small plastic bags are suitable), as they frequently attack each other. A little water-weed, such as

Elodea, in each bag helps to provide cover, oxygenation of the water, and purchase for moving around. The tadpoles which hatch from eggs have no limbs at first, but soon begin to swim sufficiently well to capture small crustaceans such as *Cyclops* and *Daphnia* after an initial short period of feeding on infusoria either already present if pond water is used or introduced with the waterweed. In due course the front and then the hind limbs appear, by which time the larvae are capable of taking *Tubifex* and similar-sized small worms, which may, if necessary, be chopped up to a smaller size. At this stage almost any kind of water-life small enough to be swallowed may be given, cultured brine-shrimps are a good stand-by provided not too many are given at a time, as any not eaten will quickly die and foul the water. Mosquito and gnat larvae can also be used. By the time the tadpoles are approaching full larval size most are capable of taking very small garden worms and water-shrimps. As the time of metamorphosis approaches the larvae should be taken out of the plastic bags and put into shallow water in a tank or bowl so arranged that they can leave the water without difficulty. This container should be furnished with a well-fitting lid, as young salamanders are excellent climbers. As soon as they leave the water they can be put into a similar tank to that prescribed for the adults, but without any water-container. The most suitable food at this time is white-worm or very small earthworms, though some species prefer small insects such as *Aphis* or *Drosophila.*

The food of adult salamanders varies: some are specialized insect-feeders and catch their prey by means of a sticky tongue. Most of the larger salamanders seize their prey in their jaws and feed readily on earthworms. Some, such as *S. salamandra*, are very fond of slugs, particularly the soft, white form frequently found in gardens. Very few will tackle hard-bodied insects or even meal-worms.

Newts

The name 'newt' is commonly, though loosely, applied to a number of salamanders, most fairly closely related, which are generally more at home in water than most of the typical salamanders. They are confined to the temperate zones of the Northern Hemisphere, and as a rule the adults repair to water each year for a breeding season lasting several months and spend the rest of the year on land. In some, such as the Asian hynobid salamanders, the eggs are laid in small sacs with a gelatinous cover, and subsequently fertilized by the male in a similar manner to most fish. In others, such as the various European newts of the genus *Triturus*, the males develop various skin appendages, particularly dorsal crests, during the breeding season and indulge in characteristic courtship dances in the presence of the females; following which, a spermatophore is deposited by the male and picked up by the female, so that the eggs are fertilized internally before being laid.

The requirements of these newts in captivity are similar to those of the salamanders proper, except that many species can be kept permanently aquatic, and this makes feeding them much easier. The water in which they are kept need not be more than a few inches deep, and provision should be made, possibly by stones, for the newts to leave the water whenever they wish. It should be remembered, if the newts are kept in ordinary glass-sided fish-tanks, that they are extremely good climbers; and when wet the smaller species, at least, can walk up a vertical glass surface with comparative ease. The tanks should therefore be fitted with well-fitting covers, which need rarely be removed if a small hole is left in the middle for the introduction of food.

It is advisable to have a quantity of water-weed in the tank, not only to provide welcome cover for the newts and encourage them to stay in the water but also to provide them with suitable attachment for the eggs when they are laid. This particularly applies to the *Triturus* species, the females of which lay their eggs singly and, when possible, carefully wrap each egg in a leaf as it is laid. *Elodea* is a most suitable weed for this purpose. The eggs should be removed to another container before hatching, and this is best achieved by taking out any weed to which eggs are attached and replacing it with fresh. Once the eggs have hatched, the larvae can be reared in the manner already outlined for the salamanders proper.

While some newts, such as the Ribbed Newt, *Pleurodeles waltl*, and the Japanese Fire-bellied Newt, *Triturus* (*Cynops*) *pyrrogaster*, have been bred in captivity with considerable success, with most species breeding is frequently achieved only during their first season in captivity (they are normally captured during the breeding season when concentrated in ponds). When the next breeding season comes around they fail to react in any way, even the males failing to produce their crests and other appendages. Recent experience suggests that this is in some way connected with diet, and that feeding the newts throughout the year on a wide range of natural foods, particularly small water-life such as aquatic crustacea and insect larvae, overcomes this difficulty. It is also possible that failure to provide any period of hibernation or semi-hibernation in captivity reduces the breeding urge, particularly in those species with a more northerly range, which normally spend several months in hibernation each year. There is considerable scope for research in this matter, taking into account such factors as temperature and light.

Both on land and in the water, newts will devour almost anything small enough to be swallowed, provided it moves and has no hard covering. In the water some species hunt their prey to a considerable extent by smell as well as by sight, and will readily swallow small pieces of raw meat, though this is not recommended as a staple food. On land most of the smaller species will take quite small insects, such as *Aphis* and *Drosophila*, catching them on their sticky tongue, but the larger species react only to large prey, such as earthworms, slugs, and caterpillars.

While some salamanders are best kept at temperatures of around 70–80°F (21·1–26·7°C) newts generally should be kept fairly cool. They nearly all feed well in water at temperatures down to about 40°F (4·4°C) and the water should not rise above 60–65°F (15·6–18·3°C), as higher temperatures may be fatal over a period. On land, 50–60°F (10–15·6°C) is suitable; below about 45°F (7·2°C) the newts start to get sluggish and stop feeding.

The permanently aquatic salamanders are those which, even in a natural state, never leave the water except perhaps in an emergency such as the drying up of the pond or stream in which they live. Most are neotonic forms in which the external gills are retained throughout life. Where the external gills are reduced or absent, normally some other means is available for obtaining oxygen from the water, such as the highly vascularized skin of the cryptobranchid salamanders. With all these permanently aquatic species there is no need to provide means of leaving the water, and the simplest form of container is a normal fish-tank. The only addition need be a bed of fine sand for one or two species which like to burrow in the bottom, and for some of the more secretive species a 'hide-away', such as a large flat stone firmly supported on a smaller one at each end, a piece of concrete or earthenware drain-pipe of suitable size, or even half a large flower-pot split vertically and placed with the hollow side downwards. It is advisable to check on the natural habitat of the species concerned, as some live in sluggish or even stagnant water, and others in streams or mountain torrents rich in oxygen. The former are quite happy in just a tank of water with a little water-weed to assist oxygenation, but the latter must have the water artificially aerated.

The most hardy of the aquatic species is undoubtedly the Axolotl, *Siredon mexicana*. It has been bred in large numbers in captivity, and much has been published about it. It can withstand a wide range of temperature and feeds readily on earthworms, water insects of all kinds, and even strips of meat. Fertilization is internal by means of a spermatophore, and the tadpoles which hatch from the eggs develop like those of normal salamanders except that there is no final metamorphosis, the larval form being retained even when the animal becomes sexually mature. The eggs should be removed from the parents and the young reared in a separate container, as axolotls are voracious feeders and snap at anything moving. In fact, one of the difficulties of keeping even adult axolotls together is that when being fed they sometimes seize each other's limbs. The lost limbs are quickly regrown, and the deprived axolotls seem to suffer no material inconvenience in the meantime, but it spoils their appearance.

The largest of all living amphibians is the Giant Salamander, *Megalobatrachus*, with two species in Japan and China. It lives in cold mountain streams with a high oxygen content, and obtains much of its oxygen supply from the water by means of a highly vascularized skin. This poses difficulty in providing the right conditions in captivity, but the giant salamanders are otherwise quite hardy, and specimens have been kept in captivity for very long periods. A closely related member of the same family is the Hellbender, *Cryptobranchus alleghaniensis*, found in many rivers and streams in the eastern Unites States. Other permanently aquatic salamanders from North America are the Blind-eel, *Amphiuma means*, the Mudpuppies of the genus *Necturus*, and the sirens or 'Mud-eels' of the genera *Siren* and *Pseudobranchus*. All these have been successfully kept in captivity, but comparatively little research has been carried out on their breeding habits.

The aquatic salamanders about which least is known are undoubtedly those generally referred to as 'cave-salamanders' by reason of the fact that they inhabit underground lakes and streams, usually in limestone areas. Several species are known from the limestone regions of the southern United States, and it is possible that further species exist as yet undiscovered in that area. A similar-looking species belonging to an entirely different family occurs in a fairly limited area in South Europe. All these cave-salamanders have elongated, slender bodies with remarkably thin legs, and most retain the larval form with external gills throughout life. Eyes are either absent or vestigial and functionless, as might be expected of animals which live in constant darkness, and the salamanders are whitish or pinkish and their skins to some extent transparent. One of the American species, the Grotto Salamander or Ozark Cave Salamander, *Typhlotriton spelaeus*, is a partial exception in that the larvae live in mountain streams and appear and behave very much like the larvae of normal terrestrial species. They have functional eyes and external gills and are strongly pigmented, but as they approach adult size they swim upstream and enter caves, following which their colour fades to almost white and their eyes degenerate into functionless dark spots under the skin. In this particular species the external gills also disappear at about the same time.

The natural diet of the cave-salamanders appears to consist largely of small aquatic crustaceans of species which are also adapted to life in caves. In captivity they feed well on fresh-water shrimps, and most will take small earthworms and insect larvae. Most caves have a constant temperature of around 50°F (10°C), and it is doubtful whether any of the cave salamanders are subjected in their natural state to higher temperatures than this, though they may well have to contend with much lower ones at times, as many cave streams are fed by cold mountain torrents. Temperatures of 40–50°F (4·4–10°C) are suitable for them in captivity. They are, of course, quite at home in complete darkness, and strong light is probably best avoided. There is no direct evidence that moderate light does harm, but the tendency in most for their normally pale skins to become darker over a period if exposed to light suggests that the normal lack of pigment leaves them insufficiently protected against the effects of strong light.

Comparatively little is known about the breeding of cave-salamanders, as the only species which has been studied in captivity to any extent is the Olm, *Proteus anguinus*, from South Europe. It has been found that the optimum temperature for this species is around 50°F (10°C), but that a fairly wide range of temperature can be tolerated. However, if breeding in captivity is aimed at it would seem most important not to let the temperature of the water rise much above 50°F (10°C). Some confusion formerly existed as to whether the Olm laid eggs or gave birth to living young, but it would now seem that it is possible for either to take place. Under the right conditions, up to fifty or sixty eggs enter the oviducts, but all except one in each oviduct dissolve to form a liquid which nourishes the remaining egg. This egg eventually develops into a comparatively large young Olm, which is born alive (the young measure about 4 inches at birth, and the adults range between 8 and 12 inches). In this way, only two young are born at a time, or only one if the process fails to work in one oviduct, but if the temperature is too high all the eggs are laid as eggs and either fail to hatch or if they do hatch the small and immature young die shortly afterwards. Further laboratory research into the whole process would obviously be most valuable.

To avoid confusion, it should perhaps be emphasized that the term 'cave-salamanders' is used here to indicate only those aquatic forms which spend their lives in underground lakes and streams. Certain non-aquatic salamanders, such as some species of the genus *Hydromantes*, are often referred to as cave-salamanders through frequently inhabiting caves or cave entrances, but these should be treated as salamanders proper.

TAILLESS AMPHIBIANS

As with the tailed amphibians, it is convenient to divide the frogs and toads into three general groups for their requirements in captivity. These groups consist of the ground-living forms, the tree-frogs, and the permanently aquatic species.

The ground-living forms cover the vast bulk of the frogs and toads, the Common Frog, *Rana temporaria*, and the Common Toad, *Bufo bufo*, being typical examples. They live generally similar lives, spending much of their time on land, but resorting to water for breeding. The breeding season in the temperate zones is usually a given time of the year for each species, but in many tropical regions may be 'triggered off' by the onset of heavy rains. Mating involves a form of amplexus and external fertilization of the eggs. The normal pattern is for the eggs to be laid in water, in a mass, a long string, or even singly. The eggs hatch out into tadpoles, which eventually develop hind-legs and then front-legs. The gills and tail are absorbed and lungs develop at or a little before the time the creature leaves the water as a miniature frog or toad. There are, however, a number of exceptions to this general pattern, some involving what might be regarded as parental care of the eggs, or even of the young. The simplest form is perhaps that of the Midwife Toad, *Alytes obstetricans*, whose male wraps the fertilized eggs around his hind limbs and carries them about until the tadpoles are about to hatch, when he enters water to liberate them. Some tropical frogs, notably certain species in the family Rhacophoridae, lay their eggs on land in moist places. Hatching is usually delayed until heavy rain covers the eggs with water or washes them into a pool or stream, but in some cases the complete development and metamorphosis of the tadpole takes place inside the egg, and the young hatch out in the form of miniature adults. Other examples could be given of departure from the normal pattern of breeding, and it is obvious that in dealing with unusual frogs, particularly tropical ones, the breeding habits of the species concerned should be checked as far as possible to ascertain their requirements in captivity.

Frogs and toads which follow a conventional pattern of breeding cannot often be persuaded to breed under laboratory conditions. The main factors affecting them are likely to be limitation of space, incorrect temperature (especially temperature changes for those which breed seasonally), and suitable water conditions. Frogs and toads are generally more choosey than are the tailless amphibians in their choice of a breeding site. Vegetation and depth of water obviously play an important part with many species, and it is possible that the chemistry of the water is also a factor. Clean pond water is better than tap-water in most districts (though care should be taken to remove any carnivorous insects likely to attack the young tadpoles) and plenty of water-weed should also be included, even if only because it provides some cover for the adults in the water and may encourage them to stay in it. Once it is possible to persuade the adults to lay eggs, the main difficulty is over. The eggs may be laid in large clumps, long strings, or sometimes a few at a time or even singly. The eggs should be removed from the adults immediately, because their active movements could easily damage the eggs or young tadpoles, while some frogs which feed under water might even eat them. The safest method is to remove the whole water-container, replacing it with another for the adults. This avoids physical damage to the eggs or harmful results from a sudden change of temperature. The time taken for the eggs to hatch and the rate of development of the tadpoles is controlled by the temperature of the water, and either too high or too low a temperature may cause excessive mortality of eggs or larvae at some stage of development. Each species has its optimum temperature, and this can only be learned by experience, although tropical species generally require a higher temperature than those from temperate regions. Most tadpoles are vegetarian, although at a certain stage of development many of them become partly carnivorous by feeding on carrion. Most of them are equipped with horny lips with which they can scrape algae from water-weed or the sides of the tank, and the presence of such algae is therefore beneficial, particularly when the tadpoles are small. Later, the tadpoles will begin to nibble the vegetation, which may be supplemented with lettuce leaves if required. Small scraps of meat may be added to the diet of many species when the tadpoles are large enough. The tadpoles quickly find these by smell and devour them readily. Care should be taken to remove any uneaten meat or lettuce before they can pollute the water, and a few water-snails in the tank as scavengers will help to avoid pollution at this stage. When the tadpoles are about three-quarters grown is the time when the purity and oxygen content of the water are of the greatest importance. Later, as the lungs develop, the tadpoles are able to surface for air and depend less on the water as their source of oxygen.

As the tadpoles approach full size they develop hind-legs, then front-legs, and finally the tail starts to be absorbed. At this stage provision should be made to enable the young frogs or toads to leave the water, or they may drown. They are not strong swimmers at this time, and the absorption of the gills which accompanies the development of the lungs diminishes their supply of oxygen from the water. Some rocks protruding above the water or a piece of wood floating on the surface may enable them to leave the water, but to avoid casualties as far as possible, a gentle slope for them to crawl out of the water is best. Some mortality may have to be accepted, as occasional tadpoles somehow fail to make the switch from gill-breathing to lung-breathing and succumb when ready to leave the water or shortly afterwards.

The newly metamorphosed frogs should be moved to a container with damp soil and some vegetation such as moss or grass. Provided the soil is sprinkled from time to time to keep it moist, no water-container is necessary, and in fact, one could be a danger, as until their legs have grown a little more, the young amphibians are still poor swimmers and might easily drown. The main difficulty over the next period is the provision of suitable food, which naturally consists of very small insects. *Drosophila*, which can easily be cultivated in large numbers, are a good stand-by, and if sufficient greenfly are obtainable they are an excellent food. The container should be covered with fine-mesh wire, as even the wingless strains of *Drosophila* may escape, and the risk of escape by the young frogs themselves increases as they grow. They are soon able to take house-flies and later blow-flies, whose larvae may be obtained commercially. In general, almost anything which moves and is of suitable size may be used as food, but with some variation between species, e.g. toads readily eat ants, but most frogs will not do so. For practical purposes, those insects which can be obtained commercially or reared in large numbers in the laboratory are the most useful. In all cases the stimulus to feeding is movement of the prey, so that only live food can be given, and for the same reason large frogs and toads should not be kept together with small ones.

Housing requirements for most adult terrestrial frogs and toads are not particularly difficult to provide, the two main factors being temperature and humidity. Full activity, including feeding, can be maintained by species from temperate regions at a temperature range of about 50–65°F (10–18·3°C). Tropical species (except those which occur in the tropics but only at high altitude) find a temperature of 70–80°F (21·1–26·7°C) more suitable, and in some cases necessary. By far the most important consideration is humidity, and for most of the frogs high humidity is necessary. A fairly large water-container in the cage is advisable, but is not strictly required if a layer of moist soil is maintained, as amphibians generally imbibe water through the skin and do not drink in the proper sense. Some species like the Edible Frog, *Rana esculenta*, enjoy sun-bathing and can remain in full sunlight for long periods, but nevertheless, require access to moisture. Some frogs and toads are better adapted to drier conditions, and certain species in a wild state live in semi-desert regions. These forms avoid desiccation by burrowing (for which some are furnished with spade-like processes on the hind-feet) and emerge only at night or after rain. A deep layer of coarse sand and one or two large rocks or stones will provide them with the necessary facilities for burrowing, but it is as well to keep the sand slightly damp or even sometimes wet, particularly at high temperatures.

The nature of the soil helps to determine the kind of food which may be given, and the natural preferences of different species reflect this. The dry-soil types of toads eat large quantities of ants and other Hymenoptera, and many of these insects are ones not normally found in really damp habitats. The long and accurately aimed tongue of the toads is better designed to pick up small insects than the short, wide tongue of the frogs, which usually ignore very small prey. On the other hand, toads are as able as frogs to deal with large forms of prey, such as large earthworms, but these and other slimy creatures like slugs and snails cannot suitably be introduced into a container with dry sand or soil, as they quickly become coated and are then virtually inedible. Some of the largest frogs and toads, such as the Bullfrogs, *Rana catesbeiana* and *R. adspersa*, and the Marine Toad, *Bufo marinus*, will attack and devour vertebrate animals up to the size of a full-grown mouse. On the other hand, certain of the burrowing toads have a comparatively specialized diet, apparently living entirely on termites.

The container for frogs and toads should be as large as possible, and overcrowding should always be avoided. The more active frogs should be given plenty of room, as obviously a creature such as the Agile Frog, *Rana dalmatina*, which can leap a distance of several feet will come to grief if confined to a small cage. It should be remembered that the skin secretions of many frogs and toads are poisonous, and may prove fatal to other amphibians kept in contact with them. A notable example is the Pickerel Frog, *Rana palustris*, of North America, which should not be kept with other species.

The tree frogs in some respects require different treatment from the terrestrial frogs and toads. There are very many genera and species of them throughout the temperate and tropical areas of the world, although most are found within the tropics. The adults spend most of their time in trees or bushes and possess various adaptations for an arboreal existence, the most common being the provision of disc-shaped expansions of the ends of the digits, which act as 'suckers' and enable these frogs to climb and leap about safely in foliage, and even to walk up a vertical pane of glass. Most are nocturnal, spending the day-time pressed close to a branch or tucked away among leaves, and becoming quite active at night. The normal food is flies, moths, and spiders, which are found in trees, and in captivity they frequently ignore terrestrial creatures, such as mealworms and earthworms. They leap to catch their prey, and their eyes seem to be adapted to comparatively long-range vision, as they often appear to overlook insects at very close range. To accommodate their leaping and climbing activities, they should be provided with a roomy cage, which is tall as well as having a large base area. It should be furnished with one or two broad-leaved bushy plants such as small laurels, and it may be found more convenient to have them planted in pots rather than set in soil in the bottom of the cage. It is a wise precaution to include a shallow water-container, although this is not strictly necessary if the foliage is sprinkled regularly with water and the floor of the cage is covered with a fairly deep layer of moist soil. The cage should be covered, to prevent escape of both the frogs and the insects on which they feed. Blow-flies are a convenient basic food, or house-flies for the smaller frogs. They may be introduced either as adults or as larvae or pupae which will be consumed when they emerge as flies. A small hole in the top of the cage, closed by a well-fitting cork, is a very useful means of introducing food without opening the cage.

The breeding procedure of the tree-frogs varies considerably between species, most of them following the normal pattern of laying eggs in water which hatch out as tadpoles, others adopting protective devices for their eggs and young. Some build mud nests in shallow water in which the eggs hatch; others construct foam nests either in the water or among branches overhanging water, into which the tadpoles drop after emerging from the eggs. The female of certain South American tree-frogs carries the eggs on her back until they hatch out either as tadpoles or as fully formed froglets, and in the genus *Gastrotheca* the females develop dorsal pouches in which the eggs are carried until they have hatched and the tadpoles have completed part or all of their development. In a few tropical American species the male remains with the eggs, which are laid on land, until they hatch. The tadpoles attach themselves to his back or sides by suction and are carried around until they drop off in water.

Very few species of tree-frog have been successfully bred in captivity, for not only is it difficult to simulate their requirements but it is also essential to gain some knowledge of their life-history before such an attempt is made. The pouch-breeding species of *Gastrotheca* offer better chances of success than most, and at least one species, *G. marsupiata*, from the highlands of Ecuador has been bred with a certain amount of success in this country.

The final group to be considered is that of the *wholly aquatic frogs and toads*. Certain genera dotted about the warmer regions of the world spend virtually all their lives in water, and obviously have to be kept in water in captivity. The most widespread group is the family Pipidae, comprising three genera in

Africa and two in South America. These are all rather grotesque, flat-bodied toads with a number of adaptations for an aquatic existence, including powerful hind-limbs with long, fully webbed toes, lidless eyes set high on the head, and 'lateral-line' organs capable of detecting water-borne vibrations. The most practical container for them is a simple fish-tank. Probably because they lack eye-lids, they often prefer to shelter from bright light, and the tank should contain either a plentiful supply of water-weed or some sort of rock shelter under which they can retire if they wish.

The best-known species is the African Clawed Toad, *Xenopus laevis*, which because of its usefulness in biological tests has been widely bred in captivity. Its various sub-species are widely spread over most of the southerly half of Africa, and it inhabits areas of water of varying size, from small pools to the margins of large lakes. Although very much at a disadvantage when out of the water, it does on occasion travel overland from one stretch of water to another. Many of the pools in which it occurs dry up periodically, and it is able to lie dormant for long periods in the baked mud until the rains refill the pool and liberate it once more. Presumably because of this, breeding normally occurs immediately after the onset of heavy rains. Amplexus is inguinal (the male clasping the female around the groin), eggs are laid which hatch into tadpoles of a rather specialized type. During the first few days of its life the tadpole remains on the bottom, but later ascends to the surface and subsequently spends most of its larval life suspended head-downwards in the water, usually congregating in rather deeper water as the time for metamorphosis approaches. The tadpoles feed essentially on infusoria, which they strain from water passed in through the wide mouth and out through the gill-openings. Eventually, the legs and lungs develop and the tail is absorbed, following which the toadlet changes to a completely carnivorous diet. Small water-insects are accurately caught in the mouth by a sort of darting movement, and in addition the sensitive fingers are used to locate food in the bottom mud and transfer it into the mouth. As the toads increase in size they take much larger prey, and fully grown toads can satisfactorily dispose of quite large earthworms. In captivity, they also learn to take small strips of raw meat, and while quite voracious feeders when food is plentiful, they can fast for fairly long periods.

In breeding clawed toads the main difficulty is to find the right conditions for triggering off the process. Thundery weather tends to act as a stimulus, and imitating heavy rain by spraying water into the tank occasionally does the trick (Vallance, 1952). The most effective method is to transfer the toads at the right time of the year (June is the best month in this country) to another tank containing water some ten to twenty degrees colder than that from which they have been removed. If eggs are laid the toads should immediately be separated, or they will eat the eggs and probably the larvae. The latter commence to eat a few days after they are born, and the necessary infusoria can be supplied by adding green pond water. Useful supplementary foods are finely mashed liver and dried and ground nettle and lettuce leaves. Plenty of water-weed in the tank is advisable, particularly as the time of metamorphosis approaches, as the young toads can easily drown at this stage if not given facilities for supporting themselves near the surface of the water.

The South American members of the family are typified by the well-known Surinam Toad, *Pipa pipa*. The adults frequent forest pools and live rather similar lives to those of *Xenopus*, but their mode of breeding is different. As the eggs are laid, the male, while fertilizing them, spreads them over the back of the female, which develops a spongy nature at this time, and into which the eggs sink. Buried in this manner in the female's back, the eggs develop through all the larval stages until finally tiny replicas of the adult push their way out through the outer covering of skin and swim off. Although less able to withstand temperature ranges than *Xenopus*, these toads are fairly hardy in captivity, but comparatively little is known about their breeding requirements.

Other wholly aquatic frogs and toads are specialized members of the normally terrestrial families, and in superficial appearance tend to resemble the pipid toads. One or two Asian species have been studied to a slight extent, but these specialized forms offer a wide-open field for detailed research.

This generalized account of the amphibians is designed as a basic guide for those who find it necessary to keep and breed amphibians as laboratory animals for biological research. Some species which are highly interesting from a zoological point of view, such as the primitive members of the family Leiopelmidae, have not been dealt with because they are unlikely to be suitable for this purpose. All the species mentioned by name are ones which it should be possible to establish in the laboratory and on which detailed information is available in existing literature.

REFERENCES

ALLISON, R. M., 'The merits of some English and Foreign male toads used for the diagnosis of Pregnancy and measures to overcome their disadvantages', *Laboratory Practice*, 1955, **4**, 277–82 (1955).

ALLISON, R. M., 'The mortality of frogs, toads and salamanders when subjected to forced hibernation', *Laboratory Practice*, 1956, 382 (1956).

ALLISON, R. M., 'Failure of enforced hibernation to inhibit the breeding of *Rana temporaria*', *Nature*, **177**, 342 (1956).

WEEKS, G., 'Refrigeration of *Rana Temporaria*', *Journal of Animal Technicians Association*, **5**, No. 3, 65–66 (1954).

KAPLAN, H. and KAPLAN, M., 'Anaesthesia of frogs with ethyl alcohol', *Proceedings of Animal Care Panel*, February 1961, p. 51 (1961).

EATWELL, A. S. J., 'Natural breeding of *Xenopus laevis*', *Journal of Animal Technicians Association*, **4**, No. 3, 58 (1953).

VALLANCE, A. F. C., *Journal of Animal Technicians Association*, **3**, No. 2 (1952).

REPTILES *

The reptiles living in the world today represent four orders, Rhynchocephalia, Crocodilia, Testudines, and Squamata. For the purpose of this chapter Rhynchocephalia and Crocodilia will not be considered. The order Squamata is divided into two sub-orders, Serpentes and Lacertilia, and these will be dealt with separately.

* The Editors regret that it has been necessary through lack of space to omit many families of Reptiles. It is hoped that those families omitted would not be important to the animal technician, and the editors sincerely hope that by this omission, the value of this important chapter has not been diminished.

Reference to the omitted families of Reptiles should be made to the author at 11 Churchill Road, St. Albans, Herts.

Order TESTUDINES

It is convenient to divide the testudines into three groups, which will be referred to as the tortoises, the terrapins, and the turtles. The tortoises are the species which live on land, the terrapins are those which frequent fresh water, and the turtles are the marine aquatic forms. To avoid confusion, it should be explained that in other countries these three terms do not always have precisely the same meaning. In America, for example, it is customary to refer to most testudines as 'turtles', the term 'tortoise' being reserved for certain specific land forms and 'terrapin' for a few fresh-water forms particularly esteemed for culinary purposes.

Tortoises

The entirely terrestrial testudines are a fairly distinctive group, although they are taxonomically separated into various genera, and some are more closely related to certain of the terrapins than to the other tortoises. They range in size from species with a maximum length of about 6 inches to the Giant Tortoises, which still survive on certain islands in the Indian Ocean and the Galapagos group, and which may reach a weight of several hundred pounds.

Different species vary to some extent in their requirements in captivity, and a knowledge of their natural habitats is an advantage. Some live in moderately temperate regions and hibernate in winter; it is possible to keep these outdoors throughout the year in this country provided they have somewhere to hibernate. Examples are the European Tortoise, *Testudo hermanni*, the so-called 'Greek' Tortoise or Mediterranean Spur-thighed Tortoise, *T. graeca*, and the Common Box Tortoise, *Terrapene carolina*, of North America. Most tortoises, however, come from tropical or semi-tropical regions and require temperatures in the range of 60–80°F (15·6–26·7°C) throughout the year. Subject to this, almost any kind of pen or cage is suitable, but as much room as possible should be allowed, as in spite of their fairly slow locomotion, tortoises like to move around and can cover a lot of ground during a day. In a pen the sides should be higher than the tortoise can reach when standing upright on its hind-legs, and wire netting should be avoided, as the tortoise might get its head and legs entangled. Many tortoises are good burrowers, and this should be remembered when designing the pen. Nearly all are diurnal, and like to tuck themselves well away at night. Both for this reason and to avoid too much exposure to rain, some sort of shelter is needed.

Many tortoises can go without water for long periods, but all drink copiously when thirsty. Drinking-water must therefore always be provided. Feeding does not present any particular problems, although there is some variation between species. All tortoises eat vegetable matter, some exclusively so. Colour and smell, which is well developed in the tortoises, play a part in the choice of food. The desert forms feed largely on the flowers and succulent leaves of cacti and other desert plants. The open-country species eat a wider range of tender leaves, flowers, berries, and other fruits. The forest tortoises prefer leaves, fruits, and certain types of fungus. In captivity, lettuce and dandelion are eaten readily by most tortoises, and a wide range of fruits, including apples, citrus fruits, tomatoes, and berries. The choice of food offered should be as varied as possible, as individual tortoises may have strong preferences, which may change from time to time. Some species include a good deal of animal matter in their diet also. The European Tortoise, *Testudo hermanni*, and the South American Tortoise, *T. denticulata*, readily eat carrion, which they can detect from quite a distance if it is sufficiently odorous, and both will also eat earthworms and various other invertebrates. These and some other species will eat raw meat in captivity, and it is probable that the occasional dead mouse, which they will tear to pieces before swallowing, helps to keep them healthy.

Tortoises often lay eggs in captivity, which with proper care and a little luck may hatch out. They should be kept at a temperature of around 75–80°F (23·9–26·7°C), but the greater difficulty is to maintain the correct degree of humidity. If kept too dry they will shrivel up, and if allowed to become too damp may be attacked by fungus. The correct humidity can really only be learned by experience, but the resultant humidity of a newspaper soaked in water and squeezed as dry as possible will be roughly that desired for the sand or soft loam in which the eggs should be buried. It is difficult at the temperature required for hatching the eggs to maintain the correct humidity of the packing material, and the best method is to place this and the eggs in a large earthenware flower-pot with a sheet of glass over the top. If the contents show signs of becoming too dry the outside of the pot should be wetted so that the moisture can soak through and restore the required humidity. The time the eggs take to hatch varies according to the species and the temperature, but is usually ten to twelve weeks in most cases.

If success in hatching the eggs is achieved there remains the difficult task of rearing the young. The main factors are temperature, which should not fall below about 70°F (21·1°C), and food. A varied diet containing considerable quantities of minerals is necessary to promote growth of the shell. The addition of a few drops of halibut-liver oil is beneficial, and there are some powdered calcium preparations on the market which will help to ensure proper shell development. Alternatively, powdered cuttlefish may be used, or a piece of the bone put into the cage for tortoises to gnaw. Failure to provide adequate minerals may result in irregular shell growth or even the death of the tortoise.

The main causes of illness in tortoises are respiratory infections, internal parasites, and eye troubles. The symptoms of what is commonly known as 'tortoise pneumonia' are wheezy breathing and a discharge from the nostrils. It appears to be contagious and can be fatal, but often reacts satisfactorily to treatment. This consists mainly of keeping the tortoise warm and free from draughts. Inhalations and antibiotics may be helpful in affecting a cure, but their use generally calls for an individual diagnosis in each case. Silver vitellin introduced into the mouth or nostrils is also reported to have been used successfully. Internal parasites, especially nematodes, are quite common in tortoises, and probably every tortoise in a wild state has its share. In captivity they sometimes increase greatly in numbers, and the obvious precautions are to keep the cage clean to avoid reinfection, and not to overcrowd. Tortoises seem to be prone to eye troubles of various kinds, and the most frequent symptoms are watering of the eyes and an inability to open one or both of them. The latter often arises in tortoises emerging from hibernation, though probably this arises simply from the eyelids having become stuck together, whereas in other cases the cause is more likely to be an infection. In all cases regular bathing of the eye with a weak solution of boracic is likely to be the best remedy.

Terrapins

These differ considerably from the tortoises in their habits and requirements in captivity, some rarely leaving the water, and others spending much of their time on land either basking or wandering around.

Their housing must take these habits into account, and for most species a large fish-tank is suitable divided into a water area and a land area. For the sun-loving species a lamp may be suspended above the land area, which may simply be a large flat rock placed in the water. Some species from temperate regions spend the winter in hibernation and will survive the winter outdoors in this country if given the right facilities. A few examples are the European Pond Tortoise, *Emys orbicularis*, the Spanish Terrapin, *Clemmys caspica leprosa*, Reeve's Terrapin, *Chinemys reevesi*, from Asia, and the Musk Terrapin, *Sternotherus odoratus*, from North America.

Terrapins in general are far more carnivorous than the tortoises, and some live entirely on an animal diet. According to the size of the terrapin, this may range over small water insects, worms, amphibians, and fish. The prey is seized in the mouth and either swallowed whole or torn to pieces with the claws of the powerful front limbs. In captivity natural food if available is best, but it may be supplemented by chopped raw meat or fish.

As in all reptiles, fertilization is internal, and in the wild state the females come ashore to lay their eggs in holes dug in soft soil, sand, or decaying vegetation. Most species have a definite time of year for laying their eggs, but there is some evidence to suggest that a few lay two or even three times a year. The period of incubation varies considerably according to the species, and is much affected by temperature and, possibly, by humidity, but usually takes two or three months, when the young, on hatching, make straight for the water. Any eggs laid in captivity should be treated in a similar way to tortoise eggs.

Food for the young should include as much calcium as possible, and the natural foods from which most of this is obtained are small fresh-water crustaceans, such as *Daphnia* and fresh-water shrimps. Woodlice seem to be an acceptable substitute, also mealworm and small earthworms. Some species will take a certain amount of vegetable matter such as lettuce right from the start. Raw meat should only be given to young terrapins very occasionally, but a fair amount can be included in the diet of larger terrapins, together with raw fish. Water-melon appears to be a particularly acceptable delicacy for the more herbivorous species. Several prepared 'turtle foods' on the market are expensive, harmless as a supplement, but inadequate as a staple diet. Given the right foods, even young terrapins should not need any vitamin supplement, but the need to supply calcium in growing terrapins is essential. Finely ground bonemeal may be mixed with meat or fish, and if pieces of cuttlefish bone are put into the water the terrapins will gnaw them. Chopped herring occasionally, or a few drops of halibut-liver oil mixed with the meat provide extra vitamins. Nearly all terrapins feed under water, and only a few are able to swallow their food on land. The food may be placed either in the water or on the land where the terrapins can take it into the water to eat it. They have good eye-sight and detect their food primarily by sight, but often use the nose to examine dead food before accepting it. Terrapins are certainly messy eaters, they tear to pieces with the claws anything too big to swallow whole, and this may result in rapid fouling of the water. To avoid having to change the water too often, large pieces of meat and fish are best chopped up to a small enough size to be swallowed. An alternative method is to have a separate small feeding tank in which the terrapins are placed at mealtime.

The main illness of terrapins in captivity, and one which particularly affects the young, is a condition known as 'soft-shell'. The shell becomes soft and pliable and the terrapin becomes lethargic, its eyes close, and it ceases to feed. Unless the condition can be cured, the terrapin dies in a week or two, probably from starvation. Evidence suggests that this condition results from an inadequate diet, and the only hope of a cure is to get the animal feeding again on the right foods. This can be done if the condition is tackled at an early stage by increasing the temperature of the water to 85–90°F (29·4–32·2°C) and doing everything possible to persuade the terrapin to feed. Penicillin and some other antibiotics can help by alleviating the condition temporarily, but will not by themselves effect a permanent cure. What appears to be a separate form of illness, with comparable symptoms, is one in which the eyes become very swollen and cannot be opened. Recent investigation suggests that this is a condition affecting the entire lymphatic system, spreading into most organs of the body. The cause is obscure, and so far no definite cure is known. These are both essentially diseases of captivity, rarely occurring in the wild, and lack of natural sunlight may be at least a factor. Some benefit seems to be derived from exposure for 5–10 minutes daily to the rays of a small ultra-violet lamp at a distance of about 2 feet. Proper diet is, however, undoubtedly the major requirement for the prevention of these diseases, as many terrapins have been successfully reared in captivity without any access to natural sunlight or ultra-violet light, and some terrapins, such as the Snapper, *Chelydra serpentina*, which are predominantly aquatic and do not have the basking habit, can also succumb to 'soft-shell' in captivity if not given proper treatment.

Sub-order LACERTILIA

There are some 2,500 species of lizards known to exist in the world today, and they vary tremendously in bodily form, habits, and habitat. This can therefore be only a general treatise, and will follow the simple form of listing some of the more important families and outlining their main characteristics and requirements.

Family Agamidae

Over thirty genera and about 300 species are included in this family, ranging from Africa and South Europe through South Asia to Polynesia and Australia. They are mostly strongly built lizards with well-developed limbs, some of them reaching a moderately large size. They are active, alert, and diurnal.

Most are terrestrial, though a few are largely arboreal. Special mention must be made of the South Asian genus *Draco*, which includes some twenty or more species of mainly arboreal lizards with long, slender tails and five or six posterior ribs greatly prolonged, movable laterally, and connected by membranous sheets of skin which enable the lizards to make gliding leaps from branch to branch or even from tree to tree. Another unusual member of the family is the Moloch Lizard or 'Thorny Devil', *Moloch horridus*, which is extremely adapted to its habitat in the desert and semi-desert regions of parts of Australia. Its food appears to consist entirely of ants. A number of the agamid lizards include a certain amount of vegetable matter in their diets, and a few, such as the various species of the genus *Uromastix* from Central and Western Asia and extreme North-East Africa, are almost entirely herbivorous. Otherwise, these lizards feed mainly on insects.

Most of the agamids can be successfully kept in captivity.

Family Amphisbaenidae

Individual specimens of various species have been kept alive in captivity for fairly long periods. In general, the best medium seems to be a mixture of soft sand and leaf-mould, kept just damp enough not to crumble. The lizards make regular runs, and surface only occasionally at night, although in a wild state they sometimes appear after heavy rains. Proper observation is extremely difficult under these circumstances, but a specimen of *Amphisibaena alba* kept by the writer for some months is still going strong, having been offered no food other than earthworms, which certainly disappear regularly. It seems almost certain that the water-uptake is through the skin, as this particular specimen has received no water other than what can be obtained from the slightly damp soil.

Family Anguidae

The best-known species in this country is the limbless Slowworm, *Anguis fragilis*, which has a wide range covering most of Europe, including the British Isles (but NOT Ireland), parts of North Africa, and South-West Asia as far as North Persia. In Scandinavia it is found as far north as Latitude 65°. It is hardy in captivity, and can withstand a wide range of temperature, being most active at around 80°F (26·7°C), but feeding (at least as far as specimens from the more northerly part of the range are concerned) at temperatures as low as 50°F (10°C). It spends much of its time burrowing in soft soil or under stones or logs, but occasionally emerges to sun-bathe (particularly in early spring, or gravid females in early summer) and frequently after rain. It prefers a slightly moist habitat, and should be provided with a layer of loam or leaf-mould, which should be sprinkled with water to prevent its becoming too dry. This is important with the young, which quickly become dehydrated if exposed to dry heat. The Slowworm will lap up drops of water with its broad tongue. Favourite foods are worms and slugs, particularly the small, white slug often found in gardens.

Another legless species often imported into this country is the so-called 'Glass-Snake' or Sheltopusic, *Ophisaurus apodus*, from the Balkans, South Russia, Asia Minor, and Morocco. It is much larger than the Slowworm, reaching a length of 4 ft and a diameter of 1½ in. It is also hardy in captivity and becomes quite tame. It lives in rocky, bushy areas, and feeds on insects, eggs, small birds and mammals, small snakes and other lizards.

With all the anguids it is as well to remember that the tail is extremely brittle and will be discarded if the lizard is roughly handled. A new tail will grow in due course, but the appearance of the lizard may be permanently spoiled.

Family Chamaeleontidae

The true chamaeleons (not to be confused with a few species, such as some of the members of the American iguanid genus *Anolis*, which are often popularly, but erroneously, referred to as 'chamaeleons' because of their ability to change colour to some extent) are found in Africa, Madagascar, and parts of South Asia, with one North African species overlapping into South Spain. They are very distinctive in appearance, with high and rather narrow bodies, opposible digits, prehensile tails, independently movable eyes, and extremely protrusible tongues. Most are high arboreal, spending most of their time in trees and bushes, though a few live more on the ground. Some species are furnished with up to as many as three 'horns' placed anteriorly on top of the head. There are about eighty species, of which the majority belong to the genus *Chamaeleo*.

Their food consists entirely of insects, which are stalked and caught on the tip of the very long, sticky tongue. This organ can be extruded very rapidly to a considerable distance, and the ability to aim it extremely accurately is undoubtedly assisted by the binocular vision made possible by the turret-like form and the remarkable mobility of the eyes.

The requirements of chamaeleons in captivity still require a lot of clarification, and most of them do not normally survive for more than a few months. In many ways, the most suitable are some of the smaller species commonly known as 'dwarf chamaeleons', which have frequently been kept in captivity for fairly long periods and bred with a fair measure of success. The reasons for this are probably several—their small size (rarely exceeding 5 in.) makes it easier to provide them with sufficient food, their young are born alive, whereas most chamaeleons are oviparous, and they can withstand much lower temperatures. The background to this latter ability is that either, like some South African species of the genus *Microsaura*, they live at a considerable distance from the Equator or, although living near the Equator, they occur only at high altitude like the Two-lined Chamaeleon, *Chamaeleo bitaeniatus*, and Marshall's Dwarf Chamaeleon, *C. marshalli*. These dwarf chamaeleons are most active in the daytime at temperatures of around 70–75°F (21·1–23·9°C)—higher temperatures should be avoided if possible—but suffer no harm if the temperature falls at night to as low as 45°F or even 40°F (4·4–7·2°C). With most other chamaeleons it is necessary to keep to a much closer temperature range, around 80°F (26·7°C).

Spacious housing accommodation is almost a necessity, and the cage should be at least 2 or 3 ft high and even much larger if possible, furnished with bushes or branches in which the chamaeleons can climb. The size of the cage is particularly important if several chamaeleons are kept together, as they have a well-developed territorial instinct, and each specimen will choose a small 'roost' for sleeping and as a base from which to operate in search of food. Any attempt by another chamaeleon to usurp or even enter on to the 'roost' will be opposed. In too small a cage this instinctive behaviour pattern is disrupted, and the chamaeleons may fail to settle down to a natural routine.

Chamaeleons require regular and plentiful food, and in a wild state their diet is somewhat varied. These requirements are not always easy to satisfy in captivity, and this is probably one of the reasons why captive chamaeleons so often fail to thrive. Flies and moths, some species of which can be bred in quantity, are useful foods, but even better are crickets and grasshoppers or, for some of the larger species, locusts. Chamaeleons drink only from raindrops or dewdrops, and seem incapable of learning to drink from any kind of vessel. The branches or vegetation in the cage should therefore be sprinkled with water every two or three days. This has, of course, the effect of increasing air humidity, and although many chamaeleons come from rain-forest areas, they do not seem able to withstand humid conditions in captivity. Whatever the reason, good ventilation of the cage is necessary. Most chamaeleons like to bask on occasion, and unless sufficient sunlight is available, a suitable lamp should be mounted in the roof of the cage.

Apart from the dwarf species, which give birth to living young, very little success has attended efforts to breed chamaeleons in captivity. The females of most species descend to the ground to lay their eggs in shallow holes which they dig in soft soil or leaf-mould. Any attempt to breed the oviparous species therefore calls for a layer of suitable soil in the bottom of the cage. The eggs or the young should be moved to a separate cage and the young, particularly of the smaller species, need plentiful supplies of quite small insects, such as *Aphis* or *Drosophila*.

Family Lacertidae

The 'true lizards' constitute a large family, with a score or more of genera and some 150 species. The main home of the family is Africa (though it does not exist in Madagascar), and a number of species also occur in Europe, where it is the dominant form of lizard, and in Asia.

The largest genus is *Lacerta*, with some thirty species in Europe, Asia, and Africa north of the Equator. Two of the three British lizards belong to this genus, the Viviparous Lizard, *L. vivipara*, and the Sand Lizard, *L. agilis*. The largest of the European lizards is the handsome Eyed Lizard, *L. lepida*, from South France, the Iberian Peninsula, and North-West Africa. The males frequently reach a length of 2 feet. Many of the Wall lizards of Europe also belong to this genus, and *Lacerta erhardii*, for example, has no less than thirty-one recognized sub-species in the south-east corner of Europe, most of them among the Greek islands.

Most of the lacertid lizards are happy in captivity with a bed of sand and a few rocks. They are mostly tolerant of fairly wide temperature fluctuations, but naturally, those from Africa require more warmth than those from farther north. They are essentially diurnal, sun-loving lizards and require a lamp in the cage in the absence of adequate sunlight. A water-dish should be provided for drinking. Food consists of a wider range of invertebrates, particularly moths, spiders, and flies. Large species, such as the Eyed Lizard, will also eat other lizards, small mammals, nestlings, eggs, succulent fruit, and occasionally tender vegetation, and many of the smaller species also eat soft fruit, such as berries. Most take mealworms in captivity.

The young of a few species are born alive: the Vivparous Lizard being an example. But most of the members of this family lay eggs which are buried in sand or soil. They are not particularly difficult to rear in captivity, provided the young are given sufficient warmth and food. Otherwise there is a tendency for the bones, especially the jaws and spine, to grow badly. This may be due to vitamin deficiency, as short periods of exposure to ultra-violet rays seem to have a beneficial effect.

Family Teiidae

Around forty genera and 150 species of teiid lizards live in the New World, and of the smaller species, probably the genus *Cnemidophorous*, with some two dozen species covering an area from the United States to South America and the West Indies, is fairly typical. It includes the Race-runners and the Whiptails, fast and active terrestrial lizards of dry, and usually, sandy places. They are not particularly easy to keep in captivity unless given spacious accommodation. They are oviparous, laying thin-shelled eggs in the sand. Food consists of insects of all kinds. A closely related genus which reaches as far north as Mexico is *Ameiva*, which includes a number of attractive, striped species, slender with long tails, some of them reaching a foot or more in length. They are also very active, and the larger species eat eggs and small vertebrates as well as insects.

Sub-order SERPENTES

Snakes lend themselves to being kept in captivity, and certainly some species should make satisfactory laboratory animals. It is necessary to emphasize the great variety of the snakes, of which some 2,500 species are known to science. These vary from tiny, worm-like snakes 6 or 7 in. long when full-grown to the Reticulated Python, *Python reticulatus*, and the Anaconda, *Eunectes murinus*, both of which have been credited with attaining a length of 30 ft.

The different species of snake are as versatile in habit as they are in form. They represent the most successful evolutionary radiation the reptiles have achieved since they declined from their period of dominance on earth, and variously occupy a wide range of habitats. There are terrestrial snakes, burrowing snakes, tree snakes, freshwater snakes, and sea snakes. In addition, there are harmless snakes and dangerous snakes, the latter including the more virulent of the poisonous snakes, as well as the larger constrictors.

Feeding

All snakes are carnivorous and feed on whole animals. These are usually caught alive, though dead animals are frequently eaten by snakes. Occasional reports of vegetarianism among snakes have yet to be proved, and certainly the snakes are designed for feeding on other animals. The long, sharp teeth are admirably suited for holding struggling prey, the jaws are capable of considerable extension, and the gastric juices are powerful enough to digest flesh, bones, and sometimes feathers and fur. Different species of snake eat different kinds of animals, and it is, of course, necessary in each case to know what is the right food to offer. The larger constricting snakes largely eat warm-blooded animals; many smaller snakes, such as the Common Adder, *Vipera berus*, eat small rodents, lizards, and sometimes young birds; others, such as the Grass Snake, *Natrix natrix*, feed mainly on amphibians; many aquatic and semi-aquatic snakes are mainly fish-eaters; many smaller snakes, such as the

American Ring-neck Snakes (*Diadophis*) and Ground Snakes (*Storeria*), consume earthworms as part at least of their diet; and other small snakes eat grubs and insects. With occasional unpredictable exceptions, each species keeps to its own particular diet, and it is no use offering a Grass Snake only mice or an Adder only fish. Some snakes feed mainly by day, others mainly by night. Food that is not eaten fairly quickly when it is put into a cage should be removed. This particularly applies to live rodents, which may easily harm or even kill a snake too sluggish or otherwise unwilling to feed. If more than one snake is kept in the same cage there is always the danger if food is left in the cage that two snakes may both seize the same morsel. Snakes often continue to swallow whatever is in their mouths, so that one snake may even swallow the other in this way. This does not apply, of course, if the food requirements of the snakes differ, but is a real danger with snakes which eat the same kind of food. Snakes react to food by both movement and smell. A live animal may be quickly seized because it is moving, and a dead or motionless animal detected by smell. Many snakes feed in darkness, on rodents in their burrows, and they rely on smell, using, in addition to the nose proper, the forked tongue to carry scent particles to Jacobson's organ, a sort of nose in the roof of the mouth.

Many snakes simply seize their prey and swallow it without further ado. The Grass Snake is an example. Some, like the Smooth Snake, *Coronella austriaca*, when feeding on lizards will throw a coil or two around the lizard to subdue its struggles. The constricting snakes, such as the Pythons, the Boas, and the Rat-snakes (*Elaphe*), kill their prey by constriction, usually seizing it in the mouth and throwing two or three coils around it or even pressing it against a solid object. When the animal is dead the coils are relaxed and swallowing commences.

The fangs of the poisonous snakes are primarily designed to kill or subdue prey, and play only a secondary defensive role. There are three groups of such snakes, in each of which the fangs operate differently. The group known as the 'back-fangs' has short, rigid poison fangs in the rear of the upper jaw. These snakes seize their prey and chew it and bring the poison fangs into play, holding on until the poison takes effect. Most of these snakes have weak venom, and some time may elapse before the prey ceases to struggle. The cobras and their allies have similar fangs in the front of the jaw, with which they can strike their prey, though in a somewhat clumsy manner compared with the viper group, in which the fangs are much longer and mounted on a movable bone so that they can be erected when the mouth is opened and quickly driven into the prey and withdrawn. Generally with these two groups, and invariably with the vipers except when small animals are being eaten, the animal so struck runs off and dies after perhaps a few minutes. The snake then tracks it down by means of the tongue.

A specialized group of vipers known as the 'Pit-vipers', which includes among others the rattlesnakes, have a pit roughly between the eye and the nostril which acts as a sensitive thermoreceptor, capable of detecting heat radiations from a warm body. The pythons have similar but smaller organs in the plates of the upper lips. If dead food is offered to these snakes they are sometimes likely to accept it if it has been warmed. Some rodent-eaters can be persuaded to eat if a mouse or rat is placed in a small box with an opening on one side just big enough for the snake to get its head through. Presumably this corresponds with the conditions in a rodent's burrow.

Housing

For most snakes, a simple, unadorned cage is best. Normal requirements are a retreat of some sort, as most snakes like to hide away when sleeping, and a water-dish or bowl. Many snakes like to bathe from time to time, and a large enough bowl is an advantage for this, but even for the water-snake it is not necessary. The Mud Snake, *Farancia abacura*, and one or two others need to be kept in a damp environment, but as a general rule the cage should be kept quite dry apart from the water-container. Most of the diurnal snakes like to bask, and sunlight or a lamp should be used. However, the temperature range of each species needs to be carefully checked, as too high a temperature can be fatal for snakes, and it should be possible for the snakes to get away from the source of warmth to a cooler part of the cage. Much of a snake's life is taken up with moving about to regulate temperature, and they will do this automatically if given the chance. Heaters should be guarded if necessary so that the snakes do not burn themselves.

For the tree snakes, which rarely descend to the ground, branches or a small bush are a practical necessity. Burrowing snakes require sand or soil as the case may be, and the permanently aquatic snakes must naturally be kept in water.

Dangerous snakes

Obviously some special precautions need to be taken when dealing with these. The larger constrictors should be handled carefully, as apart from their constricting powers, they can give a nasty bite. Individual snakes may, of course, become tame; but the fewer risks taken the better. The cage in which these powerful snakes are kept must also be quite strong, as they can exert a considerable pressure in a confined space.

With the dangerously poisonous snakes the golden rule is to handle them as little as possible. In any case, the use of tongs or other instruments for holding the snakes always runs the risk of causing some damage unless carefully used. The ideal is to so arrange the cage that all the necessary jobs of feeding and cleaning can be carried out, as it were, at a distance. Food can be introduced through a small hole high up in the cage, with a well-fitting door or lid. Cleaning and the removal of uneaten food is possible with long-handled instruments which can also be operated through a small port. The best arrangement for a water-container is a sunken pool with a drain-pipe attached to a flexible hose outside the cage. By raising or lowering the free end of the hose, it can be used for both filling and emptying the pool. A cock, operated from outside the cage, will avoid any possibility of even small enough snakes escaping through the pipe. A few species of cobra known as the 'Spitting cobras' are capable of ejecting their venom accurately over a distance of several feet, and it can be dangerous if it gets into the eyes. If there is any chance of this happening, goggles should be worn. The treatment of snakebite cannot be dealt with in this article, but obviously the correct methods should be learned and the proper sera in respect of the specimens kept should be held in readiness. If adequate precautions are taken they should never be needed, but accidents can happen.

Sloughing

All snakes slough or cast their old skins at regular intervals, usually every two or three months. Up to a week prior to sloughing the snake becomes lethargic and ceases to feed, the colours become drab, and the eyes take on a milky appearance. Both sight and the sense of smell become impaired. A day or two before the skin is cast the milkiness largely disappears from the eyes, as a result of a lubricating fluid being secreted between the old and new skins. Eventually, the snake seeks out some rough object, such as a stone or branch, on which it rubs the sides of its head until it has lifted the old skin away from the lips. The skin is pushed back beyond the head and the snake then proceeds to crawl out of the old skin, which it can do easily if it is able to entangle the skin in branches or vegetation. If the snake is healthy the skin should come off entire, including the covering of the eyes. It sometimes happens, particularly with snakes in poor health, that patches of old skin remain on the body, or even that the snake fails to cast its skin altogether when the time comes. If this happens, the old skin will dry on to the body, usually shrinking a little in the process. It must then be removed by hand, which is made easier if the skin is first soaked in water or preferably liquid paraffin. If the covering of the eyes does not come off when the snake skins it may also fail to do so at the next sloughing, and thus several coverings can build up on the eye, with the result in extreme cases that the snake may not be able to see. These coverings must be removed with forceps, which is a tricky operation. Many snakes like to immerse themselves in water for long periods as the time for sloughing approaches, and it is a good thing to have a large enough bowl of water in the cage at this time.

Diseases and pests

Snakes in captivity can suffer from a number of well-recognized diseases, and at least one external parasite can be very troublesome. Fortunately, all these complaints are fairly amenable to treatment and can be prevented by proper care. The most difficult disease to deal with is a respiratory infection commonly known as 'snake pneumonia'. The symptoms are lethargy, loss of appetite, and in severe cases a watery discharge from the mouth or nostrils. If a snake suffering from this disease is examined the mouth and throat will usually be found to be full of thick mucus, the presence of which may impair breathing and cause the snake to 'wheeze' noticeably. The extent to which the disease is infectious is not altogether clear, and it seems probable that it is acquired only by snakes already in poor health. It often makes its appearance in snakes which have been subjected to extremes of temperature, and there is more than a suspicion that it can be transferred by the Snake Mite, which will be discussed below. An infected snake should be quarantined and kept free from draughts but with good ventilation at a regular temperature of around 75–85°F (23·9–29·4°C) or perhaps slightly higher for some tropical species. Sulphanilimide powder dropped on to the floor of the snake's mouth frequently clears up the infection in a matter of days, repeating the treatment every other day until the symptoms have disappeared. The amount administered should be only enough to dust the inside of the mouth lightly, as overdoses of sulphanilimide taken into the system may have detrimental effects. Silver vitellin is also reported to have been used with some success in cases of pneumococcal pneumonia in reptiles, a few drops of a 15 per cent solution being administered into the mouth or nostrils; so also is Achromycin, giving about 50 mg for each 3 ft of snake, in tablet form.

Another disease affecting the mouths of snakes is known as 'mouth-rot' or 'mouth-canker'. It is evidenced by a whitish, cheesy deposit inside the mouth, particularly in the groove along each side of the lower jaw-bone. If unchecked, erosion of the flesh and even of the bones of the jaw can take place. The manner in which the complaint is acquired is not altogether clear, but healthy snakes in clean surroundings are unlikely to acquire it. If treated early the condition can be cleared up by removing the deposit and lightly dusting the affected parts with sulphanilimide powder or painting them with dilute tannic acid. The general health of the snake must be considered if a recurrence is to be avoided.

Snakes kept in damp or dirty surroundings easily acquire skin sores, consisting of small pustules between the scales filled with a watery or cheesy deposit. The condition can become extensive in some cases, practically covering the snake's body. Treatment consists of dusting the sores with sulphanilimide powder, but the main necessity is to rectify the housing conditions, and sometimes this is in itself sufficient to bring about an improvement. The disease is likely to arise in overcrowded conditions, and probably comes from contact with faecal matter. When damp, this gives off a certain amount of ammonia, which seems to be the irritant that may start the complaint. Small snakes in particular are very susceptible to such dirty conditions, and may even die quite quickly if exposed to them. For this reason, snakes packed together in bags or boxes for transport should be carefully examined on arrival and immediately washed clean of any faecal matter. If the sores can be cleared up the scales may grow into shape again over a period, and each successive sloughing will result in a better appearance.

Intestinal disorders are common in snakes in captivity, and almost certainly arise from bad feeding or housing. The practice of feeding snakes on raw meat must be deplored in this respect, as it easily gives rise to impaction of the faeces. This may sometimes be resolved by anal irrigation with warm water or castor oil, aided if necessary by gentle massage, but in severe cases is very difficult to remove. Cages should also be kept clean enough to ensure that food does not become smeared with faecal matter before ingestion. Regurgitation of food may be a sign of either stomach trouble or too high a temperature.

Intestinal parasites are unusual in snakes, while the most troublesome external parasite of snakes is undoubtedly the Snake Mite, *Ophionyssus natricis*, which under certain circumstances build up rapidly to very large numbers among captive snakes. First indications may be the appearance of tiny white spots on the snake's body, and closer examination will show the mites embedded between the scales or moving about on the skin. A magnifying glass may be needed to detect the smaller ones, though when gorged they are easily visible. In any infested snake there are almost certain to be some embedded in the rim of the eye. They irritate the snakes, lower their condition, and probably act as vectors for some snake diseases. The eggs are laid in cracks and crannies in the cage, or in debris, such as pieces of cast skin. The mite can be kept under control provided consistent precautions are taken. Newly arrived snakes should be examined and quarantined to ensure that the mite is not introduced with them. Facilities for bathing may help the snakes in ridding themselves of the mite, which can drown if immersed for long enough, and many mites can be removed or killed if the snake is 'rubbed down' with liquid paraffin. Some insect powders are very effective if sprinkled in the cage, though care must be taken to avoid any containing DDT. 'Pulvex' can be recommended. The powders should not be allowed to get into the drinking-water or come into contact with the snakes' food, and they are best used either inside the snakes' sleeping box or under sheets of thick paper laid on the floor. A recent development in America is the use of the Sorptive Dust SG67, a silica aerogel causing desiccation of the mites, but shown under test to be harmless to snakes. It is now being marked in the United States for this purpose under the trade name DRI-DIE 67.

A word should be said about forced feeding. Almost any complaint from which snakes can suffer is likely to lead to loss of appetite, and starvation may even precede the illness in causing death. Newly captured snakes, particularly of the more nervous species, often refuse to feed until they have become used to captivity. Although snakes are able to go without food for long periods, lack of food brings about a lowering condition, making the snake more vulnerable to disease, in addition to the risk of eventual starvation. Although forced feeding is not normally to be recommended, it may sometimes enable a snake to be kept going under such circumstances until normal feeding is resumed. To avoid physical damage or shock to the snake, care and patience are necessary for this operation. In some cases it is sufficient to open the mouth of the snake slightly and insert the food, and the snake will automatically proceed to swallow it. In other cases the piece of food can be gently pushed into the throat and massaged into the gullet until it is swallowed by the snake. If these methods fail it is probably best to administer a liquid mixture of beaten egg and perhaps some finely chopped raw meat by means of a rubber and glass syringe of the eye-dropper variety. All this involves opening the snake's mouth to start with, and this must be done with extreme care to avoid shock and damage to the jaws or teeth. The snake should be held firmly but gently behind the head with one hand, and the other used to open the mouth with a smooth wooden spatula. Some dexterity in introducing the food or syringe will then be needed, and an assistant may be necessary.

FISH

The fish commonly kept in the laboratory may be divided, according to their natural habitat, into two main types—marine and fresh-water. Each of these groups contains both cold-water and tropical fish. Freshwater fish are more popular because they are easier to maintain than the salt-water types.

HOUSING

Fish are usually kept in glass aquaria, but aquaria may be constructed of any material which is waterproof, is not easily corroded by water, and is not poisonous to fish.

All-glass aquaria may be undesirable because they are of uneven thickness and are therefore liable to crack and cannot be repaired. Furthermore, the uneven glass distorts the appearance of the fish and makes observation of them difficult. Tanks are usually made from sheets of glass held in position in a metal frame with putty. Aquarium putty (sold under various trade-names) is specially designed for its purpose; it remains soft, and so allows for the different rates of expansion and contraction of glass and metal, and, unlike ordinary putty, is not toxic to fish. For tanks of up to 1 cu ft in volume (6 gallons capacity or 18 in. × 10 in. × 10 in. deep) glass may be used and the frame may be of pressed steel. For larger tanks of up to 20 gallons capacity, $\frac{3}{8}$-in. brass angle is suitable for posts and top frame, and $\frac{3}{4}$-in. angle base frame; from 25 to 40 gallons, $\frac{3}{4}$-in. posts and top frame with 1-in. bottom frame; beyond that, 1-in. angle iron throughout. The base itself should be of slate or reinforced plate glass.

Aquaria are covered to minimize the loss of water by evaporation, to reduce temperature variations, to keep out dust, and to prevent some species (e.g. *Xiphophorus*) from jumping from the tank. Lids are usually made from non-ferrous metal. The electric-light bulbs commonly used to illuminate the tank may be supported in the lid. A sheet of glass may be used as a lid, but in this case the metal edge of the tank must be protected with rubber strips to preserve it from the corrosion which would otherwise occur due to condensation of water between the lid and the frame. A lid must be so shaped and positioned on the tank that the water which condenses on its underside drips back into the tank and does not run away to the exterior. This will not be necessary if a lid and glass cover are both used.

To find the capacity in gallons of any rectangular container, multiply the length, height, and width in inches together, and divide by 231, e.g. an aquarium 20 in. × 16 in. × 15 in. would be

$$\frac{20 \times 16 \times 15}{231} = 20{\cdot}7 \text{ gallons}$$

WATER

Water, the environment of fish, is all important. Water may be hard or soft, acid or alkaline. Fortunately, the common species will tolerate considerable variations in the constitution of the water, although they are happiest with near neutral conditions. Most tap waters are safe to use after they have been left standing for two or three days. This period of 'maturation' permits the escape of dissolved gas and allows the growth of organisms, which the fish may eat, in the water. Before it is used, the water should be tested on litmus paper, which will turn red in acid, blue in alkaline, or remain unchanged in neutral conditions. Acidity may be reduced by adding sodium bicarbonate to the water, and alkalinity by adding tannic acid or a filtrate of peat moss. Hard water may be softened by diluting it with rain water or with ice formed from condensed water round the ice box of a refrigerator (NOT water from ice trays) or with distilled water to which had been added sea salt in the proportion of three level teaspoonfuls to the gallon. The water may become hard if soluble, chalky or alkaline rocks (e.g. limestone, marble) are placed in the aquarium. If rocks are used for decoration, smooth, naturally worn rocks or well-scrubbed coal or slate are suitable; but they should be free from cracks in which food particles could collect and decay. It is not necessary for the fish to cover the floor of the aquarium with gravel. The floor of an aquarium may be covered with gravel if some medium in which plants may root is needed.

HEATING

Tropical fish must have a warm environment. Heat may be applied from small gas jets placed under the tank, which is protected by asbestos, but electrical heating is more commonly used. A heater may be made from a coil of wire placed in a hard-glass test-tube which can be made water tight with a rubber bung through which is passed insulated wire leading to a simple thermostat. One such heater of about 100 watts submerged in an average 10–12-gallon tank would provide sufficient heat, though often two heaters are fitted in case one should break down. The average water temperature for tropical fish is 75°F (23·9°C). It should be remembered that, though most fish will tolerate large, *gradual* temperature changes (often of ±20°F (11°C)), the bacterial count of the water increases and the oxygen content decreases as the temperature rises, with obviously harmful results.

LIGHTING

Artificial lighting is not essential for fish, but it is essential if the fish are to be seen clearly and if the natural light is inadequate. If an aquarium contains plants these will normally require artificial light. Two light bulbs of 40 or 60 watts each switched on for 12–15 hours daily is usually recommended, but the situation and contents of an aquarium determine the amount of artificial light that is needed. Algae develop when there is too much light; plants die when there is too little. Experience soon indicates the amount of light needed. As a guide, 'very poor' light is a position where a newspaper cannot be read easily by daylight.

PLANTS

It is now accepted that plants are not essential for keeping healthily stocked aquaria. All the necessary interchange of gases occurs quite naturally at the water surface. It is therefore important that tanks should expose a large surface area of water to the air. Each inch of fish (tail excluded) should have 24 in.^{2} of surface area of water; thus it comes about that a deep tank will not necessarily hold more fish than a shallower one of the same width and breadth.

Plants are useful in forming a refuge for fry (immature fish) and some types of eggs. Aquatic plants are grouped roughly as floating, rooted, or bunched, each group needing a different treatment. The floating plants, e.g. Crystalwort (*Riccia*), Fairy Moss (*Azolla*), and the Common Duckweed (*Lemna*) need no planting; they float on the surface, giving excellent hiding places for fry, and acting as a light filter, which is often a better method of preventing the growth of algae than is altering the amount of light.

The rooted plants have definite root bases similar to common terrestrial plants. Many reproduce by strawberry-like runners, and when planted the gravel or compost should cover the roots but not the crown or junction of the roots and leaves. Examples of rooted varieties are the many types of *Cryptocorynes*, the *Vallisneria* (Eel Grass, Tape Grass), and *Sagittaria* (Arrowhead) species, both resembling wide-bladed grasses, and the various types of *Echinodorus*, of which the popular Amazon Sword plant is one.

Bunched plants, as their name implies, are planted in bunches and have scarcely any roots. At times adventitious roots grow from leaf nodes, but these are few and are far from essential. The common Canadian Pondweed (*Anacharis*, formerly *Elodea*) is one of the best known, and others are Fanwort (*Cabomba*) and Milfoil (*Myriophyllum*). Some plants, usually the rooted types, appear to die off when transplanted, but they almost invariably recover. Dead leaves, along with other debris, should be removed from the aquarium regularly. Many water plants flower.

AERATION

Aeration is primarily important as an aid to oxygenization, by its *agitation* of the water. Aeration is obtained by a small electric pump which forces air through a tube to the bottom of the aquarium, sending up a spray of small bubbles. These bubbles contain oxygen, which is absorbed by the water, mostly at the surface. Aeration is particularly valuable to turn on at night in aquaria partly depending in the daytime on the oxygen from plants. The plants make no oxygen at night, but give off carbon dioxide. Aeration is also valuable in hot weather when the natural oxygen content of water is low. When overcrowding of fish is unavoidable the ordinary capacity of the aquarium may be doubled by the use of aeration.

Air is composed approximately of 4 parts of nitrogen to 1 part of oxygen. Water takes up oxygen more readily than nitrogen, and these two gases are absorbed from air into water in the ratio of 2 parts of nitrogen to 1 part of oxygen.

The amount of oxygen, as a part of air, which pure water is capable of absorbing by volume at different temperatures is shown in the table below:

Water at 50°F (10·0°C) 7·8 parts per 1,000 by volume
Water at 60°F (15·6°C) 6·9 parts per 1,000 by volume
Water at 70°F (21·1°C) 6·3 parts per 1,000 by volume
Water at 80°F (26·7°C) 5·7 parts per 1,000 by volume
Water at 90°F (32·2°C) 5·0 parts per 1,000 by volume

It will be seen how rapidly the oxygen content diminishes as the temperature increases.

FILTRATION

Filtration is a method of removing unwanted dirt and organisms from the aquarium water. The action of the filter also keeps the water circulating, and thus speeds up the interchange of oxygen or carbon dioxide at the surface, as does normal aeration.

Most types of aquarium filters have to be attached to a pump, which forces air down a tube into the aquarium water, the air bubbles rise up a second tube, taking small amounts of water with them, the water then falls into a perforated container which holds the filter medium, usually glass wool. Activated carbon is often added to the medium to further purify the water; the clear water then drips through the filter's perforations back into the aquarium.

SNAILS

Water-snails eat algae, but they excrete much waste material into the tank. They are therefore of doubtful value, and may well be excluded from the tank, with the possible exception of the Japanese Burrowing Snail (*Melainia*), which aerates the compost, and the very large Apple Snail (*Ampullaria*), which rapidly produces infusoria invaluable for rearing fry. Infusoria are any animal organisms that are of suitable size to feed to young fish before they are old enough to take small *Daphnia.*

FOOD

It is generally conceded that living foods produce the best results, but dried prepared foods are easily bought and stored, and most fish will enjoy the change. Crushed water-snails, chopped earthworms (*Annelida*), and finely chopped or scraped raw liver of chicken or beef are excellent foods. The liver of the mouse is excellent, since it is very easily mashed into suitably sized particles. Dried foods should be fed in small pinches, and should be eaten in 5 minutes. Any uneaten food must be removed from the tank before it begins to decompose.

Daphnia (Water Flea) is an acceptable food, as also are White Worm (*Enchytraeus*), Brine-Shrimps (*Artemia*), and Micro-Worm (*Anguillula*). Cultures of White Worm may be bred on a bread-and-milk medium. Micro-Worms do well in a thin oatmeal and water medium.

Brine-Shrimp eggs are sprinkled on the surface of a jam-jar of water in which one teaspoonful of common salt has been dissolved. The jar is then floated in the aquarium, and the eggs hatch in 24–36 hours. The shrimps should be washed free of salt before they are used as food.

Some fish need more vegetables than they can get from the vegetation in the tank, and lettuce and boiled spinach can be supplied in these cases.

BREEDING

Most cold-water fish need a lot of space for breeding, but some smaller varieties, such as Goldfish (*Carassius*), can be bred in an aquarium. The tropical fish either bear live young (are ovo-viviparous) or lay eggs (are oviparous).

The famous Guppy (*Lebistes*), the Platy, and Swordtail (*Xiphophorus*) are examples of the family *Poeciliidae*, which are live-bearers. At sexual maturity the anal fin (which is just posterior to the anus or vent) of the male develops into an organ of reproduction, the *gonopodium.* The sperm or milt of the fish pass along a groove in this organ, which is placed touching, but not into, the female. The fertilized eggs are retained in the female for about four weeks, the period varying slightly with temperatures and from species to species. The young fish hatching from the eggs are from $\frac{1}{8}$ to $\frac{1}{4}$ in. long. They do not require microscopic food, but can be fed immediately on dust-fine powder food, Micro-Worms and Brine-Shrimps.

The egg-laying or oviparous fish are far more numerous, and may be classified according to the position taken up by the eggs. The main groups are: (i) the demersal eggs, which are heavier than water and sink; (ii) the sticky eggs, which adhere to plants or gravel; and (iii) the non-sticky eggs.

The Zebra-Fish (*Brachydanio*) is an example of the non-sticky, demersal type of egg. When ready to breed the female appears more rounded when viewed from above or below, and should be placed in a breeding tank with the male in the evening; spawning should take place next morning. Like many fish, the Zebra is an avid egg-eater. To prevent egg-eating the water should be about 3–4 in. deep, and the bottom of the tank should be covered with small marbles or glass rods between which the eggs may fall and be hidden. The parents should be removed after spawning, and the eggs should hatch in 48 hours. Like most fry, they hang, by suction, to the glass sides of the tank for a few days. No food should be given until they are free-swimming, as they exist on their yolk-sac until that time. Uneaten food may foul the water.

At this stage the fry are fed microscopic food, or 'infusions'. Infusions are cultures of tiny infusoria which can be made by placing boiled hay, banana skins, manure, or dead leaves in water for a few days, after which protozoa and other organisms will have multiplied. These cultures can easily 'go bad'. Alternative foods are the yolk of a hard-boiled egg made into a 'mist' by stirring it in water, and the infusoria created by the *Ampullaria* snail. Small amounts of Complan* are also ideal for feeding to fry. As the fry grow they can be changed to a diet of Micro-Worms and Brine-Shrimps. Fry should be fed small amounts at frequent intervals; not less than four feeds per day.

The Tiger Barb (*Puntius*) and the Goldfish are examples of the sticky-egg type. Barbs are sexed in the same way as Zebras, but the males are more highly coloured. The male Goldfish develops small white tubercles on the gill-plates in the breeding season. Pairs may be mated as described for Zebras, but clumps of bush plants, such as Milfoil, should be placed in the tank to receive the eggs. Since gravel can be a hindrance, the plants can be tied with cotton and weighted down with a stone. The pairs should be removed after spawning, and the eggs and fry treated in the same manner as the Zebras.

Most common fish can be bred in this way. Sexing some species is impossible without internal examination, but the plumpness of females and the colour of the males is indicative of the sex.

The *Anabantids* (e.g. *Betta*—the Siamese Fighting Fish) have an air-breathing apparatus called a labyrinth organ. They blow nests of bubbles, into which the fertilized eggs are blown and guarded by the male. The most dangerous time for Anabantids is at about three weeks of age, when the fry go to the surface to fill their air-breathing organs. At this time great care must be taken to ensure that the air is at the same temperature as the water. Many die at this time, and a glass cover fitting tightly over the tank will greatly help the survival of the young fish.

HANDLING

Fish are covered by a protective film, and should never be handled unless it is essential to do so, as once the film is broken the fish are more susceptible to infections. If a fish must be picked up, use a soft-net, keep it wet, and hold the fish to stop it falling or jumping, and thus injuring itself.

* Glaxo Laboratories Ltd.

CHAPTER TWENTY

Animal Genetics

Offspring resemble their parents to a greater or lesser degree. These resemblances are due to the inheritance of parental characters by the offspring. Inheritance is not a simple process: sometimes the characters of one parent are noticeably predominant in an offspring, sometimes those of the other. Moreover, the inherited characters can be modified by external factors over which the animal has no control. For example, body size may be determined as much by the availability of food during growth as by inheritance. There are therefore two main factors which determine the growth and development of an organism, the inherited or *genetic* factors, and the external or *environmental* factors. The science of inheritance is called *genetics*.

GENES AND CHROMOSOMES

Inherited characters are controlled by *genes*. The gene can be considered as the basic unit of inheritance, one gene controlling one character, though this is an oversimplification as will be shown later. Each animal possesses a considerable number of genes. The genetic make-up of an animal, which determines its inheritance, is called the *genotype*. The genotype is then conditioned by the environment to form the animal's adult characteristics, which is called the *phenotype*.

Genes occur in linear order along the *chromosomes*. Chromosomes occur in the nuclei of all cells, but are only visible when the cell undergoes division. A group of mouse chromosomes is shown in Fig. 1. The number of chromosomes in each cell is usually constant within a species, but differs between species. For example, the number in the mouse is 40, in man 46, and in the Chinese hamster 22. This characteristic number is called the *diploid* number of chromosomes and is given the symbol $2n$.

Two types of cell are exceptional in that they do not possess the diploid number of chromosomes. These are the gametes, i.e. the egg and the spermatozoon. Instead the gametes possess one-half of the diploid number, which is known as the *haploid* number of chromosomes ($1n$). Thus, in the mouse the spermatozoa and eggs carry the haploid complement of twenty chromosomes, each chromosome being different from the others. At fertilization the union between the gametes restores the diploid number of forty chromosomes in the embryo. But for each chromosome in the egg there is a similar one in the spermatozoon. In the mouse embryo after fertilization there are thus twenty pairs of chromosomes, and in the human embryo twenty-three pairs. The two members of each pair are called *homologous* chromosomes. The cells of the embryo divide in a manner that results in all cells retaining the diploid chromosome complement, hence all adult cells are diploid. This type of cell division is called *mitosis*. The special type of cell division which occurs just

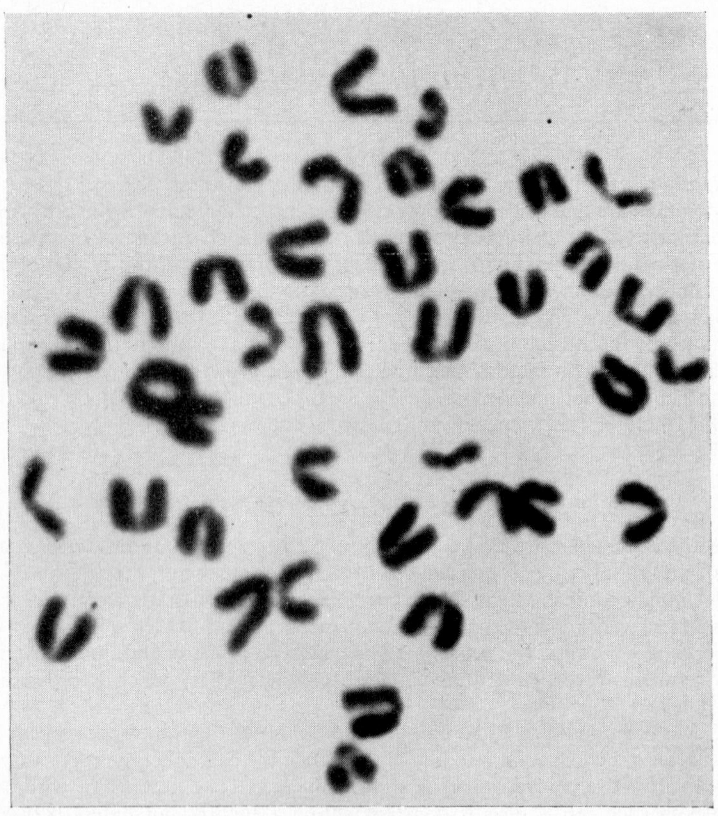

FIG. 1. Diploid set of mouse chromosomes. This mitosis is at metaphase. Each chromosome is split into two chromatids, which are attached at the centromere. The mitotic spindle has been removed chemically. $2n = 40$.

before the gametes are formed, and which results in the eggs and spermatozoa having the haploid complement of chromosomes, is called *meiosis*. We will discuss these types of cell division later.

Both the haploid egg and the haploid spermatozoon carry a complete set of genes. After fertilization the embryo thus contains two sets of chromosomes on which are carried two sets of genes, one set paternal and the other set maternal. Each character in an animal is therefore controlled by two genes, one paternal and the other maternal. The two genes then interact in determining the expression of the character they control, and the type of interaction can be very varied.

To take an example, consider coat colour in rabbits. An offspring may receive the gene for *Himalayan* from the mother and a similar gene from the father. An individual with two identical genes is said to be *homozygous* for that character. On the other hand, the father may have donated the gene for *Himalayan* and the mother the gene for *albino*; the offspring would thus have two different genes for coat colour and be *heterozygous* for that character. Heterozygous animals, or *heterozygotes*, have two different genes for the one character. Which of the two genes will predominate over the other in expression depends on which is *dominant* and which *recessive*. As the gene for *Himalayan* is dominant to albino, the heterozygote will be *Himalayan* and the recessive gene *albino* will not be apparent in the individual. Genes which control the same character, e.g. *Himalayan* and *albino* in the above example, and which occur at the same point on a chromosome (see below), are called *allelomorphs* or *alleles*. Many but not all allelic genes show dominance or recessivity. Sometimes both genes can interact and produce an intermediate or unusual expression, and others may dominate on different parts of the body. The character we have chosen is also influenced by the environment, for a low temperature during development can influence the expression of the gene *Himalayan*.

Genes which govern a particular character always occur at the same place on the chromosome carrying them. It is often necessary to define this place on the chromosome, and the term *locus* is used. Alleles thus occur at the same locus. Genes for other characters will occur at different loci. Three or more allelic genes may occur at a particular locus, e.g. the genes controlling *Agouti* coat colour in mice, or the *ABO* blood groups in man. Except in unusual circumstances, any particular individual can, of course, possess only two of these genes, for most individuals are diploid. Examples of unusual inheritance will be given later.

We must now study what happens to the chromosomes during mitosis or meiosis. From what has been said above, it is clear that a knowledge of the behaviour of chromosomes during cell division is needed for our understanding of inheritance.

CHROMOSOMES AND CELL DIVISION

Normal cell division–*Mitosis*

Mitosis occurs in cells throughout the body, and each daughter cell inherits a diploid complement of chromosomes similar to those of the mother cell. The events of mitosis are represented diagrammatically in Fig. 2. The non-dividing cell consists of a nucleus and cytoplasm. In the first stage of cell division (called *prophase*) chromosomes appear in the nucleus. Each chromosome is long and thread-like, and is composed of two identical halves called *chromatids*. During the second stage, *metaphase*, the chromatids shorten, and remain attached to each other at a particular point known as the *centromere* (Figs. 1 and 8). The chromosomes then become arranged at the mid-point of a structure called the *mitotic spindle* (Fig. 3). The nuclear membrane has by now disappeared. In the succeeding stage, *anaphase*, the centromeres divide and the chromatids move apart along the spindle to form two identical groups of chromatids at each end of the mitotic spindle (Fig. 3). During the last stage, *telophase*, the spindle disappears, each group of chromatids become included in a nuclear membrane, and two daughter cells are formed by division of the cytoplasm. The chromatids then duplicate themselves in the nucleus. Each daughter cell thus possesses a chromosome complement identical with that of the mother cell.

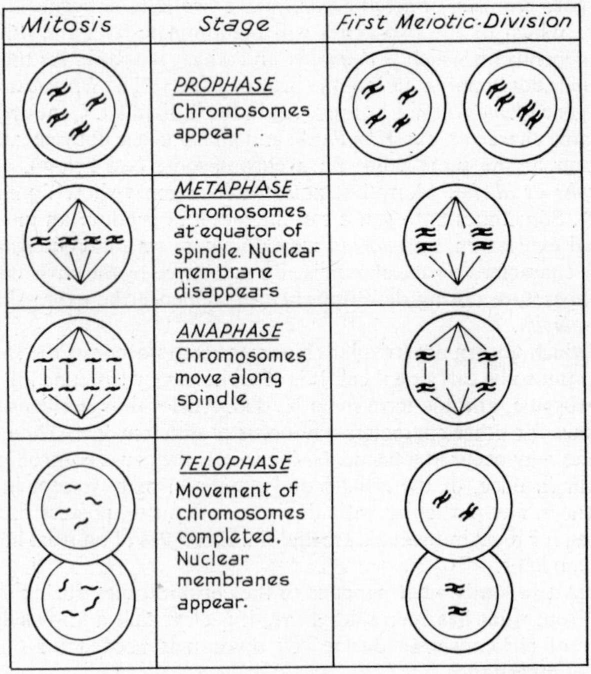

Mitosis	Stage	First Meiotic Division
	PROPHASE Chromosomes appear	
	METAPHASE Chromosomes at equator of spindle. Nuclear membrane disappears	
	ANAPHASE Chromosomes move along spindle	
	TELOPHASE Movement of chromosomes completed. Nuclear membranes appear.	

FIG. 2. Diagrammatic representation of meiosis and mitosis.

Reduction division–*Meiosis*

If the diploid number of chromosomes is to remain constant in a species the two gametes which contribute equally to the new generation must carry the haploid number of chromosomes. The reduction in the number of chromosomes takes place during the two meiotic divisions which immediately precede the formation of the gametes in the testis or ovary. Meiosis is divided into the same stages as mitosis (Fig. 2). During prophase of the first meiotic division a series of events takes place which are very different from events during mitosis. Each chromosome becomes paired with its corresponding homologous chromosome (Fig. 4). There are thus four chromatids, i.e. two pairs, alongside each other (Fig. 6). Some of these chromatids may exchange a segment with one of the chromatids of the homologous chromosome (Fig. 6). This

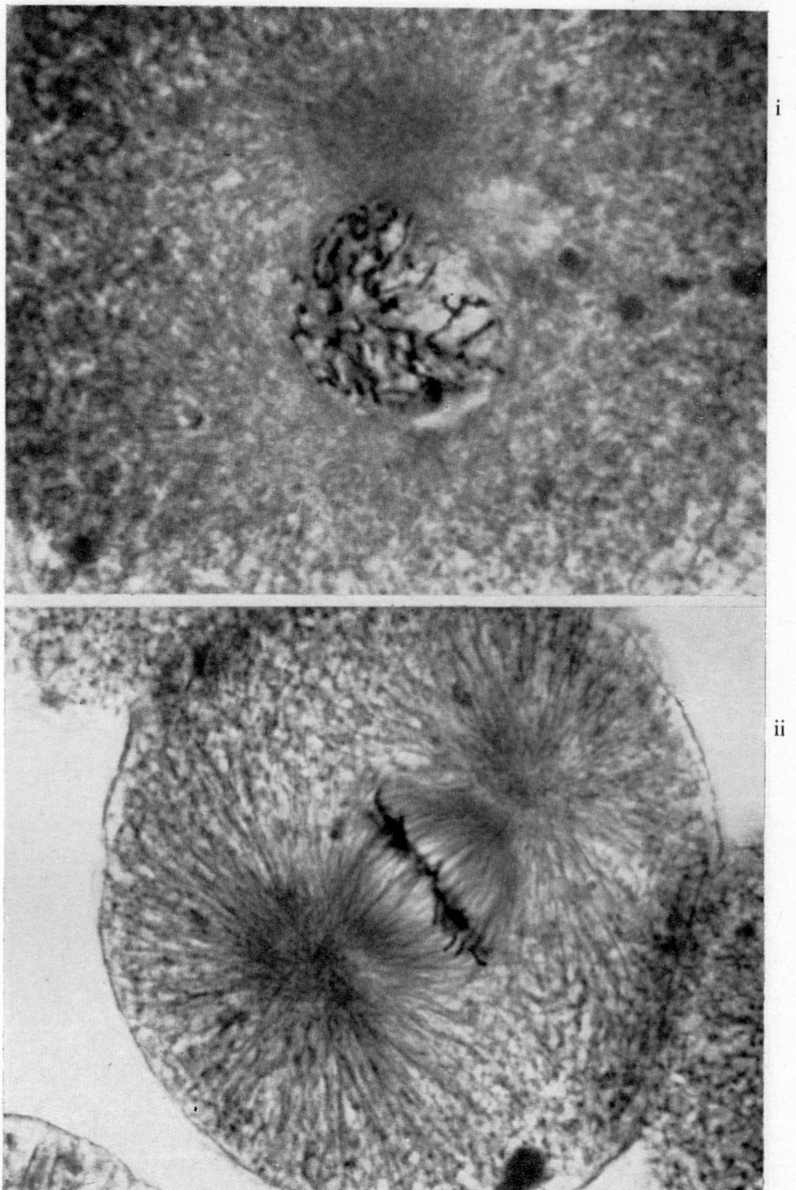

FIG. 3(*a*). Successive sequencies of the stages of mitosis.

(*a*) Mitosis in fixed cells of the whitefish blastula.

(i) Prophase, the chromosomes are becoming distinguishable. (ii) Metaphase, the chromosomes are shortened and attached to the mitotic spindle. (iii) Early

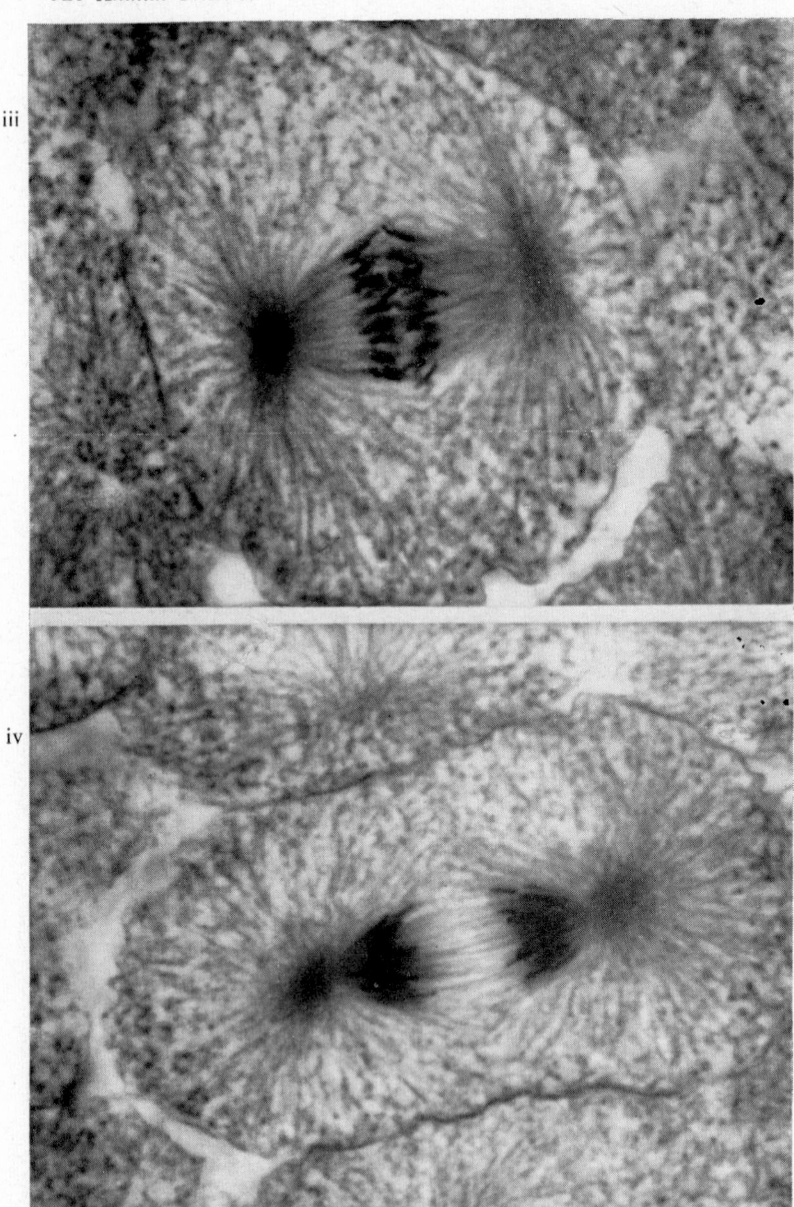

Fig. 3(*a*) iii and iv

anaphase, the chromatids are just beginning to move apart along the spindle. (iv) Late anaphase, the chromosomes are at the end of the spindle. (v) Telophase, the spindle is breaking down and the cell has almost divided into two. (vi) Two

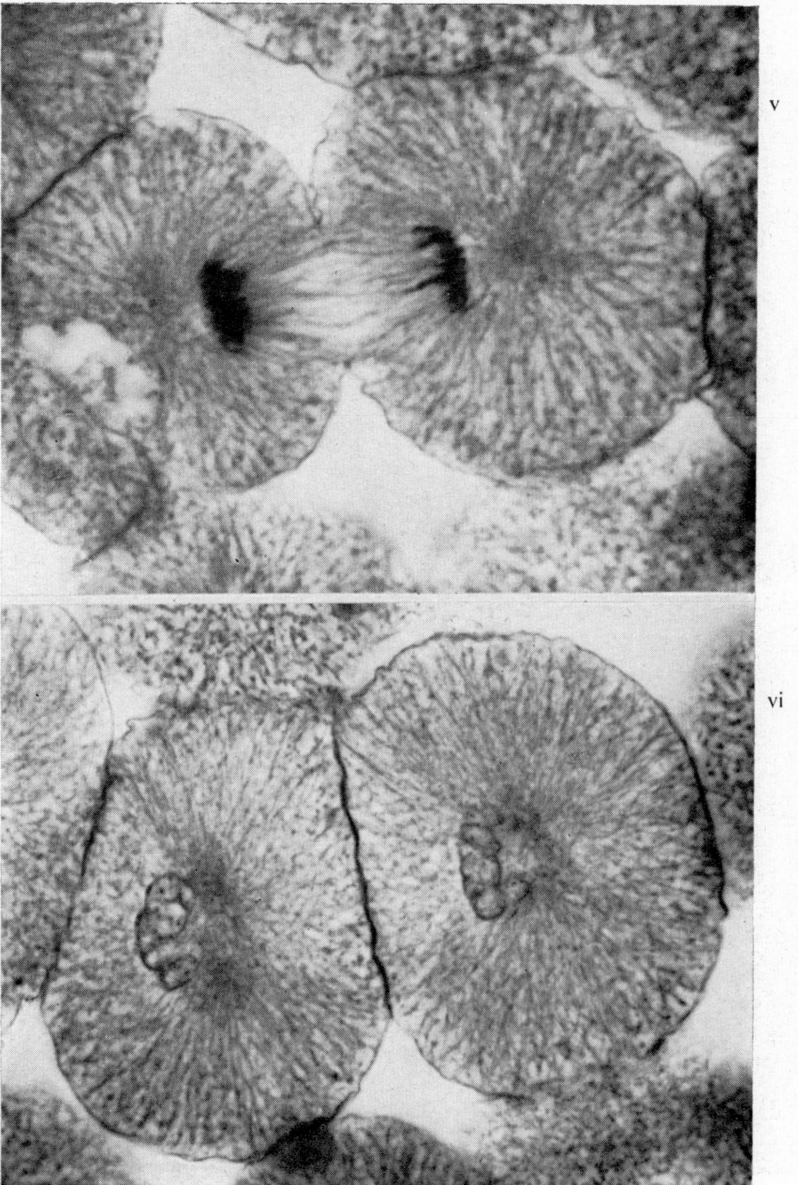

FIG. 3(*a*) v and vi

daughter cells have formed, the spindle has almost disappeared, and a nucleus has been reconstituted in each daughter cell. These pictures are published by courtesy of Mr M. R. Young.

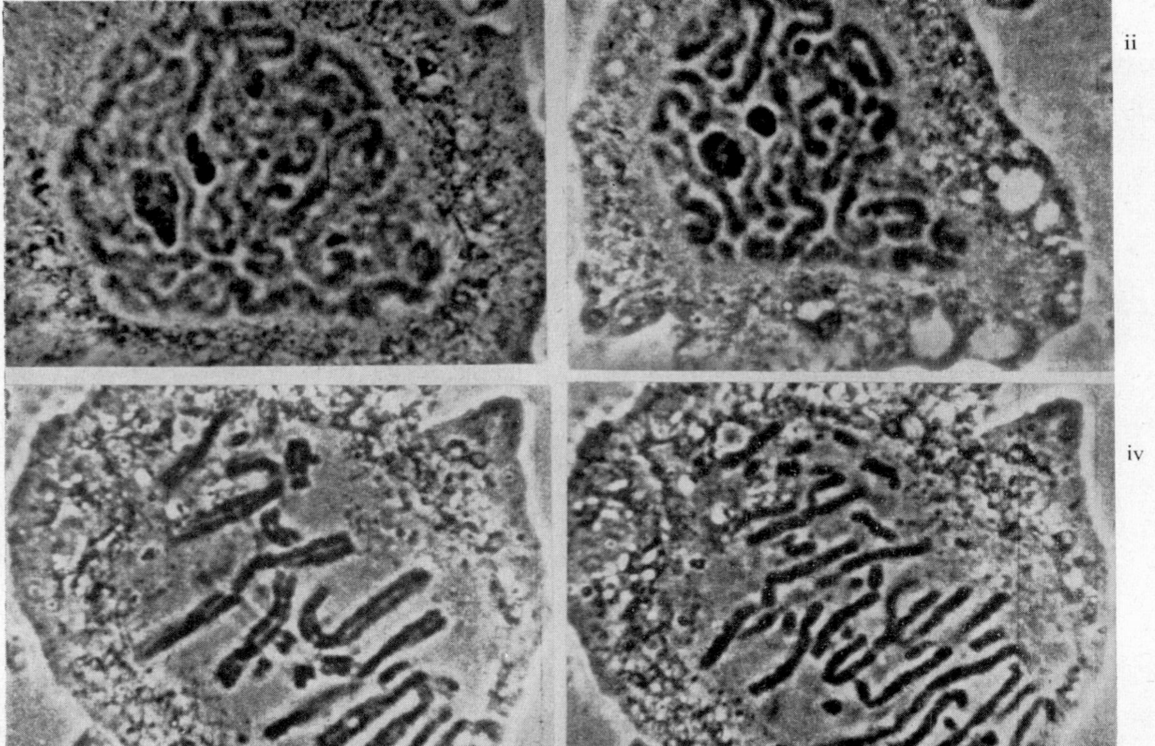

FIG. 3(*b*) i – iv

(*b*) Mitosis in a living endosperm nucleus.

(i) and (ii) Prophase. (iii) Metaphase. (iv) to (vii) Anaphase. (viii) Telophase. Note that at metaphase each chromosome is composed of two chromatids. These pictures were taken by time-lapse photography, and are published by courtesy of Dr A. Bajer and Films of Poland.

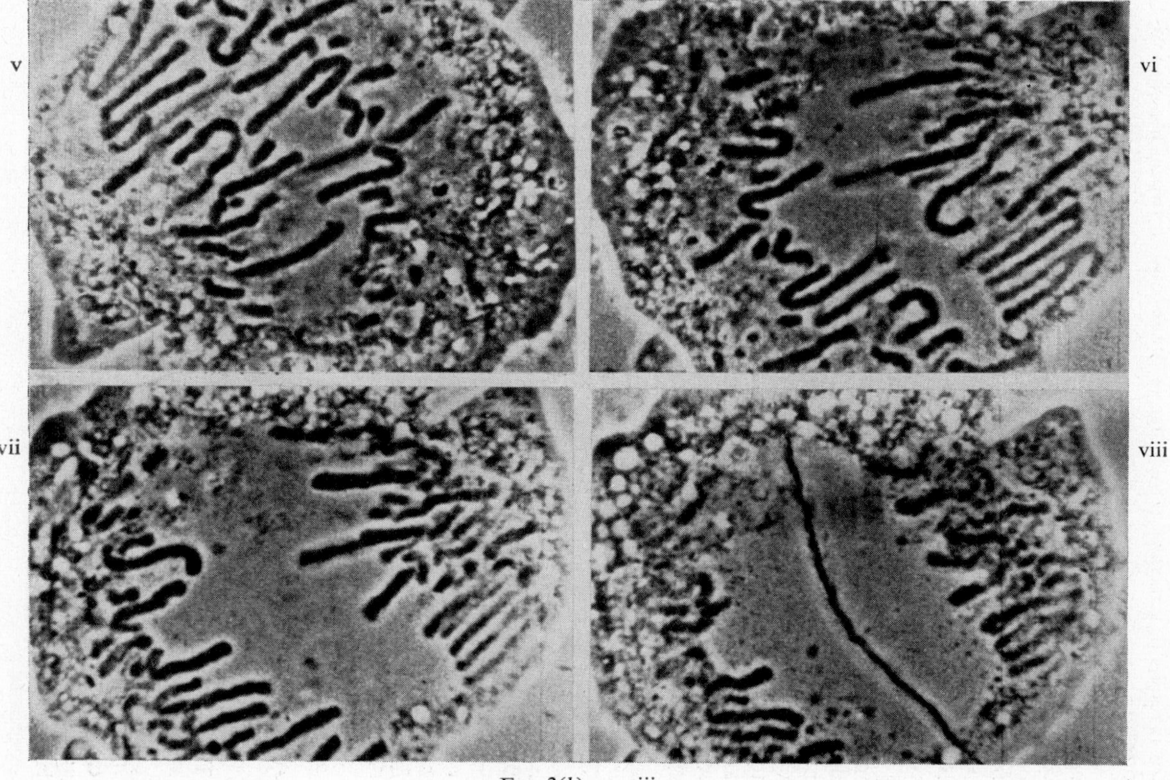

FIG. 3(*b*) v – viii

exchange of material between two chromatids is most important genetically, for it results in a reciprocal exchange of genes between two of the chromatids.

The two chromosomes remain in close association during metaphase. In anaphase another event occurs which has no counterpart in mitosis: the centromeres do not divide and the homologous chromosomes, each consisting of two chromatids, move apart to the opposite poles of the spindle. The failure of the centromeres to divide thus results in the two daughter cells each possessing one-half of the number of chromosomes present in the mother cell. In other words, the daughter cells are haploid, though each chromosome is composed of two daughter chromatids.

The second meiotic division follows immediately after the first. This is like a mitosis: the daughter chromatids which had formed during the first meiotic division separate at anaphase during the second division. Each of the two cells produced after the second meiotic division thus contains the haploid number of chromatids, and these haploid cells then differentiate into eggs or spermatozoa. Meiosis therefore halves the number of chromosomes in the gametes, and fertilization restores the diploid number in the embryo. Meiosis and fertilization are thus complementary, and their relation in the chromosome cycle is shown diagrammatically in Fig. 5.

THE INHERITANCE OF CHARACTERS

We can now use our knowledge of chromosome behaviour in studying the inheritance of genes. It is necessary to know the types of offspring, and the ratios of these types, produced from various crosses, i.e. homozygote × homozygote, homozygote × heterozygote, heterozygote × heterozygote. We will first study the inheritance of a single gene in such crosses, then two genes, and finally more complex situations.

(*a*) INHERITANCE OF CHARACTERS CONTROLLED BY A SINGLE GENE

The laws governing the inheritance of a single gene were laid down by Mendel, and are often referred to as *Mendelian*.

For simplicity, genes are given a symbol, usually one or a few letters. The initial letter of a dominant gene is usually written in capitals. A gene that causes waving of the coat in mice is called *Rex*, symbol *Re*. Homozygous *Rex* animals are *ReRe*, homozygous *non-Rex* are *rere*, and heterozygotes are *Rere*. How are the parental genes distributed among the offspring when two animals are crossed?

Suppose we cross two animals, a male mouse homozygous for *Rex* and a female homozygous for *non-Rex*, i.e. *ReRe* × *rere*. Both paternal chromosomes carry *Re*, hence all spermatozoa will carry the gene *Re*. Likewise, all of the mother's eggs must contain *re*. At fertilization all of the offspring will

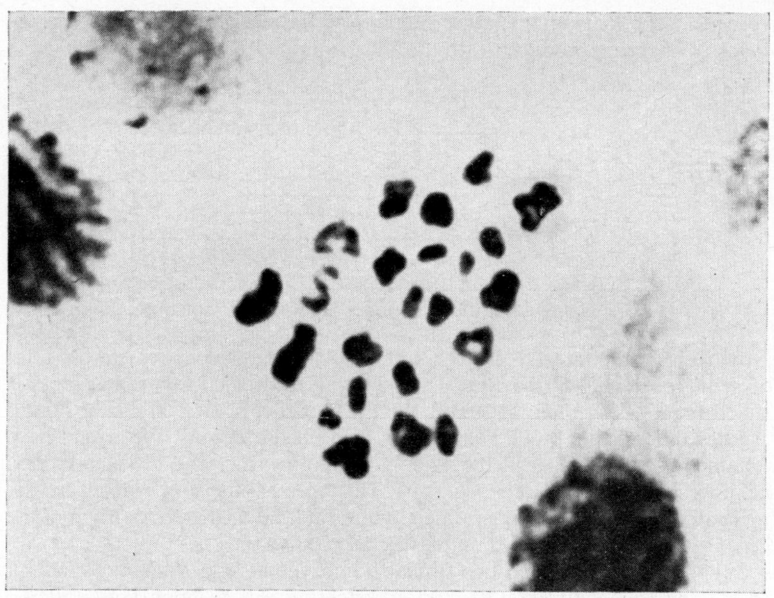

FIG. 4. Meiotic prophase in the human spermatocyte. The diploid number of chromosomes in man is 46, hence there are 23 pairs of prophase chromosomes during meiosis. Published by the courtesy of Dr C. E. Ford.

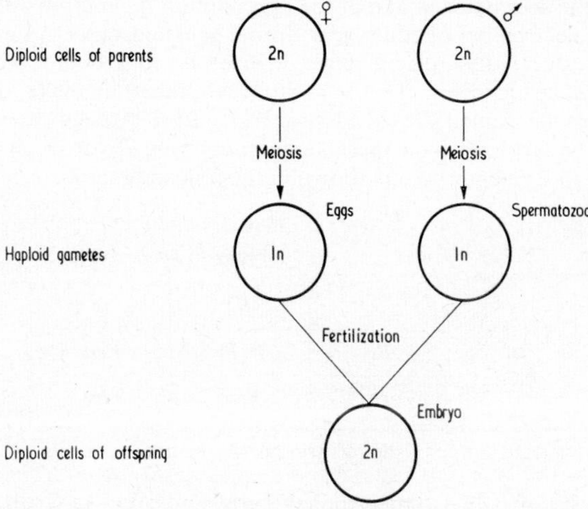

Fig. 5. Relationship between meiosis and fertilization, and the constancy of the chromosome complement.

receive a gene *Re* from the father and *re* from the mother. We can forecast the types of offspring produced with the aid of a table:

rere ♀ × *ReRe* ♂

		Types of spermatozoa	
		Re	*Re*
Types of eggs	*re*	*Rere*	*Rere*
	re	*Rere*	*Rere*

The genotype of the offspring is shown in the squares.

Now *Re* is dominant to *re*, hence all of the offspring, which are heterozygotes of genotype *Rere*, will be *Rex*. The offspring from such a parental cross is called the F_1 *generation*, or simply the F_1. Crossing of two dissimilar homozygotes thus results in all F_1 offspring being heterozygous. We would have obtained the same result if the male had been *non-Rex* and the female *Rex*, i.e. if we had made the *reciprocal cross*. The types of offspring from all crosses we shall describe are not determined by the paternal or maternal origin of the genes unless they are carried on the *sex chromosomes*.

Suppose we had crossed two similar homozygotes, e.g. *ReRe* × *ReRe*. All sperm and eggs will carry *Re*, and all offspring will be homozygous *ReRe*. Likewise, the cross *rere* × *rere* will produce all *rere* offspring. The crossing of two similar homozygotes thus results in all offspring being similar to the parents.

Suppose we cross two heterozygotes, i.e. *Rere* × *Rere*. The gametes, being haploid, will contain either *Re* or *re*. In the great majority of systems studied one-half of the spermatozoa and of the eggs contain *Re* and the other half *re*; there is no selective loss of either gene during gametogenesis. A heterozygote therefore produces dissimilar gametes in contrast to a homozygote, which produces similar gametes. The separation of parental alleles during the formation of the gametes is called *gene segregation*. An egg carrying *Re* or *re* can then be fertilized by a spermatozoon carrying *Re* or *re* giving rise to *ReRe*, *Re re* or *rere* embryos, as shown in the following table:

Rere ♀ × *Rere* ♂

		Types of spermatozoa	
		Re	*re*
Types of eggs	*Re*	*ReRe*	*Rere*
	re	*Rere*	*rere*

The genotype of the offspring is shown in the squares.

In the majority of such systems studied there is no selective fertilization, i.e. spermatozoa carrying *Re* have an equal chance of fertilizing eggs carrying *Re* or *re*. There is thus random fertilization, and the four types of F_1 offspring shown in the diagram occur with equal frequency.

Now, *Re* is dominant to *re*. In the cross outlined above the expected ratio of offspring will be 1 *ReRe* : 2 *Rere* : 1 *rere*. But phenotypically *ReRe* is indistinguishable from *Rere*. The F_1 ratio expected will thus be 3 *Rex* : 1 *non-Rex*, though two of the *Rex* mice will be heterozygous for *non-Rex*. Sufficient offspring must be classified to establish a 3 : 1 ratio, for small numbers can lead to mistaken conclusions. An example of a 3 : 1 ratio is given in Table 1. It is most important to note that a cross between two *Rex* animals has produced some non-*Rex* offspring. Another important fact to note is that genes maintain their integrity between generations, even when recessive and not visible. Thus *re* segregates normally in the F_1, though absent phenotypically in either parent.

TABLE 1. 3:1 SEGREGATION IN THE MOUSE

Two animals heterozygous for the gene *pink-eye*, *p*, were crossed. The 216 offspring were classified and the results were:

Pp ♀ × *Pp* ♂

Types of offspring expected	1*PP* : 2 *Pp*		:	1 *pp*
	3		:	1
Numbers of offspring expected	162		:	54
Numbers of offspring found	161		:	55

(Data from CARTER, T. C. and FALCONER, D. S., *J. Genet.*, **50**, 399, 1952)

The remaining crosses can be dealt with briefly. The offspring ratio for crosses between homozygotes and heterozygotes depends on whether the dominant or recessive gene is homozygous:

Rere ♀ × *ReRe* ♂

		Types of spermatozoa	
		Re	*Re*
Types of eggs	*Re*	*ReRe*	*ReRe*
	re	*Rere*	*Rere*

Rere ♀ × *rere* ♂

		Types of spermatozoa	
		re	*re*
Types of eggs	*Re*	*Rere*	*Rere*
	re	*rere*	*rere*

The cross *ReRe* × *Rere* results in all offspring being phenotypically *Rex*, whereas *rere* × *Rere* results in one-half *Rex* and one-half *non-Rex*. In contrast to matings between heterozygotes, no new genotypes are found in the F_1 of these crosses. We can thus summarize the types of offspring and the frequency of these types resulting from various crosses. Crosses between similar homozygotes result in offspring identical with the parents, whereas those between dissimilar homozygotes result in heterozygous offspring. Heterozygous pairs produce offspring in the ratio 3 dominant : 1 recessive, two of the dominants being heterozygous. Lastly, crosses between homozygotes and heterozygotes result in two genotypes in the offspring, these genotypes being similar to those of the parents.

Mendelian ratios can be used to determine whether animals are homozygous or heterozygous for particular genes. For example, if a homozygous *Rex* male

(*ReRe*) is mated to a homozygous non-*Rex* female (*rere*), all offspring will be heterozygous *Rex* (*Rere*). If the male was heterozygous, the cross would be *Rere* × *rere* and one-half of the offspring would be non-*Rex*. Likewise, if an animal was heterozygous for a recessive gene, e.g. albino, *Cc*, matings to an *albino* homozygote, *cc*, would result in one-half of the offspring being *albino*. Mendelian ratios are also used to test whether a character is controlled by a single gene. If an *albino* homozygote is crossed to a *non-albino* homozygote the cross would be *cc* × *CC* if *albino* is controlled by a single gene. All offspring will thus be heterozygotes *Cc*. The heterozygotes are now crossed back to the recessive parent, i.e. *Cc* × *cc*, and one-half of the offspring will be *albino*. This type of cross back to one of the parents is called a *backcross*, and backcrosses should always give rise to a 1 : 1 ratio if the character is determined by a single recessive gene.

(*b*) INHERITANCE OF TWO CHARACTERS

The inheritance of two characters depends on the relationship between the two genes and their position on the chromosomes. The two genes could be either on different chromosomes or at different loci on the same chromosome. When on the same chromosome, genes are known as *linked* genes. We will first study inheritance of two characters in the F_1 when the genes are on separate chromosomes.

(i) Segregation of genes carried on separate chromosomes

Consider two genes in the mouse, *Agouti* coat colour, *A*, which is dominant to *non-Agouti a*, and *Rex*, *Re*, a coat-waving gene and its recessive allele *re*. A male heterozygous at both loci will be *Aa Rere*. What kinds of spermatozoa will be produced by the male? If we ignore all the chromosomes except those carrying these two genes the chromosome constitution of the animal can be represented as follows:

$$\frac{Re}{\overline{}\;re} \qquad \frac{A}{\overline{}\;a}$$

After meiosis each spermatozoon will carry the haploid number of chromosomes, and hence contain one of the genes at each locus. Suppose a spermatozoon carried *Re*: it will also have an equal chance of containing *A*, and be *Re A*, or *a* and be *Re a*. Likewise a spermatozoon carrying *re* could also carry *A* or *a*, and be *re A* or *re a* respectively. Thus the double heterozygote *Rere Aa* can produce four types of spermatozoa *A Re*, *A re*, *a Re*, *a re*. Moreover, the four types of spermatozoa are produced in equal numbers. Likewise, a female of identical genotype will produce four types of eggs in equal numbers. How can we forecast the types of offspring in the F_1 from a mating of two such individuals?

The simplest way is to construct a table in the same way as for the inheritance of a single gene. As before, the four types of spermatozoa are written along the top of the table, and the four types of eggs down the side.

The expected genotypes can then be filled in by combining the different types of gametes:

Aa Rere ♀ × *Aa Rere* ♂

		Types of spermatozoa			
		A Re	*A re*	*a Re*	*a re*
	A Re	*AA ReRe*	*AA Rere*	*Aa ReRe*	*Aa Rere*
Types of eggs	*A re*	*AA Rere*	*AA rere*	*Aa Rere*	*Aa rere*
	a Re	*Aa ReRe*	*Aa Rere*	*aa ReRe*	*aa Rere*
	a re	*Aa Rere*	*Aa rere*	*aa Rere*	*aa rere*

Each type of spermatozoon and egg is produced in equal numbers, and each spermatozoon has an equal chance of fertilizing any egg. Each genotype in the above table will thus be produced with equal frequency. Some of the genotypes in the squares are similar, and can therefore be grouped together. When this is done the following genotypes and phenotypes occur:

GENOTYPES	No.	PHENOTYPES	No.
AA ReRe	1	*Agouti Rex*	9
Aa ReRe	2		
Aa Rere	4		
AA Rere	2		
AA rere	1	*Agouti non-Rex*	3
Aa rere	2		
aa ReRe	1	*non-Agouti Rex*	3
aa Rere	2		
aa rere	1	*non-Agouti non-Rex*	1

The phenotypic types resulting from a cross between two double heterozygotes thus appear in a 9 : 3 : 3 : 1 ratio. An example is shown in Table 2.

TABLE 2. 9:3:3:1 SEGREGATION IN THE MOUSE

Two animals, both double heterozygotes for the genes *a*, *non-agouti*, and c^{ch}, *chinchilla*, were crossed. The 361 offspring were classified, and the results were:

Aa Cc^{ch} ♀ × *Aa* Cc^{ch} ♂

Phenotype of offspring	*AC*		Ac^{ch}		*aC*		ac^{ch}
Expected ratio of offspring	9	:	3	:	3	:	1
Expected numbers of offspring	202·5		67·5		67·5		22·5
Numbers of offspring found	219		60		62		20

(Data from CARTER, T. C. and FALCONER, D. S., *J. Genet.*, **50**, 399, 1952)

If one of the two genes was homozygous in the above cross, e.g. *ReRe*, then the cross would be *ReRe Aa* × *ReRe Aa*. The Re locus would thus be fixed, and the genes *Aa* would segregate in the 3 : 1 ratio described earlier for a single gene segregation.

The results of all of the crosses described so far, and the following deductions drawn from them, were discovered by Mendel, who worked with the garden pea. From these Mendelian ratios we can deduce that genes retain their individuality from generation to generation, and that they are not contaminated by other genes. Secondly, we know that allelic genes segregate away from each other during the formation of the gametes and freely recombine at fertilization. In the following parts of this chapter, however, we will describe circumstances which cannot be explained in Mendelian terms.

(ii) Segregation of two genes carried on the same chromosome

Genes that occur on different parts of the same chromosome are known as *linked* genes, and they segregate quite differently from those carried on different chromosomes. Suppose we have a male which is heterozygous for two genes, *Aa Bb*, both genes being carried on the same chromosome. Also, suppose that the two dominant genes came from his father and the two recessives from his mother. At meiosis this particular pair of homologous chromosomes will thus be:

$$\frac{A \quad B}{\overline{\ a \quad b\ }}$$

When the chromosomes separate at meiosis, most spermatozoa will carry either *AB* or *ab*. The types *Ab* and *aB* will be formed only if there is an exchange of genetic material between homologous chromosomes during meiotic prophase. This process is called *crossing-over* and is expressed diagrammatically in Fig. 6, showing a crossing-over between the locus *A* and *B*. After a single crossing-over two of the four chromatids have a new association of genes: *A* is linked to *b* on one chromatid, and *a* is linked to *B* on another. The other two chromatids, *AB* and *ab*, are similar to the original parental chromosomes. The frequency of these new associations of genes will clearly depend on the frequency of crossing-over.

Ab and *aB* are called the *crossover classes* or *recombinants*. Suppose that a crossing-over had occurred in approximately 10 per cent of spermatozoa in a male. Since the recombinants are produced in approximately equal numbers, 5 per cent of the spermatozoa will be *Ab*, and 5 per cent *aB*. The remaining 90 per cent will be parental types, i.e. 45 per cent *AB* and 45 per cent *ab*. If this male was backcrossed to a female that was homozygous for both recessive genes, i.e. *aa bb* (in which crossing-over would not affect the ratio of the gametes, all of them being *ab*), then the ratio of the offspring would be: 45 per cent *Aa Bb*, 45 per cent *aa bb*, 5 per cent *Aa bb*, 5 per cent *aa Bb*. An example is shown in Table 3.

The frequency of crossing-over is approximately proportional to the distance between two genes. A low percentage of recombinants indicates strong linkage, i.e. the genes are close together. An example is the close linkage between *pink-eye* (*p*) and *short-ear* (*se*) in the mouse. Genes which are a great distance apart on the same chromosome will show little or no linkage; if the chromosome is a long one the genes might appear to be unlinked because two or more crossovers may occur. If two genes are present on the same chromosome, e.g. *A* and *B* in the diagram above, they are said to be *linked in coupling*, and they will tend to associate together until separated by a crossover. If *A* and *B* are on different chromosomes they are *linked in repulsion*, and will tend to segregate away from each other. The distance between linked genes is usually

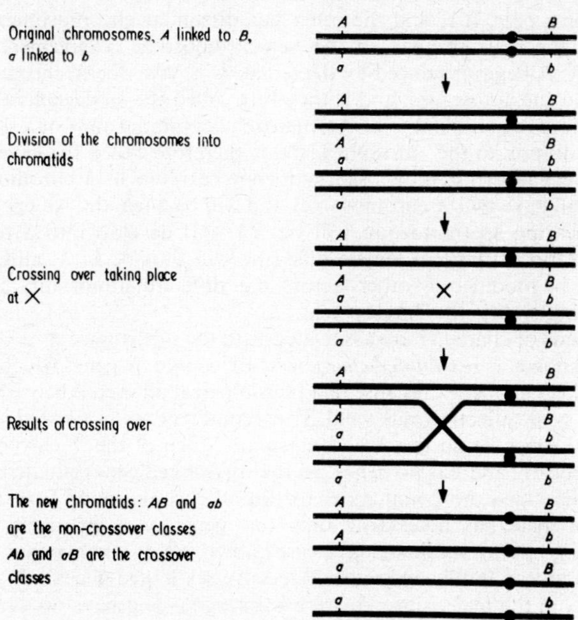

FIG. 6. Diagrammatic representation of crossing over between two loci.

measured in units of crossing-over, one unit being 1 per cent of crossing-over. Linkage analysis of hundreds of pairs of genes in the fruit fly *Drosophila melanogaster* and other species has been a major tool in demonstrating the arrangement of genes in linear order along the chromosomes.

TABLE 3. LINKAGE BETWEEN TWO GENES IN THE MOUSE

The two genes are *a*, *non-agouti*, and *we*, *wellhaarig*, which causes waving of the coat. A double homozygous female was crossed to a double heterozygous male, and the offspring were classified. The results were:

aa wewe ♀ × *Aa Wewe* ♂

Type of eggs	*a we*			
Types of spermatozoa	*A We*	*A we*	*a We*	*a we*
	Aa Wewe	*Aa wewe*	*aa Wewe*	*aa wewe*
Offspring classified	332	32	44	317

Total number of offspring = 725

Conclusion: *A* is linked to *We*, and *a* to *we*

(Data from HERTWIG, P., *ZIAV*, **80**, 220, 1942)

(c) SEX CHROMOSOMES, SEX DETERMINATION, AND SEX LINKAGE

The sex of an animal is determined by a pair of chromosomes called the *sex chromosomes*, the remaining chromosomes being called the *autosomes*. There are two types of sex chromosome: the *X* and *Y*. One sex has an identical chromosome pair, *XX*, and the other has dissimilar chromosomes, *XY*. In mammals the male sex has an *XY* sex-chromosome complement and the female *XX*. All eggs produced by the female will have one *X* chromosome in addition to the autosomes, and is therefore called the *homogametic sex*. The male produces equal number of spermatozoa containing an *X* or a *Y* chromosome in addition to the autosomes, and is therefore called the *heterogametic* sex. If an egg is fertilized by a spermatozoon carrying an *X* chromosome the embryo will have two *X* chromosomes and will be a female. An egg fertilized by a *Y*-bearing spermatozoon will be *XY* and develop into a male. The expected ratio of male to female offspring will thus be 1 : 1, although this ratio may be modified by other factors, e.g. differential mortality of the two sexes during embryonic development.

Two kinds of character are associated with the inheritance of sex. The first of these, known as *sex-limited characters*, are caused by genes which can only have an effect in one sex because of a physiological difference between the two sexes, e.g. genes affecting milk yield. The second type are *sex-linked characters*. Sex-linked genes lie on the *X* chromosome. Much of the *Y* chromosome is inert genetically and has no genes, so that sex-linked genes situated on the *X* chromosome have no counterpart on the *Y* chromosome. This condition, where the male has necessarily only one gene for a character, is called *hemizygosity*. Many sex-linked genes are known, e.g. two in man are red-green colour blindness and haemophilia. Recessive sex-linked genes will appear as dominants in the male, since they are hemizygous and have no allele on the *Y*-chromosome. The frequency of sex-linked recessive conditions in a population is thus much higher in males than in females, where the recessive is hidden by the dominant allele. There is obviously less chance of a female being homozygous for a recessive gene than there is of a male being hemizygous.

Sex-linked characters exhibit a type of inheritance not displayed by genes borne on the autosomes. Suppose a male mouse with the dominant gene *Bent-tail*, *Bn*, is crossed to a female with the homozygous recessive, *bn*. Representing the cross diagrammatically, with *x* and *y* as the sex chromosomes:

$$bn_x bn_x ♀ \times Bn_x\text{-}_y ♂ x$$

Types of Spermatozoa

		Bn_x	-_y
Types of eggs	bn_x	$Bn_x bn_x$	$bn_x\text{-}_y$
	bn_x	$Bn_x bn_x$	$bn_x\text{-}_y$

The offspring will be 2 *Bnbn* females, 2 *bn* males. The female offspring have inherited the characteristics of their father through the inheritance of his

X-chromosome, and the male offspring have inherited the recessive characteristics of their mother because there is no allele on the *Y*-chromosome. This phenomenon is called *criss-cross inheritance*, and is characteristic of sex-linked characters.

We have assumed that no genes are carried on the *Y* chromosome. This is not strictly true: in some species a portion of the *Y* chromosome contains genes and behaves with the corresponding part of the *X* chromosome in a manner similar to relationship between two autosomes. The similar portions of the *X* and *Y* chromosomes are called the *pairing segments* of the sex chromosomes. Genes on the pairing segments follow the normal rules of Mendelian inheritance, whereas those on the *non-pairing segment* of the *X* chromosome follow the laws of sex-linked inheritance described above. Very rarely, genes occur on the non-pairing segment of the *Y*-chromosome. One of the few known genes of this type is *bobbed* in *Drosophila*, and at least one is known in man. Such genes are obviously confined to males and are transmitted from father to son.

INTERACTION OF GENES: GENE MUTATION

In Mendel's original experiments the allelomorphic genes that he studied were either dominant or recessive. This relationship does not always hold, and all gradations are known between cases which show complete dominance and those where the heterozygote is strictly intermediate, e.g. the pink F_1 hybrid from a cross between red and white snapdragons. Other types of relationship occur. One of the loci governing coat colour in mice is the *Agouti* locus. The gene *A*, *Agouti*, which produces the coat colour typical of wild mice, is dominant on the back of the animal. An allele a^t, *tan*, is dominant white on the belly but recessive to the gene *Agouti* on the back. Heterozygotes Aa^t are thus *Agouti* on their back and white on their belly. A third allele at this locus, *a*, *non-Agouti*, is recessive to both *A* and a^t. In cattle the gene *R* is incompletely dominant to *r*. Thus *RR* cattle are red, *rr* white, and *Rr* roan. Incomplete dominance of this type will upset the expected 3 : 1 Mendelian ratios when two heterozygotes are crossed, and will instead give a 1 : 2 : 1 segregation of red : roan : white respectively.

Genes at different loci, i.e. non-allelic genes, also interact with each other, and such interactions can also lead to modifications of the expected segregation ratios. A well-known example of this is the gene *albino*, *c*, which when homozygous completely masks the action of other genes affecting skin and hair colour. The masking effect of one gene by another gene which is not an allele is called *epistasis*. But since *albino* is recessive, the coat colour in mice heterozygous for albino, *Cc*, will depend on the other genes controlling coat colour. Epistasis will also modify Mendelian ratios. For example, if two heterozygotes *AaCc*, *Agouti non-albino*, are crossed the segregation in the F_1 should be 9 *Agouti non-albino*, 3 *Agouti albino*, 3 *non-Agouti non-albino*, 1 *non-Agouti albino*. Since albino is epistatic, however, the ratio will actually be 9 *Agouti non-albino*; 3 *non-Agouti non-albino*, 4 *albino*.

Some characters are dependent on two or more genes acting together. In fowls the presence of the two non-allelic dominant genes for rose comb (*R*) and pea comb (*P*) results in a new type of comb called the walnut comb. Two or more non-allelic genes which interact to produce a character not formed by either of them alone are called *complementary factors*. Genes which have no known effect unless they occur with other genes are known as *modifiers*. The extent of white as opposed to coloured hair in roan cattle is due to modifying genes. It is probable that many phenotypic characters are the result of the interaction between several or many genes. The coat-colour genes in rabbits and rodents can be taken as examples. In mice three alleles at the *agouti* locus, A, a^t, and a confer agouti or non-agouti coat colour with a white belly; at the *black* locus B and b confer a black or brown coat respectively; and *albino*, c, is epistatic to both loci. A mouse of genotype $AA\ BB$ is agouti black, $AA\ bb$ is agouti brown and lighter in colour, $aa\ bb$ produces a deep chocolate brown, and $aa\ BB$ is black. In addition to these, other genes act by making the coat slightly lighter, spotted, or striped. Some of these genes, all non-allelic, are *pink eye*, p, *dilute*, d, which are both diluting genes, *pied*, s, a spotting gene, *Tabby*, T, which causes striping, and many others.

It is also probable that most if not all genes influence several characters, although their effect on one particular character may be most evident. When a gene causes changes in two or more characters the gene is called *pleiotropic*. Thus, the dwarfing gene in mouse, *dw*, also affects the development of the gonads. Most genes probably also exert effects on such fundamental characters as viability and fertility.

Genes which have marked effects on viability can clearly lead to a disturbance in the Mendelian ratios, the most extreme examples being those which cause the death of the embryo. The dominant gene A^y, which causes yellow coat colour and obesity in mice when heterozygous is lethal to the embryo when homozygous. Thus, matings between two heterozygous yellow mice produce yellow : non-yellow offspring in the ratio 2 : 1 because the homozygous yellows die before birth. Lethal genes often have some effect in the heterozygote, although not always so. In man *sickle cell anaemia*, a disease of the red blood cell, is caused by a gene which is lethal when homozygous but which has only slight effects in the heterozygote.

GENE MUTATION

Occasionally the nature of the gene becomes changed, and this process is called *gene mutation*. The mutated gene then reproduces itself in its new form, and the new and old forms are then allelic. Mutation occurs very infrequently, though some genes have a higher mutation rate than others. Most mutations are recessive and may differ from the original gene in various ways: morphological, physiological, or biochemical. Many mutants produce striking alterations in the phenotype, e.g. *white eye* in *Drosophila* and *Yellow* in mice, while others may have only small effects. Estimation of mutation rates is difficult; dominant mutations can be scored easily, but recessives can only be identified when homozygous. Recessive mutations can thus segregate for many generations in heterozygotes before being detected in a homozygote. The process of mutation is generally reversible, and *back mutation* occurs when a mutant gene reverts back to its original form. The rates of forward and back mutation are usually different.

Many if not all of the allelic genes which occur in animal populations have arisen by mutation. With many genes it is obviously impossible to know which is the normal and which is the mutant allele. Usually the dominant is referred to as the normal or *wild-type*, and the recessives as mutants. The term mutant is thus used loosely. Mutation can be induced by various experimental means, e.g. X-rays, ultra-violet light, and various chemicals, and these techniques have been used to produce new genotypes. The effect of such agents is usually non-directed, i.e. mutations of various sorts occur randomly at various loci. Ionizing radiations induce many dominant lethals; indeed, the incidence of induced dominant lethals is often used as an indication of the mutagenicity of an agent.

CHROMOSOMAL ABERRATIONS

Another type of genetic change which can profoundly influence the phenotype of an animal is a change in the chromosome number or structure. Most animals are diploid ($2n$), i.e. they contain two complete sets of chromosomes. Occasionally, one or more extra sets occur, and this phenomenon is called *polyploidy*. Polyploidy is commonest among plants, and many cultivated plants are tetraploid ($4n$) and larger than the diploid variety. Recently a boy was found to have many triploid ($3n$) cells, but polyploid mammals seldom develop to birth. Mouse, rat, and human triploids have been found in the early and middle stages of gestation, but are usually retarded or dying.

The loss of one chromosome from a set can also occur, and such an individual is called a *monosomic*, i.e. $2n - 1$. An individual with one extra chromosome is called *trisomic*, i.e. $2n + 1$ (Fig. 7). Until recently monosomic or trisomic mammals were unknown. Now it is clear that various intersex states in humans are the result of an anomalous sex chromosome constitution through the loss or gain of a sex chromosome. Moreover, trisomics other than those involving the sex chromosomes have been found in man, and the extra chromosome may play a significant role in idiocy or other mental defects (Fig. 7).

Changes may also occur within a chromosome. Part of a chromosome can be lost, and this is called a *deficiency* or a *deletion*. A classical example is *Notch* in *Drosophila*, which causes a notch in the wing margin. It is inherited as a sex-linked dominant in the female, but is lethal in the male. Deficiencies of various sorts have recently been found in humans, and also in certain cancerous cells. X-rays or other irradiations can induce deficiencies. Conversely, a chromosome may have an extra small piece in it, and this is called a *duplication*. A piece of a chromosome may break from its normal place and become attached to an unrelated chromosome. This type of rearrangement is known as a *translocation*. Like deficiencies and duplications, translocations can be inherited, and the extra piece can function like an extra chromosome.

QUANTITATIVE INHERITANCE AND THE SCIENCE OF ANIMAL BREEDING

Mendelian genetics was clearly sufficient to explain clear-cut qualitative differences such as those we have already described, but was more difficult to reconcile with the *quantitative* differences between animals in which all gradations occur between two extremes, e.g. height, weight, intelligence,

(*a*)

(*b*)

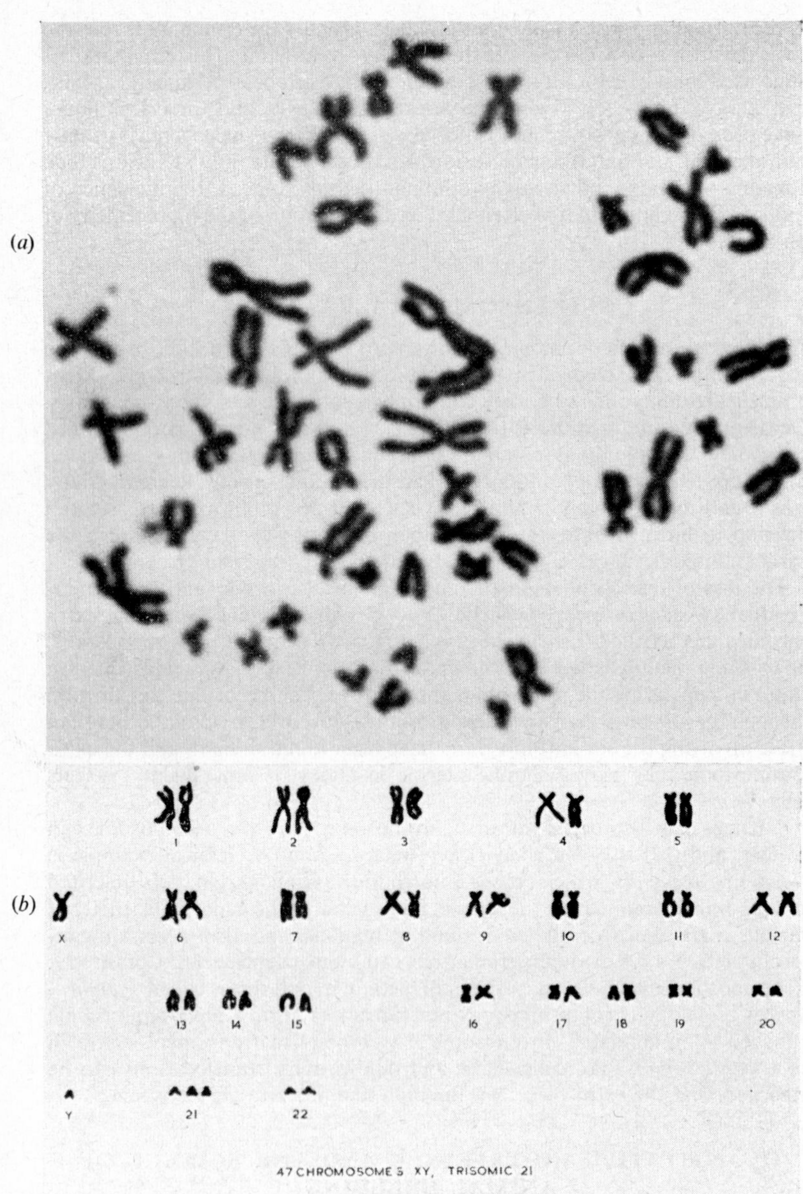

FIG. 7. (*a*) Metaphase chromosomes obtained from a cell obtained from a Mongoloid idiot.

(*b*) When the chromosomes are arranged in homologous pairs, an extra chromosome, no. 21, is found. This person thus possessed 47 chromosomes instead of the diploid 46. Published by the courtesy of Dr C. E. Ford.

fertility, etc. But it was shown that the inheritance of quantitative characters could be accounted for in Mendelian terms because the continuous variation of these characters was due to the joint action of many genes, each having a small effect but acting cumulatively. One of the first examples to be interpreted in this way was the colour of grain in wheat which appeared to be influenced by three genes, each acting additively. In a cross between the race with three dominants (*AA BB CC*) and that with the three recessive (*aa bb cc*) the F_1 were all intermediate (*Aa Bb Cc*). The parental forms appeared again in the F_2, together with five intermediates between these types. The F_2 was much more variable than the F_1, and the three genes were clearly segregating to give a wide variety of phenotypes in the F_2. When characters are controlled by many genes the classification of the different genotypes becomes extremely difficult or impossible, for the variation within a population is almost continuous. In such complex situations alternative methods of approach are needed, as it may be impossible to identify more than a few of the genes involved. Various methods have been evolved for the study of such *quantitative* characters.

A type of breeding now used extensively is *inbreeding*. This involves crossing two closely related animals, e.g. father × daughter, brother × sister, etc. Father–daughter or brother–sister crosses are perhaps the most used in inbreeding. After several generations of inbreeding the inbred offspring closely resemble each other and certain characters are fixed in the population. For example, a skin graft from one animal to another will only be accepted by the recipient if the donor is of similar genotype, i.e. an identical twin or a member of the same inbred line. Why do inbred animals of the same line resemble each other so closely? Consider one locus, e.g. *A*, in a mating system of brother × sister. If the brother and sister are both *AA* all the offspring will be *AA*, and this locus will be *fixed*, i.e. all members of the inbred line will be *AA*, as long as brother–sister mating continues. Mutation of *A* to *a* will interfere with this system, but gene mutation occurs very rarely, and we can neglect it here. Alternatively, the brother and sister might have been homozygous *aa*. Then *a* would become fixed in the inbred line. If both brother and sister were heterozygous *Aa* the offspring would occur in the ratio 1*AA*: 2*Aa* : 1*aa*. With continued brother × sister mating we will ultimately select either *AA* or *aa* animals for crossing, and one of the two genes will be fixed. Thus, inbreeding increases homozygosity in a population. With continued inbreeding, homozygosity occurs at more loci, and the more predictable is the genotype of the offspring. After many generations of inbreeding each inbred line possesses its own characteristics, i.e. those that have been fixed.

But if there are any harmful genes in a stock which become homozygous by inbreeding, then there will clearly be a deterioration of the stock, and this is known as *inbreeding depression*. Such harmful genes are sufficiently frequent to make continuous inbreeding difficult: many inbred lines become homozygous for deleterious or lethal genes and may become undersized, infertile, or die out completely. Many inbred lines have been successfully established, however, presumably free of such deleterious genes. In some cases it may well be that a residue of heterozygosity persists in an inbred line, and that this small amount of heterozygosity is necessary for survival.

If two inbred lines are crossed the resulting hybrid is superior in vigour to either of the two parental lines. This type of cross is known as *outcrossing*, and the superior quality of the offspring is known as *heterosis* or *hybrid vigour*. Heterosis is probably due to the increased heterozygosity of the hybrid as compared with the homozygosity of either parental line. The two parental inbred lines are unlikely to be homozygous for the same harmful recessive genes, and the presence of such genes in the hybrid will thus be masked by the normal alleles from the other line. For example, if one inbred line has a recessive gene *a* and the other has *b*, both genes causing poor growth, then the cross between the two lines will be *aa BB* × *AA bb*, and all F_1 offspring will be *Aa Bb*. If the F_1 offspring are crossed the F_2 and successive generations show a decline in heterosis. This would be expected if heterozygosity was the cause of heterosis, since breeding the F_1 would result in segregation at the heterozygous loci restoring homozygosity in the F_2 and later offspring. Heterosis has been applied especially to maize breeding, where the crossing of two inbred lines produces an F_1 which is extremely sturdy and productive. Recent evidence has also indicated that hybrids are superior to their inbred parents for various types of assay.

Identical twins are of considerable value in studying the expression of quantitative characters. The two members of a pair are genetically identical, since they arise from the same fertilized egg. Identical twins are thus invaluable in studies where animals of the same genotype are required, and can be used experimentally in the same way as inbred animals. Unlike inbred animals, however, twins can possess many heterozygous loci. Human twins who were separated early in life are yielding valuable information on the expression of genetic traits in different environments. Because they are of identical genotype, differences between separated twins in intelligence, aptitude, etc., yield much information on the effects of environment on these characters.

ANIMAL BREEDING

The foremost genetical approach to animal breeding is by *selection* of the breeding colony for the desired characters. Genetics is of use in animal breeding only so far as these characters are heritable, and the aim of the geneticist is to increase the frequency of the desired genes in a population. In turn, increasing the frequency of these genes depends on the efficiency with which individuals carrying a favourable genotype can be selected. The well-known breeds of farm animals have resulted partly from the selection of breeding types over many generations and partly from the improved control of environmental conditions, e.g. feeding, hygiene, etc.

One of the most used methods at the present time is the grading up of livestock by the selection of one parent. The male is clearly the best parent to select, since he can be mated to many females, especially by techniques of artificial insemination. Many beef herds, especially in the United States, have been graded up by the use of selected sires. When breeding animals for beef the phenotype can be used as an indication of the genotype of a bull. But selection of the male on his phenotype is obviously of little use in upgrading dairy cattle, and a good estimate of his genotype is obviously required. Examination of the pedigree is helpful, but at best such information can be only an approximate guide. Even with beef breeds the phenotype often gives little indication of the genotype due to effects of dominance, epistasis, and other phenomena. The best method of determining an animal's genotype is to test it by mating the male to large numbers of females in widely differing environments, and then measuring the performance of his daughters. This technique is known as *progeny testing*. By use of statistical techniques the character being studied, e.g. milk yield, can then be analysed in terms of sire, dam, and environment, and the most suitable sires can be selected. Progeny testing is being widely used in agriculture at present.

Other techniques of animal improvement are being widely practised, and are mentioned briefly. Some breeds have been upgraded by crossing to other breeds. The import of a few sires into a country is used to improve the native breeds, the crossing being usually accompanied by a rigorous selection programme. For some characters, selection through both parents is practised, e.g. increasing the egg production of hens. Wide crosses, e.g. horse × donkey, are invaluable in certain areas of the world.

Another use of genetics in animal breeding is the elimination of undesirable genes from populations. Selection against an undesirable dominant is simple, and these genes can be removed in one generation. Recessive genes are much more difficult to eradicate, since heterozygotes cannot be detected without test-mating and phenotypic selection will remove only the homozygotes.

CHAPTER TWENTY-ONE

Mammalian Reproduction

Breeding season

The reproductive organs of mammals living in the wild state are active and produce mature cells only at a certain period each year, which is usually referred to as the breeding season. In some species, e.g. rodents, this period occurs during spring and summer, and in others, e.g. goat and sheep, it takes place in the autumn. The limits of the breeding season are controlled by external factors, such as the amount of light per 24 hours and the availability of a good food supply. The production of mature germ cells (spermatozoa and eggs) is dependent on the secretion of hormones (gonadotrophins) by the pituitary gland, which, in turn, is controlled by the external factors mentioned above.

When animals are brought into laboratory conditions the breeding season becomes extended and the production of germ cells may occur throughout the year (as in small rodents). The ferret is an example of a species whose breeding season has not been altered by domestication, although, as is well known, the onset of activity and mating in this species can be brought forward several weeks by providing extra periods of illumination during autumn.

Production of germ cells

The means whereby mammals increase their numbers is characterized as sexual reproduction, in which new individuals develop from germ cells produced by the female, after these have united with male germ cells.

Male germ cells (sperms or spermatozoa) are produced in numerous compartments or tubules in the testes; they travel through small ducts (vasa efferentia) to a larger duct (vas deferens) which leads directly to the exterior of the body through a copulatory organ (penis). In some animals the lower end of the vas deferens is enlarged for sperm storage (ampulla). The prostate, seminal vesicles, and bulbo-urethral glands provide secretions to suspend and activate the spermatozoa.

Female germ cells (eggs or ova) are produced as individual cells in the ovary. Each ovum develops within a Graafian follicle which enlarges as the egg matures and finally ruptures to release it (ovulation). The release of ova occurs periodically, and generally at a definite time in relation to the internal physiological rhythm known as the di-oestrous cycle. After release the ova travel down the oviduct or Fallopian tube into a larger organ, the uterus, where the young animal develops; the terminal part of the female tract is the vagina.

Di-oestrous cycle

This cycle refers to a series of changes taking place in the reproductive tract of the female leading to a condition of 'heat' or oestrus when mating occurs. As mentioned above, this latter event takes place at or about the time of ovulation. Certain species of mammals have clearly defined periods of oestrus of short duration which recur throughout the year at regular intervals; that is to say, in a cyclic manner. The laboratory rat, whose oestrous cycle has been most carefully studied, is one such species, and it is described as being polyoestrus (having many cycles). Certain other mammals, whether they become pregnant or not, have only one period of heat per year, e.g. the British fox. The vixen (female) comes into oestrus and ovulates spontaneously in January, and whether mated or not she remains in anoestrus (out of heat) until the following year. Such animals are monoestrus (having one oestrus per breeding season).

Table I relates to changes that take place in the external and internal reproductive organs during the course of a typical oestrous cycle. The separation of the five stages is based on the results of microscopic examination of the vaginal smear. This division allows for two stages of oestrus (Stages II and III) in order to show more precisely the time of ovulation in the cycle of vaginal changes.

There are several methods of taking vaginal smears: (1) a drop of saline is pipetted into and out of the vagina, and then transferred to a slide; (2) a small sterile cottonwool plug is moistened with saline, inserted into the vagina with forceps, and then wiped on a slide; (3) a nickel–chromium wire loop, previously heated in a flame and cooled in saline, is used to remove some cells from the vaginal wall by gentle scraping.

Ovulation

The shedding of mature eggs or ova from the ovary is known as ovulation, and may be either spontaneous or induced. Spontaneous ovulation occurs when the Graafian follicle ruptures independently of mating; it takes place at each oestrus, and so the length of the oestrous cycle is related to the time between ovulations. Induced ovulation takes place only after the stimulus of mating, the act of copulation causing the ovaries to release mature eggs. In some rodents evidence of mating may be detected by the presence of a cornified plug in the vagina. In others a vaginal smear will reveal the presence of spermatozoa. Mature eggs may remain in the ovaries for anything up to ten days and then degenerate as the next crop matures.

Embryonic development

The spermatozoon unites with the egg (ovum) and thus the hereditary factors from mother and father are brought together. The fertilized egg (zygote) undergoes subdivision (cleavage) into a number of smaller cells as it passes along the Fallopian tube, and is now referred to as an embryo. A cavity forms within the embryo as it enters the uterus, so that it takes the form of a hollow,

TABLE 1

THE DI-OESTROUS CYCLE OF THE RAT AND ASSOCIATED CHANGES IN THE REPRODUCTIVE ORGANS

STAGE	EXTERNAL GENITALIA		VAGINAL SMEAR				Mating behaviour	Ovary	Uterus
	Vulval area	Vaginal wall	Epithelial cells	Cornified epithelial cells	Leucocytes	Mucus			
I. Pro-oestrus (Fig. 2)	Becoming swollen	Dry	+++	+	±	−	±	Follicles maturing	Becoming distended (fluid)
II. Oestrus (early) (Fig. 3)	Maximum swelling	Very dry and corrugated	−	+++	−	−	+	Follicles mature	Fully distended
III. Oestrus (late) (not illustrated)	Swelling receding	Slightly moist	±	+++	±	−	±	Ovulation	Less distention
IV. Met-oestrus (Fig. 4)	Not swollen	Moist	+	++	++	−	−	Corpora lutea (C.L.)	Fleshy and pink
V. Dioestrus (Fig. 1)	Not swollen	Moist	+	−	+++	little, stringy	−	C.L. regressing, follicles growing	Resting condition

− = none. ± = occasional. + = few. ++ = many. +++ = very many.

often expanded, structure which is called a blastocyst. Attachment occurs with the wall of the uterus and a specialized organ (placenta) is formed, partly from maternal and partly from embryonic tissue. The placenta contains separate embryonic and maternal blood vascular systems. Blood from the mother, passing into the placenta, brings food materials and oxygen for the nourishment of the embryo, and carries away any waste products produced by the embryo. All these substances pass between the maternal and embryonic blood systems, through one or more layers of cells which constitute the 'placental barrier'. This barrier prevents cells passing from mother to embryo, but viruses and some bacteria can get through.

Embryonic development involves first the formation of the various tissues, and then the organs; finally, the main process occurring is growth of all parts. In the later stages the new individual is referred to as a foetus.

Gestation and parturition

The period between fertilization and implantation of the blastocyst (pro-gestation) is usually very short (4–6 days in small rodents), but may be doubled when conception results from a post-partum mating, owing to the inhibiting effect of lactation on implantation. This period may last for many months in those animals (e.g. badger, stoat, seal) that exhibit the phenomenon known as delayed implantation. Gestation or pregnancy is the term applied to the period occupied by the development and growth of the young animal within the uterus of the mother, i.e. from implantation of the blastocyst to the birth of the young animal. During this period enlargement of the uterus takes place and there is considerable development of the mammary glands. The unborn animal (foetus) lies within a sac formed by the amnion and chorion; these sacs contain a considerable body of fluid which protects the foetus from shock and injury.

Towards the end of pregnancy the unborn animal usually takes up a position with its head towards the mother's rear. The birth of the young animal is brought about by strong contractions of the uterine muscle assisted by voluntary contractions of the abdominal muscles. Later the membraneous sac containing the foetus enters the vagina, after which the sac ruptures and part of the amniotic fluid escapes.

For some time the newly born animal remains attached to the uterus by the umbilical cord and the placenta. This cord is normally severed by the mother, and she may eat the placenta and the membraneous sac while cleaning the young animal. Parturition is the term applied to the emptying of the uterus and the termination of pregnancy.

The gestation period varies considerably with different species, and the young are born at varying degrees of development, i.e. guinea pigs are born fully furred, with eyes open and quite mobile, rats are born naked and blind.

In the rat a slight discharge of blood may be observed 10–12 days before parturition.

Lactation

From birth to puberty few changes take place in the mammary glands, but with the initiation of oestrous cycles there is some development in the tissue

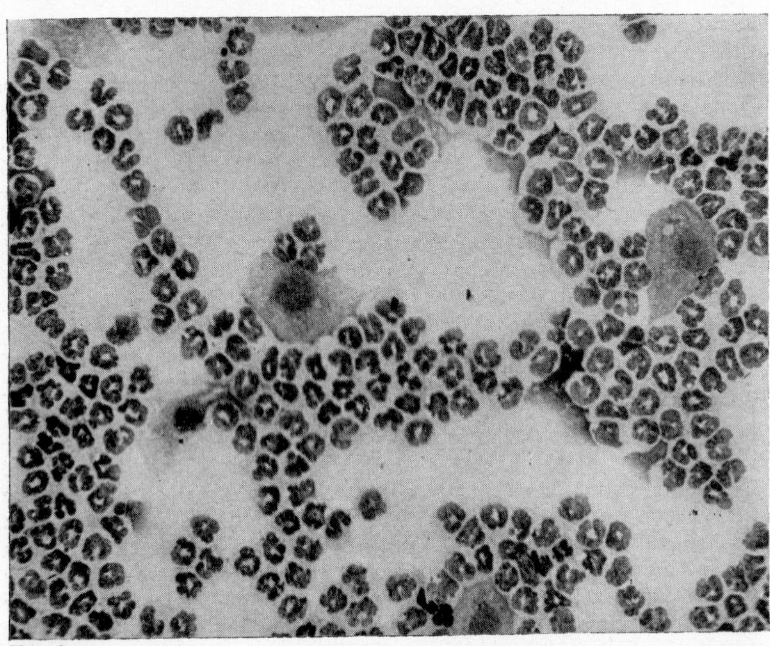

FIG. 1

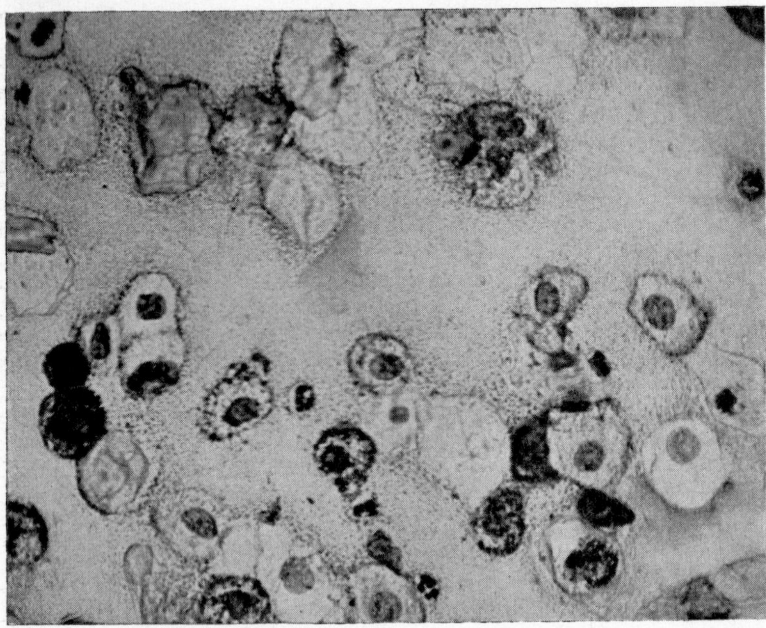

FIG. 2 (see page 346 for captions)

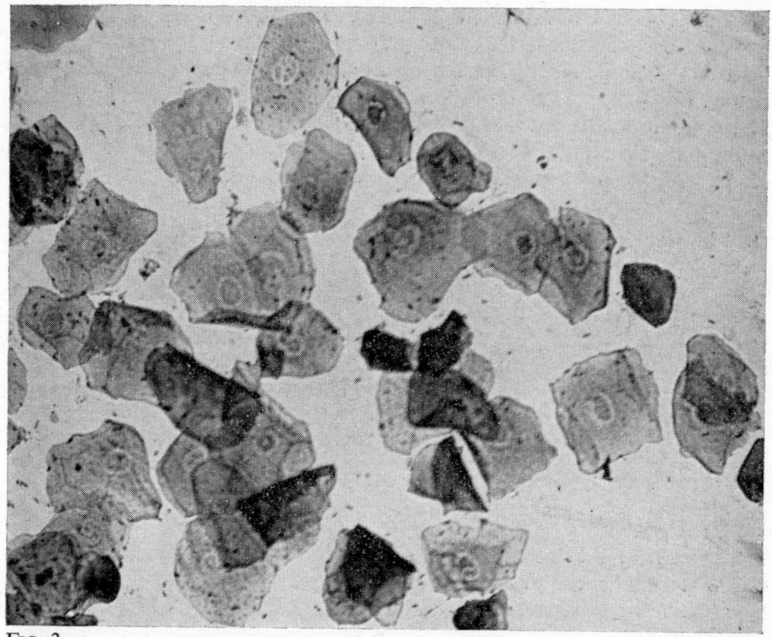

FIG. 3

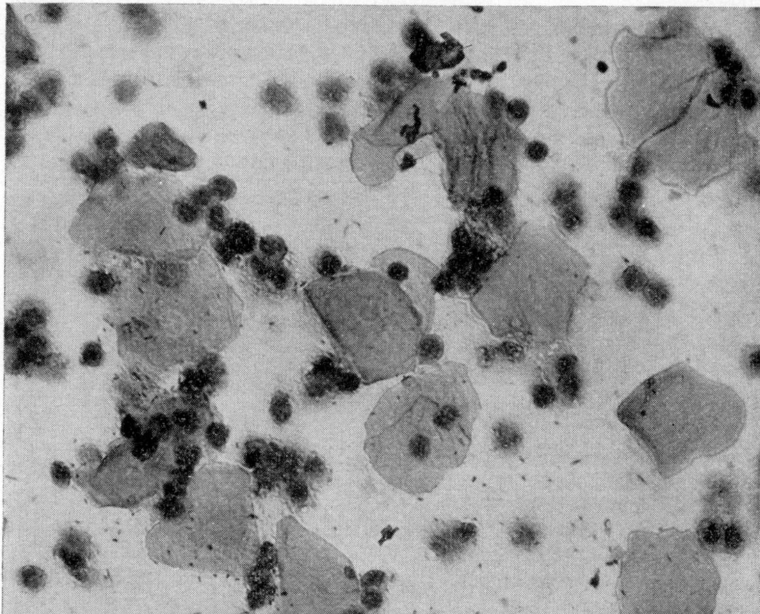

FIG. 4 (see page 346 for captions)

elements directly concerned with milk secretion (alveoli). The major growth, however, occurs in the latter half of gestation, and the alveoli secrete a sticky fluid rich in globulin; this, mixed with milk formed later, constitutes 'colostrum', which is important for the new-born animal as a source of antibodies. The amount of alveolar cell growth occurring in the last part of gestation determines the amount of milk secreted after parturition. Growth of alveoli ceases with the onset of the regular secretory process. After parturition the rate of secretion increases rapidly to a peak and then slowly declines. The process of suckling is essential for the maintenance of secretion. The duration of lactation depends on the state of development of the young at birth and their rate of subsequent growth; thus the lactation period in the rat is a good deal longer than that in the guinea pig.

The composition of milk of different animals varies, and is related to the degree of development of the young at birth, the relative size of the young to the mother, and the immediate post-natal energy requirements of the young. Thus, rat's milk contains 14·9 per cent fat, 2·8 per cent sugar, and 9·2 per cent protein; by contrast, cow's milk contains 3·8 per cent fat, 4·8 per cent sugar, and 3·1 per cent protein.

Hormonal regulation

The growth of the secondary sex organs and the development of the sex characteristics depend on the hormones secreted by the gonads, i.e. androgen by the testes, oestrogen and progesterone from the ovaries. Progesterone is derived from the corpus luteum, which develops out of the ovarian follicle after the ripe ovum is shed. Progesterone prepares the uterus for the reception of the fertilized egg and maintains pregnancy. The placenta also secretes progesterone at parturition and allows a post-partum oestrus to occur in some species. The development of the mammary glands and lactation is also dependent on ovarian hormones. Finally, the secretion of all testicular and ovarian hormones, as well as the production of spermatozoa and ova, depend on the action of gonadotrophins secreted by the pituitary gland.

FIGS. 1–4 THE DI–OESTROUS CYCLE OF THE MOUSE

Reprinted from Journal of Animal Technicians Association, **12.** No 1

1. LEUCOCYTES: Present at all stages of the cycle except when the female is on heat
2. At the approach of oestrus the leucocytes disappear and the smear consists mainly of EPITHELIAL CELLS with marked nuclei
3. OESTRUS: The leucocytes have disappeared. The smear consists of CORNIFIED CELLS ONLY. The female is on heat
4. POST-OESTRUS: The return of the leucocytes among the cornified cells. The female is no longer on heat

(See also Table I, p. 342)

TABLE 2

SYNOPSIS OF REPRODUCTION IN LESS COMMON SPECIES

Animal	Breeding season	No. of oestrus periods	Mechanism of ovulation	Intervals	Length of gestation	Average litter size	Recurrence of oestrus after parturition	Age to wean young
Hedgehog	April–May July–August	Repeated	Spontaneous	—	35–40 days	5	Next breeding season	40 days
Wood rat	No definite season	Repeated	Spontaneous	7 days	32–33 days	2	End of lactation	21 days
Bank vole	No definite season	Repeated	Spontaneous	5 days	20 days	4	Post-partum	21 days
Field vole	No definite season	Repeated	Spontaneous	5 days	20–21 days	3–4	Post-partum	14 days
Orkney vole	No definite season	Repeated	Spontaneous	5 days	20–21 days	3–4	Post-partum	14 days
Swamp rice rat	No definite season	Repeated	Spontaneous	7 days	27 days	4–8	Post-partum End of lactation	19 days
Chinese hamster	No definite season	Repeated	Spontaneous	4 days	20 days	3–7	Post-partum *	24 days
Root vole	No definite season	Repeated	Spontaneous	5 days	20–21 days	4	Post-partum	19 days

* Less regularly than in mice. Only the gentlest of females will permit mating at the post-partum oestrus. (Personal communication, G. Yerganian, Children's Cancer Research Foundation Inc., Boston, Mass.)

TAB

SYNOPSIS OF REPRODUCTI

By Dr A. S. Parkes, F.R.S.,

Re

Animal	Breeding season	No. of oestrus periods in a breeding season	Duration of oestrus	Mechanism of ovulation
Monkey (Macaque)	Probably no definite season	Repeated	—	Spontaneous
Cat	No definite season	Repeated	7–10 days	Sometimes spon taneous but gen erally after ma ing
Dog	Two seasons a year	One	7–13 days	Spontaneous
Ferret	March to July	—	Very prolonged in absence of male	Only after matir
Rabbit	No definite season	—	Very prolonged in absence of male	Only after matin
Guinea pig	,, ,,	Repeated	6–12 hours	Spontaneous
Rat	,, ,,	,,	10–20 hours	,,
Mouse	,, ,,	,,	10–20 hours	,,
Cotton rat	,, ,,	,,	2–3 days	,,
Golden Hamster	February to October	,,	12 hours	,,
Horse	February to August	Repeated	3–15 days	Spontaneous
Cow	No definite season	,,	12–30 hours (in summer) 6–8 hours (in winter)	,,
Sheep	August to March (varies with locality)	,,	30–36 hours	,,
Goat	September to February	,,	2–3 days	,,
Pig	No definite season	,,	3–4 days	,,

Notes

Breeding season. In some animals there is a definite restricted season out of which breeding never occurs. Slight seasonal fluctuation in reproductive activity, as seen in the rat and mouse, does not amount to a true restricted breeding season.

Oestrus and ovulation. The period at which an animal will mate, and at which ovulation

LABORATORY ANIMALS

ɔfessor E. C. Amoroso, F.R.S.

1

nterval between ovulations in ınmated animal during breeding season	Length of pseudo-pregnancy	Length of pregnancy	Usual size of litter	Recurrence of oestrus after young born
28 days	No special pseudo-pregnancy	24 weeks	1	2–3 months
15–21 days	No special pseudo-pregnancy	64–66 days	3–6	4th week of lactation
—	60 days	60 days	3–6	Next breeding season
—	42 days	42 days	6–10	End of lactation or next breeding season
—	14–16 days	30 days	5–10	3rd week of lactation
14–16 days	No special pseudo-pregnancy	62 days	3–5	Post-partum oestrus, then regular cycle
5 days	12 days (after sterile mating)	22 days	6–10	Post-partum oestrus, then not till end of normal lactation
5 days	,, ,,	19 days	6–10	,, ,,
7 days	?	27 days	4–8	,, ,,
4 days	9–10 days (after sterile mating)	16 days	7–9	End of lactation
20–25 days	No special pseudo-pregnancy	11 months	1	Early in lactation
3 weeks	,, ,,	9 months	1	,, ,,
16–17 days	,, ,,	5 months	1–2	Next breeding season
2–3 weeks	,, ,,	5 months	1–2	,, ,,
3 weeks	,, ,,	4 months	8–15	End of lactation

the shedding of eggs from the ovary into the top of the female reproductive tract, normally takes place.

Pseudo-pregnancy. If pregnancy does not occur after oestrus some animals have a period of pseudo-pregnancy, during which changes similar to those seen in pregnancy occur in the uterus and mammary gland.

CHAPTER TWENTY-TWO

Breeding of Common Laboratory Animals

INTRODUCTION

Any system for breeding laboratory animals must take into account the habits and nature of the species concerned; for example, does oestrus occur seasonally or is it a feature of a regular cycle; can males and females be confined in one cage or pen without fighting; is it safe for an adult male to be left in the cage together with its unweaned offspring?

Among laboratory animals, the female will permit mating only when she is in oestrus (or 'on heat', or 'in season'). In some species the male as well as the female exhibits a breeding season, e.g. ferrets; in other species the male remains fertile and capable of mating throughout the year, though the female comes into oestrus and permits mating only for a short period once or twice a year, e.g. dogs; in other species the male remains fertile throughout the year and the female exhibits a continuously recurring oestrus cycle, e.g. rats.

Some animals are content to live and breed peaceably together in a pen or cage, e.g. mice and guinea pigs, while others must be housed separately once they are adult and have not been reared together, or they will fight among themselves, e.g. adult male mice and Chinese hamsters.

When a female of a species having a continuously recurring breeding cycle becomes pregnant the oestrus cycle is suppressed and does not recur until towards the end of lactation, except in those species in which there is a *post-partum oestrus*, which occurs within a few hours of parturition. Females having a post-partum oestrus will permit mating at this time if the male is present. When such matings occur, the females are pregnant while still suckling a litter. In guinea pigs it is desirable that they are permitted to mate at the post-partum oestrus to avoid the long delay which intervenes before the restoration of the regular breeding cycle at the end of lactation. Mice also mate at the post-partum oestrus, and when this has occurred care must be taken to wean the previous litter at nineteen days of age so that the female may be free to care for the new litter.

THE ESTABLISHMENT OF A COLONY

In the beginning, clean, vigorous stock animals must be obtained, and it is customary to import such stock from existing, flourishing colonies. If good-quality stock animals are not obtainable, then such stock as can be had must be improved. Stock may be improved slowly by the same methods as must be applied to the maintenance of a colony—namely the careful selection of breeding stock and the rigorous application of hygenic measures. Advice on sources of new, reliable stock is available from the Medical Research Council's Laboratory Animals Centre at Carshalton.

CHOOSING A BREEDING SYSTEM

Having obtained a nucleus of disease-free animals, they must be isolated from other animals who may carry, and be capable of transmitting, disease. The new colony must therefore be *closed*, that is, closed to the introduction of new stock. This is a minimum requirement for the sake of a healthy stock of animals.

The choice of a breeding system depends on the demands which are likely to be made on the colony and the size of the colony. A large colony can, and probably ought to, accommodate more than one breeding system; a small colony must use one system only if it is to run economically. Before deciding upon a system, the uses to which the animals are to be put must be considered and the following questions answered:

Will the demand be for—

(i) suckling young, with or without their dams; will these dams be lost or will they be available to produce further litters;

(ii) weanling and young animals, or adults;

(iii) closely inbred animals of known relationship with one another, or large numbers of randomly bred animals?

Whatever system is adopted for the production of experimental animals—and the larger the number of such animals needed, the simpler that system will have to be—it should not be forgotten that the section of the colony which is to provide future breeding animals must be closely recorded.

Breeding colonies often consist of two parts, a 'master' colony, which is carefully recorded and controlled, and a 'production' colony, which is arranged to produce large numbers of animals with the minimum of record keeping. The master colony supplies breeding animals for the production colony, and no new breeding stock is laid down from the production colony.

When breeding small laboratory animals the aim is, usually, to produce either an inbred colony—which is achieved by mating brothers and sisters—or a random-bred colony—which is achieved by mating animals which are not closely related.

Inbreeding (Pure Lines)

If pure lines are maintained it must be accepted at the outset that accurate and extensive records must be kept, and that every animal in the colony must bear an identification mark so that it may be distinguished as an individual. There must be no doubt about the validity of records or the identity of animals. If there is ever uncertainity on either point, then the only safe course is to slaughter the animals concerned.

A line is said to be pure when it has been brother–sister mated for twenty consecutive generations. At this stage the animals in a well-run colony will be nearly homozygous and almost identical genetically. Brother–sister mating must continue to be practised after the twentieth generation because: (i) there will probably be some residual heterozygosity in the colony, and (ii) mutations will occur which, in themselves, create new heterozygosity.

If only one pair of animals is kept from each generation a 'single line' strain is produced. But more than one pair of animals must be kept to guard against complete loss of the line and to produce animals for experimental purposes. Any one pair could give rise to, say, five pairs of offspring, and it would be possible to continue breeding these five pairs as five single lines (i.e. retain one pair from each generation from each of the five lines). These would be called 'parallel lines'. Within a few generations there could be considerable genetic variation demonstrable between animals arising from the five separate lines. Therefore it is not valid to assume that all the animals in a colony are genetically similar merely because they all arise from brother–sister matings. The colony may, in fact, consist of a number of small colonies of differing genetical make-up which have arisen from parallel-line breeding programmes. This situation is illustrated in Fig. 1, where each circle represents one brother–

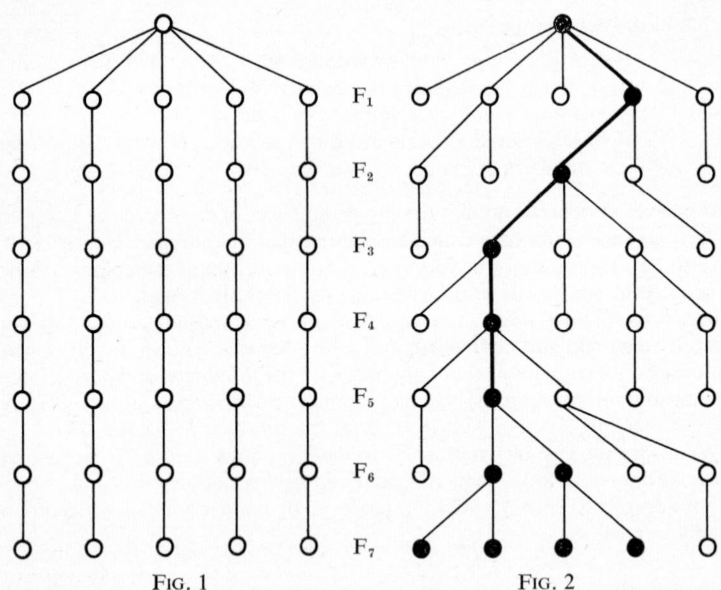

FIG. 1 FIG. 2

Breeding plans to illustrate parallel lines (Fig. 1) and a pure line (Fig. 2).

sister pair. Parallel lines must not be permitted to continue if the colony is to remain genetically uniform. Obviously, parallel lines will always be present in the colony (if only to produce a sufficiently large number of animals), but one only of the lines must be allowed to expand for the continuation of the colony and all other lines must be halted. It is customary to curtail all but one of a group of parallel lines after three generations. Such a breeding plan is shown in Fig. 2, where the line would continue through one of the solid circles in generation F_6 and one in F_7.

Random Breeding

Here the aim is to retain all the variables originally present in the colony and to keep these variables equally and evenly distributed throughout the colony. Closely related animals, e.g. brothers and sisters or cousins, are never mated, and thus close inbreeding is avoided. Care has to be exercised to ensure that new stock is not laid down exclusively from a few sections of the colony while other elements are allowed to die out. To guard against such unintentional selection it is necessary to devise a mating scheme which includes all sections of the colony equally, and which precludes the chances of close inbreeding or of repeating similar crosses in successive generations. The scheme should be one that is easy to understand and operate, and does not require complicated record keeping.

It is advisable to have not less than eight sections in a colony, and preferably more. All the sections should contribute equally to the breeding stock of the next generation, and none of the sections be allowed to die out. In a large colony each section will consist of many breeding pairs to allow enough animals to be produced for experimental purposes. This also allows some selection of good breeding stock within each section. In a small colony each section may consist of a single breeding pair, and in this case it is advisable to maintain the colony by not less than twelve pairs, because this is the smallest colony which could be expected, with careful management, to retain a considerable degree of heterozygosity over a number of years.

It should be appreciated that a random-bred colony will not remain genetically stable however carefully and thoroughly the matings are controlled. The colony remains closed (i.e. no new animals are brought in) and, in time, a closed colony loses some of its heterozygosity, the animals becoming more and more similar genetically. The smaller the colony and the less the control kept over matings, the quicker will be the loss of heterozygosity.

Harems

A common method of breeding animals is to allow groups of one male and from two to twenty (depending upon the species) females to run together in one cage or enclosure all their breeding lives. Mating takes place when the females come into oestrus, and the young are born and reared in the presence of the other members of the harem. Not all species will breed under these conditions, but the method is commonly used for mice and guinea pigs. Both these species have a *post-partum oestrus*, which occurs within a few hours of parturition, at which they will mate, so that the female is pregnant while she is suckling the previous litter. The number of females in any harem group is limited by the capacity of one male, the size of cage or pen available, or the habits of the animal; e.g. mice like to build nests in corners, and as each box has only four corners, it is advisable to group mice into harems of one male and not more than four females.

A common variation on this method is to mate the animals on the harem system, but to segregate the pregnant females just before parturition and permit them to give birth to their young in the absence of the male, but in the presence of other pregnant or parous females. Unless the females are segregated to litter in individual cages, it is not possible to keep detailed records of the breeding performance of the individuals in a harem. When females are allowed to litter together only the performance of the group as a whole may be recorded. For some purposes such limited records may be all that is required, but even these records are rendered useless if the harem groups are split up or the males moved round from group to group.

ADVANTAGES OF THE HAREM SYSTEM. Little space is required for the production of relatively large numbers of animals. The number of young from each female in any period will be maximal if advantage is taken of the animals' willingness to mate at the post-partum oestrus.

DISADVANTAGES OF THE HAREM SYSTEM. Some new-born animals may be smothered (overlaid) by older, unweaned young or adults. This disadvantage is more often observed in mice if the elder litter is not weaned, than in guinea pigs, presumably because the new-born guinea pig is a relatively robust creature. Overcrowding can contribute to losses from this cause. It is difficult to keep records of individual breeding performance unless the females are isolated for parturition.

Monogamous Pairs

Here one male and one female are mated and remain together all their breeding life, the female, in some cases, littering in the presence of the male and mating at the post-partum oestrus. The system is commonly used for the production of *inbred* strains of rats and mice. Inbred strains are those in which only brothers and sisters, or—less commonly—father and daughter or mother and son are mated.

Complete records of individual breeding performance must be kept, or such a colony cannot be utilized to best advantage.

The disadvantage of the system is that a large number of males has to be maintained and, because the pairs are housed individually, much space, labour, and equipment is needed even for a small colony. Record keeping must be accurate and detailed, and therefore takes much time.

The selection of breeding stock

It has been said that there is more art than science in choosing good breeding animals. However this may be, it is certain that the choice should not be left to the eye alone.

Breeding stock is never taken from the first litter from any female. This litter is often smaller, both in number and size, than the second litter, which is usually the best litter from any female. Further, delaying the task of selecting the parents of the next generation until a female throws a second litter provides an opportunity to consider the progress which has been made by the members of the first litter.

The selection of breeding stock begins in a negative manner by the rejection of obviously unsuitable litters or of apparently suitable litters thrown by unsuitable parents. The grounds for eliminating litters are:

(1) Visible physical abnormalities or deformities.

Note should be made of any litter exhibiting any deformity, and a careful watch kept among related litters for the appearance of the defect.

(2) Litters which are—

(i) small in number and/or size;
(ii) comprised of *one sex only*;
(iii) 'uneven', that is, comprised of animals of widely different sizes and body weights.

(3) Litters, either of whose parents have developed some abnormality, e.g. a tumour; or, when brothers or sisters, in a previous litter, have developed some abnormality.

The grounds for eliminating a female are:

(i) showing any sign of viciousness, such as unwillingness to permit her litter to be handled by a technician with whom the animal was familiar;
(ii) producing litters showing any of the signs listed above;
(iii) the maiming of any of the litter by biting; any neglect of the litter, such as a habitual failure to return straying young to the nest or failure to keep the young clean.

In fairness to the animals it must be remembered that females who neglect their litters usually do so because they are unable to suckle the young, and failure in lactation may arise from inadequate nutrition of the mother.

The process of elimination outlined above cannot be done without reference to records which have been kept in some detail.

The second step is to choose new breeding stock from the remaining litters, and here the experience of hand and eye is needed. The animals should appear clean, no 'gummy' eyes, running noses, dirty paws or tails, and no sign of diarrhoea. The fur should be thick, sleek, and glossy, not 'staring'. The animals should appear alert, active, and inquisitive; and, when handled, should feel to be firmly fleshed, not thin and fragile.

The number of animals set aside as new stock should be in excess of the number thought to be needed. This allows for losses from accidents and from animals which prove to be infertile. Furthermore, if there is an unexpected increase in the demand for animals there will be sufficient stock available to help meet that demand. Excess stock can be killed later, but it is a wise precaution to carry as large a stock as possible through to rearing a first litter.

Where several lines, or sub-strains, are carried in one colony care must be taken to allow space and cages for new stock from each line. If this is not done a line may inadvertently be squeezed out of existence merely because there is no accommodation for it. On the other hand, there is little point, in an ordinary breeding programme, wasting time, space, and trouble encouraging a weak line to thrive when its place could be better filled by the expansion of a more vigorous line.

The Chapter on p. 446 gives worked examples showing the calculations applicable to the problems of producing definite numbers of animals of specified ages.

LABORATORY ANIMALS BREEDING DATA FOR FEMALES

SPECIES	Average age generally paired	Average weight at pairing	Mean duration of pregnancy	Average number born	Age of litter at weaning	Average weight of young at weaning	Period between parturition and next possible mating
SEXES WHICH USUALLY LIVE TOGETHER							
MOUSE (*MUS musculus* L.)	6 weeks	18–20 gm	19–21 days	8–11	21 days	10–12 gm	Post-partum oestrus
RAT (*RATTUS norwegicus*)	70–80 days	150 gm	20–22 days	9–11	22 days	35–40 gm	Post-partum oestrus
COTTON RAT (*SIGMODON hispidus*)	6 weeks	80–100 gm	27 days	4–8	22 days	35–40 gm	Post-partum oestrus
GUINEA PIG (*CAVIA porcellus* L.)	12 weeks	500–550 gm	65–72 days	3–4	14 days	180–200 gm	Post-partum oestrus

FEMALES TO MALES AT OESTRUS ONLY							
GOLDEN HAMSTER (*MESOCRICETUS auratus*)	6 weeks	100 gm	16 days	5–7	21 days	40 gm	End of lactation (28 to 32 days post parturition)
CHINESE HAMSTER (*CRICETULUS griseus*)	8–12 weeks	35–40 gm	20–21 days	4–8	25 days	6–8 gm	Post-partum oestrus
RABBIT (large) (*ORYCTOLAGUS cuniculus* L.)	9 months	2,500–3,000 gm	32 days	6–8	8 weeks	1,500 gm	3rd week of lactation
RABBIT (small)	6 months	1,500–2,000 gm	30 days	6–8	6 weeks	1,000 gm	3rd week of lactation
FERRET (*MUSTELA putorius furo* L.)	9–12 months	750–800 gm	42 days	6–10	8 weeks	450 gm	End of lactation or next breeding season
DOG (*CANIS familiaris*)	14 months	Variable	60 days	4–8	8 weeks	Variable	Next breeding season
CAT (*FELIS catus* L.)	7–9 months	2,500 gm	64–66 days	3–6	6 weeks	700–800 gm, but variable	4th week of lactation
MONKEY (*MACACA mulatta*)	F. ♀ 3 years	4·5–9 Kg	24 weeks	1 only	3 months	700–800 gm	3 months
MONKEY (*MACACA mulatta*)	M. ♂ 4 years	6·75–11 Kg					

See following notes for further details

RECORDING

Breeding-stock animals must be marked individually in such a way that there can be no doubt as to any animal's identity, and there must be no duplication of numbers within any one generation of any one strain in any one room. Though it is desirable that no identity number should ever be duplicated among living animals, this is possible only in slow-growing animals big enough to be tattooed with long numbers. In a large colony of small, fast-growing animals, such as mice, this is next to impossible.

The information recorded is determined by the breeding system in use and the requirements of the establishment, but some recorded information is essential for the conduct of a successful breeding programme. Each animal must be so recorded that its ancestry may be traced back to the origin of the colony.

The minimum of information needed about any breeding female is:

(i) its identity, parentage, and date of birth;
(ii) date of mating and identity of male used at each mating;
(iii) dates of birth of litters, and the number and sex of the offspring;
(iv) the fate of the litters (e.g. to breeding stock or issued for experiments);
(v) the cause of death (post-mortem record cards should be attached to the animal's stock card).

And about any breeding male is:

(i) its identity, parentage, and date of birth;
(ii) the dates on which it is mated and the identity of the females;
(iii) the cause of death.

This amount of information is adequate to record the breeding performance and to trace the ancestry of any animal.

Records are often kept in books, but it is becoming increasingly popular to keep a separate record card for each animal. Cards may be sorted and compared (and lost!) more easily than the pages of a book. The information which is needed most frequently—the animal and its cage number—should be written boldly in the top right-hand corner of the card. Each litter is best recorded on a separate card, which is later filed behind the mother's stock card, on which is entered only a précis of the information about each litter. This is desirable for two reasons: (i) litter cards must be kept in the animal room near the relevant cages, and thus are liable to become damaged and soiled, and (ii) they carry much day-to-day information which need not be entered fully on the mother's permanent record card. The litter card should show:

(i) the mother's identity and cage of origin;
(ii) date of birth of litter;
(iii) numbers and sex born;
(iv) date of weaning;
(v) number, sex, and body weight of weanlings (individually or as a group);

(vi) the fate of the litter, e.g.

to breeding stock (quote identity numbers);
to reserve stock (quote identity and/or cage number);
to experiment (quote name of experimenter and nature of experiment, and identity numbers if known).

Space should be allowed for day-to-day notes on the progress of the litter.

CATS

It is now firmly established that cats can be bred successfully under laboratory conditions and litters produced in each month (Lamotte and Short,[16] 1966). Provided the female is on heat, there will be no difficulty in mating her even in cramped conditions. Queens may be moved to other buildings to be mated, but toms will take some time to adjust to new surroundings.

The female will indicate the onset of oestrus by calling, rolling, and a posture of flattening the back, and elevating the hind legs, accompanied by a treading movement.

One of the most important requisites for reproduction is an adequate diet with sufficient amounts of animal protein, calcium, iodine, and Vitamin A (Payne, Seamer, and Short,[17] 1966; Scott and Dickinson,[18] 1956). Another important requisite is the provision of warm, well-lit, clean, and comfortable accommodation, though an expensive building is not necessary. One of the most important aids to success in cat breeding is the selection of trained, interested, dedicated animal technicians. The whole success or failure of a cat colony can be dependent on these people. The cat is polyoestrus, showing recurrent oestrus at fourteen-day intervals lasting about three to six days, but if the cat is not mated, can last up to ten days. Oestrus behaviour and calling may occur one or two days before the female will accept the male. After accepting him, mating may occur several times a day for 3–4 days.

Oestrus behaviour in a female may continue after she will no longer accept a male. Three or four services should be sufficient for a female at her first mating, and two should suffice at subsequent matings.

The average length of gestation in the cat is sixty-four to sixty-six days, but actual length of individual cats vary from fifty-seven to seventy days, and there is a variation in days from litter to litter. This wide variation is probably because cats can mate on three or four consecutive days.

Pseudo-pregnancy

This is not nearly so common in cats as in bitches. A sterile mating will cause this condition, which usually lasts about 23 days after copulation. Scott *et al.*[1] had the impression that sterile matings resulting in pseudo-pregnancy can occur in a cat colony when a male is running with the females, especially during the winter months, even when the male is known to be fertile.

A close watch should be kept on lactating females for signs of oestrus. Fertile mating can often be effected at this time, inducing three pregnancies in a year. An average litter size of about four kittens per litter can be expected.

The average length of lactation is seven weeks, and weaning can be started during the fourth or fifth week after birth. Mother's milk must be substituted during the fourth week and kittens encouraged to drink.

RABBITS

Pregnancy in the rabbit usually lasts some thirty-one days. Several factors influence the duration, such as the size of the litter and the weight of the female. Young born between twenty-eight and thirty-four days after mating usually survive. But these extremes are rare. Rabbit milk is yellow and viscous, and contains a much higher percentage of total solids than cow's milk. A formula for a semi-synthetic milk for the hand-rearing of suckling rabbits is given in the chapter on Pathogen-free Animals.

Different breeds will reach sexual maturity at different ages; factors influencing this are size of breed and nutrition. In general, the larger the breed, the longer it takes to reach sexual maturity. Small breeds take about five months, and large breeds about seven months. The average fertile mating to infertile in most rabbitries is about 60–40.

Young can be weaned at six weeks of age or earlier, but whether the young are weaned or not, the mother should be mated at about four weeks after parturition. If the litter is weaned at six weeks the doe will have two further weeks with them and two weeks before the next litter is born.

The doe should always be taken to the buck's cage for mating. Taking the buck to her may result in fighting. Some bucks refuse to mate in strange surroundings. Mating will usually occur in a few moments and should always be observed.

Does may be palpated for pregnancy at about twelve days. Practice is needed to become proficient at this early stage. The following method is recommended: place the rabbit on a table, hold the ears with the right hand, slip the left hand under the tummy of the rabbit, and with the tips of the fingers gently press the abdomen just in front of the pelvis. The embryos if they are present can be felt, and at twelve days will be the size of small marbles. To palpate successfully both the animal and the handler need to be completely relaxed.

Young rabbits from mothers who die or from litters of high numbers can be fostered. Fostering should be carried out within ten days of birth and should be within three days in age of the litter to which they are being fostered. When transferring the young into the nest of the foster doe it is advisable to remove the doe for a short while and keep her occupied with a titbit. It may also be necessary to mask the smell of the new young in her nest as suggested by Sandford[2] (1957). The doe's sense of smell should be changed by rubbing a little 'Vick' ointment or paraffin on her nostrils and forelegs.

Pseudo-pregnancy

It is only when the doe ovulates, but the eggs are not fertilized, that pseudo-pregnancy occurs. The stimulation which causes the doe to shed eggs is usually the behaviour of the buck, but may be caused by does riding each other.

Pseudo-pregnancy continues for fourteen to sixteen days; during this period the doe will behave as though pregnant, i.e. the mammary glands are stimulated to activity and the uterus increases in size. This change is caused by hormones from the corpora lutea. Pseudo-pregnancy may be prevented by letting the doe be mated twice within a period of 5 hours.

The Moult

The rabbit coat will have attained its normal, adult state at the age of 6–8 months, according to the breed. Moulting is a continuous process, but most rabbits exhibit one main and conspicuous moult each year—usually in the autumn, but the timing is variable and there are many exceptions. A normal moult starts at the head and proceeds backwards and downwards. The flanks (the area round the tail) and the belly are the last places to clear of moult.

The severity of loss of hair and the speed of its replacement by new hair vary considerably. The process may be fast and be completed within a fortnight or so, or it may be slow and extend over several weeks. A second, main moult (double moult) may occur immediately after the first has been completed. A few rabbits will exhibit further moults in one year.

The type of hair, its rate of growth, and the duration of the moult are determined by the inheritance of the rabbit, but superimposed on these characteristics are the effects of diet and environmental temperature. Hair is composed of protein, so the growth of a complete new coat increases the requirement for dietary protein.

Other types of moult occur, e.g. in the pregnant doe the hair of the chest and belly loosens due to the effect of hormones secreted during pregnancy. This effect also occurs in pseudo-pregnancy.

GUINEA PIGS

The period of gestation in this animal is determined, to some extent, by the litter size. When the litter conceived at a post-partum mating contains only a single offspring the period of gestation may be as long as seventy-two days. Rowland[3] (1949) reported gestation periods ranging from sixty-two to seventy-two days; the average, when the litters were conceived at post-partum mating, was sixty-eight days. Bruce and Parkes[4] (1948) found that the percentage of fertile post-partum matings was 74 per cent, but it is suggested by Rowland that the occurrence of post-partum mating is adversely affected if the proportion of females to males exceeds 12–1, but it is considered not unlikely that floor area may be a factor in this connexion. Thirteen animals may be kept in a floor area of about 25 sq ft. If the number of animals housed together is increased, say to thirty-six females and three males, then one male will become dominant and will attempt to mate all the females, to the exclusion of the remaining males. By utilizing the post-partum oestrus, the average number of young weaned from each sow in a year has been reported to be as high as 18·9.

COTTON RAT

Short[5] (1957) recommended the monogamous-pair system of breeding cotton rats, but suggests that it is possible to rotate one male among several females. Sexually mature females of this species can be extremely pugnacious, and may inflict fatal damage to the male. To prevent this occurring early pairing at six weeks of age is recommended.

The provision of a nest box or plenty of nesting material is important when young are due to be born so that the mother may retire from the presence of humans.

HAMSTER (GOLDEN)

The hamster has a four-day oestrus cycle, Orsini[6] (1961). The obvious indication of this cycle is the appearance of the post-oestrus discharge on the morning of *day-2*; i.e. the morning subsequent to the night in which the female is in heat.

This discharge may appear protruding from the vaginal orifice when the animal is picked up, or may surge forth from the apparently sealed vagina in response to slight pressure at the sides of the vaginal orifice. This post-oestrus discharge is extremely regular in its occurrence on each fourth day in the cycling animal. It is a thick, opaque-white, mucous which is very viscous, adhering to the finger and 'stringing out' from the vagina from 2 to 8 in. as the finger is removed. Later in *day-2* the inner portion of this vaginal discharge thickens and becomes less stringy, beginning to resemble a whole waxy plug. On *day-3* many of the animals will show a distinct waxy plug; this is non-mucous and can be expressed. It is not always present, and may be lost prior to examination. On *day-4* there is no special discharge characterizing the day; if the waxy discharge of day-3 has not been extruded it may be found at this time or the vagina may be moist with serous fluid. *Day-1* is the day the animal comes into heat, i.e. the day prior to the recurrence of the post-oestrus discharge, which first manifests itself as a slight, translucent secretion and is not always present on day-1. Oestrus itself extends from the evening of day-1 into the morning of day-2.

HAMSTER (CHINESE)

Post-partum oestrus

Yerganian[7] (1962) reported post-partum oestrus in this animal but added, 'only the gentlest of females permit such absence, no matter how brief, from their newborn. Advantage could be taken of females who allow handling and absence from their newborn at post-partum as a method of selecting future stock with greater docility.'

Management

Sexually mature females of this species are extremely pugnacious, and if the sexes are not separated immediately after mating the females inflict considerable damage to testes and tails of the males. Mating generally takes place during the late evening, so it is necessary for someone to observe the mating and remove the male. To overcome this inconvenience Yerganian[8] (1958) described a reversed-lighting scheme of 11 hours of darkness starting at 7.0 a.m. By 9.0 a.m., 2 hours later, adult females were entering their peak of oestrus and matings are conducted routinely at this time. Females displaying hostility towards their mates could be readily separated before inflicting injury. Sufficient light with which to work during the dark period could be provided by low-voltage bulbs (15–20 watts). A reversed-lighting scheme may be set to any timing cycle that is convenient for individual workers, so long as mating trials are conducted within 2 hours of the lights being turned off.

Mating

When the sexes are put together repeated copulatory attempts by the male and lordosis in the female can be witnessed within seconds. When mating is effective the male retains his hold and forces the female to fall to one side. If the male has been successful the female may suddenly turn and attempt to bite his scrotum. Both sexes will rest for a while before repeating the mating pattern. In time the male fails to respond to the female's desire to continue the mating. The female will then become aggressive and attack the scrotum or tail. This is the time to part the pair.

MONKEYS

Monkeys may be bred in the urban animal house without special feeding, Short and Parkes[9] (1949). One of the secrets is the age of the male. Males mature later than females, and should be at least four years of age; from six to fifteen is their full reproductive age. Van Wagenen[10] (1950) reports that the first menstruation occurs in females around the second birthday, while the onset of the growth and development of the testes is seen about the third birthday or early in the fourth year.

Of primary importance is the day-to-day record of the menstrual history of each female. The same person should make a daily inspection of all females in the colony at the same time each day for the purpose of observing the perineum of each animal (the perch on which she sits may give a clue), and a record of bleeding or not bleeding should be made. With the menstrual records at hand the females may be allotted to the male cages on the eleventh or twelfth day of their cycle. The first day of menstruation is day-1 of the cycle.

Monkeys could be mated at noon each day and removed at noon on the first or second day; this places the animals together at a time when there is no immediate competition for food.

Rectal palpation is necessary to determine early pregnancy, because the Macaca mulatta almost always menstruates once after conception (implantation bleeding). The bleeding is often delayed for a few days; that is, the conception cycle is longer than the menstrual cycle characteristic at the same time for that particular monkey, and the implantation bleeding is also longer.

Together these two signs enable one to diagnose pregnancy around the twenty-third day, but palpation of the uterus is needed to confirm.

Pregnancy in the Macaca mulatta is around six sexual cycles, about 164 days. Young have been born between 147 and 180 days.

FERRETS

The ferret has a well-defined breeding season of about six months of the year which in Great Britain extends from March to August. Signs of the onset of the season are seen first in the males (hobs) about the middle of January. The testes begin to enlarge and descend, and by the end of February they are eager to mate. The onset of oestrus in the females (gills) is easily recognized by the swollen state of the vulva. The extent of the swelling is variable, and in the absence of mating this swollen condition will persist throughout the breeding season, because ovulation does not occur spontaneously.

As a guide, mating should occur at about the fourteenth day from the onset of the swelling of the vulva. Copulation may vary from 10 minutes to 3 hours, but the average is about 1 hour. The vulva should be examined about seven days after mating and should be shrinking rapidly if ovulation has occurred. If the condition remains unchanged the female should be remated.

The gestation period is forty-two days, and there appears to be little variation whether the litter is large or small. Females may be palpated for pregnancy at three weeks. Pseudo-pregnancy occurs after an infertile mating and will last forty-two days.

The young can be weaned from the mother at six weeks old. Mothers resent interference with their nests during the first fortnight after giving birth, and it is not practicable to count the number in the litter at birth.

Females will come into oestrus about sixteen days after weaning. It is not uncommon in the early part of the season for a female suckling a small litter to come into oestrus during the early part of lactation. According to Grinham[11] (1952), when this condition is observed the female should be mated lest the oestrus condition interferes with the normal course of lactation.

DOG

Bitches come into season for the first time when they are about eight months old, and the season recurs every six to eight months, usually in the spring and autumn. (Basenjis are unique among dogs in having only one season each year, in the autumn.) Bitches are not full grown at eight months old, so it is customary not to mate them until their second season, when they are more nearly mature.

The approach of the season (or oestrus period) is marked by a slight swelling of the vulva. The bitch ovulates spontaneously, and this swelling, which becomes more pronounced during oestrus, subsides at the end of the season whether or not she has been mated. The endometrium is shed at the beginning of oestrus, causing the bloody discharge from the vagina. The duration of the season is twenty-one days.

Most bitches will accept the dog only from about the tenth day until the end of the season, though the dog remains fertile and capable of mating throughout the year. Most successful matings take place from the tenth to the fourteenth day of the season.

The gestation period is sixty-three days, though variations of several days in the length of this period are not uncommon. Pups are usually weaned at the age of six weeks, and should not be left with the bitch longer than eight weeks, when she will have stopped lactating and will be worried by the continued presence of the pups.

Pseudo-pregnancy

This may occur in unmated or unsuccessfully mated bitches. Its duration is the same as that of pregnancy (sixty-three days). During this time the mammary gland develops as in pregnancy, and the bitch may prepare a bed in which to litter. At the end of pseudo-pregnancy this behaviour ceases and the mammary gland regresses to its normal state.

MICE

Most mouse breeders accept the traditional idea of a four to five-day oestrus cycle which is both regular and spontaneous. Bruce[12] (1962) reports this is only true if a male is present. If female mice are housed singly the cycle is longer (five to six days) and more irregular: if they are housed in small groups there is a natural suppression of oestrus, with an increase in the number of spontaneous pseudo-pregnancies: if they are housed in large groups (thirty females per box) they may become anoestrus for weeks on end.

The introduction of a male initiates a new cycle, so that when grouped females are paired with stud males oestrus is synchronized and the majority of females mate on the third night of pairing. Only if females have been maintained singly prior to the introduction of the male are matings spread fairly over the first four or five nights.

Ross[13] (1961) outlines three systems of management to ensure matings on particular days.

(i) Exposure of female mice housed in a stock box to caged males for two days prior to pairing, thus achieving peak matings on first and second night after pairing with stud male.

(ii) Confining the stud male behind a perforated metal division in a small box for two days before release to the females, resulting in a peak mating on the first night after release.

(iii) The use of intact males caged, or castrated males free, to ensure short regular oestrus cycles, thus increasing the number of females on heat available for pairing on demand.

It is desirable when breeding mice to select for high productivity, that is for the largest number of healthy offspring per female in a given time.

Carter[14] (1951) suggested that the measure of productivity be the number of mice born and weaned divided by the number of prenatal days. This index will be less than 1–0, and occasionally less than 0·1.

Lane-Petter *et al.*[15] (1959) proposed using Carter's index multiplied by 100 as a measure of productivity, and calling this index *Q*. It was suggested that *Q* should be called the index of productivity and to calculate it cumulatively from the date of the first exposure to the chance of mating to the birth of successive litters up to the last.

RATS

The requisites for successful rat breeding are a healthy, vigorous, highly productive, properly recorded stock, possessing good mothering ability; and draught-free buildings with temperatures of around 72°F (22·2°C) ±2° and relative humidity of 55–65 per cent. This degree of humidity will prevent ringtail. Lighting is important, and 12–14 hours of light per day is necessary. Rats are easy to handle if treated properly, and a careful, quiet, sympathetic technician is most desirable.

The mating systems mostly used are monogamous pairs and harems. The first is operated by putting a pair of animals together for the whole of their productive life. This includes the female's gestation and lactation periods and will take advantage of the post-partum oestrus.

The second method is putting one or two males with a number of females (e.g. two males with ten females), removing females when pregnant and replacing them after their litters have been weaned. Five to six litters is the maximum for each female.

Sexual maturity is reached at about 72 days, but it is advisable to delay mating until about 90–110 days of age. The gestation period is 21–23 days and the length of the oestrous cycle is 4 days. The stages of this cycle are described in the chapter on Mammalian Reproduction.

In mesh cages, females with litters require sufficient bedding to enable them to cover up their young.

Observed Matings

It is often necessary to know whether or not rats have mated so that the parturition date or the number of days duration of the pregnancy may be known. The following method for observing matings is simple and adequate for most purposes. A pair of rats is mated in a screen-bottomed cage. Paper (preferably black paper) is spread under the cage 1 or 2 inches clear of the floor grid. The following morning the paper is inspected for the presence of a 'mating plug' (copulation plug). The plug is a hard, white or yellowish, irregular-shaped mass about $\frac{1}{2}$ inch long and $\frac{1}{4}$ inch in diameter. Sometimes the plug has not been shed and may be seen in the vagina. The absence of a plug is no proof that the animals have not mated, for rats will eat the plug if it does not fall through the mesh. If a screen-bottomed cage is used and the paper is inspected carefully for whole and for part-eaten plugs only about one in fifty matings will pass unrecorded.

Ten to twelve days after the plug a discharge of slightly blood-stained mucus may be seen in the vagina. This is replaced, usually on the fourteenth day, by a thick, brownish-red discharge, which diminishes, and, finally, disappears over the next few days.

Most rats have a gestation period of twenty-two days, but strains do show variations. Observed matings will quickly reveal the length of the gestation period in a particular colony.

Fostering

Newly born rats are usually left undisturbed for the first two days of life; however, most rats will permit a known technician to investigate the litter when the young are an hour or two old. If this has to be done it is easiest to remove the doe from the cage first. The litter may then be examined and returned, tidily, to the nest, after which the doe is returned to the cage. Fosterlings may be added to a litter at such a time. They are usually marked by snipping off the tip of the tail, as earmarking is impossible in the very young rat. Some experimental procedures require litters to be transposed between two mothers, and in this case the foster litter is wrapped in the nesting material of the original litter. Such cross-fostering may be carried out with impunity. Fostering by adding pups to an existing litter is most successful where there is little difference in age and size between the original litter and the fosterlings. When a whole litter is to be fostered and the original litter sacrificed a new-born (or hysterectomy-derived) litter may be given to a doe which was suckling a litter of her own of up to four days of age.

It is often recommended that fosterlings are left in their new nest for a period of up to half an hour before the foster-mother is readmitted to the cage, so that the new pups may take on the odour of the original litter. It has also been suggested that the doe's own sense of smell should be confused by dabbing her fore-paws or head with some non-toxic, strong-smelling substance, such as oil of cloves or even perfume. It is doubtful if either procedure enhances the chance of survival of the young. It is far more important that the fosterlings are vigorous and that the manipulation is done by someone who is familiar to the does and who causes them the minimum distress, for rats will tolerate a surprising amount of interference from a human being to whom they are accustomed.

POULTRY BREEDING

Poultry is an important laboratory species especially in the field of genetics, virology and biology, so it is essential that the animal technician has some knowledge of breeding, egg production and incubation.

SYSTEMS OF BREEDING

Cross-breeding

Cross breeding refers to the mating of two distinct breeds. It is normally practised by commercial breeders for the production of laying stock and table birds.

It results in the combination of many different genes, and since desirable characters are usually dominant, they express themselves in the progeny of first crosses. Consequently the progeny of cross breeding are frequently superior to their parents because they inherit and express the more desirable qualities possessed by sire and dam.

Line-breeding

This term denotes the restriction of mating to one line by descent, the progeny having common ancestors. It is a form of inbreeding designed to avoid the mating of very closely related individuals, e.g. mother and son, father and daughter.

Inbreeding

Inbreeding is the mating of closely related individuals, e.g. brother to sister, father to daughter, or mother to son. Inbreeding brings together both desirable and undesirable characteristics. Many undesirable characteristics are recessive, but during inbreeding these characteristics combine and find expression in the progeny and can be eliminated only by selection.

Out-crossing

Out-crossing is the mating of individuals from different families of the same breed. It is the most commonly practised system of breeding because it is the simplest. Its prime purpose is to avoid the ill effects of inbreeding, but bringing new blood from different strains every year can lead to trouble.

METHODS OF MATING

Flock Mating

Flock mating means mating a number of males with a flock of females in the same pen. The number of males varies with the size of the flock. 100 females to six heavy breed males or five light breed males. The males should have been reared together. Free range or large pens is essential for this type of mating.

Pen Mating

This means the mating of one male with a number of females, ten to twelve females with one male.

Artificial Insemination

This method has been carried out successfully all over the world. Experience in handling both males and females is essential. The semen is milked from the male into a glass tube and injected, by the aid of a syringe and rubber tube, into the oviduct of the female.

FACTORS AFFECTING FERTILITY

Many factors affect fertility, for example, constitutional vigour, which is inherited. There is a wide difference between the fertility of individual birds. Birds giving high fertility in their first year will maintain it in their second year if they are kept in proper breeding condition, e.g. well fed, lean, hard, and active. On the contrary, underfeeding is opposed to good fertility. The feeding must not be unbalanced or deficient, as the condition of the stock will be adversely affected in direct proportion to the degree of imbalance. Other factors are environment, including temperature and space.

Duration of Fertility

After the removal of the male from the breeding pen fertility is maintained for a week, by the tenth day only about 50 per cent of eggs are fertile, and after twenty days it is further reduced to 15 per cent.

Feeding

Little need be said about diet for poultry, as reliable proprietary foods are marketed by the leading foodstuff manufacturers. The food can be obtained as pellets, mashes, or crumbs for chick-rearing, growing, laying, and breeding stock.

Antibiotics are added to some commercial diets, and this may be of some importance to the research worker. For the routine feeding of laboratory fowls not kept for breeding or egg production, Diet S.G.1 (Short and Gammage, 1959) is quite adequate.

Those wishing to make up their own poultry diets should consult The Ministry of Agriculture Bulletin, *The Nutritional Requirements of Farm Livestock*, No. 7 Poultry (1963).

For further reading, *Modern Poultry Husbandry*, by Robinson (1961); *U.F.A.W. Handbook*, chapter 46, 3rd Ed.

Selection of Eggs for Hatching

The egg should weigh between $2\frac{1}{16}$ and $2\frac{1}{4}$ oz, have a clean, strong shell, and have uniform shell thickness and shape. Thin-shelled eggs are undesirable, as the thin shell inhibits the normal interchange of respiratory gases and the growing embryos use calcium from the shell. Eggs are 'candled' (held before a strong light) to reveal defects which would reduce hatchability. The defects are:

(1) a tremulous air space;
(2) an air space which moves as the egg rotates;
(3) the presence of blood spots;
(4) 'mottling' of the shell, appearing as lighter areas of shell.

Storage of Eggs for Incubation

Eggs may be stored at 50–55°F (10–15°C) in a draught-free container with 8 per cent humidity for up to eight days, but it is better to start incubation within a day or two of laying. The eggs are stored on their sides or with the broad end upwards; never with the narrow end uppermost. If eggs are stored for more than 5 days it is advisable to turn them once.

Artificial Incubation of Eggs

This is done in a specially designed apparatus called an incubator. The factors which must be controlled are:

(1) ventilation—composition and rate of flow of air;
(2) temperature;
(3) humidity;
(4) turning of the eggs.

VENTILATION

There are two critical periods in incubation—the 4th day and the 18–19th day. At these periods the majority of embryonic mortality occurs. Poor ventilation results in the accumulation of carbon dioxide which may poison chicks at any time during incubation, or once they have broken into the air-space of the shell and start to breathe. Most losses due to poor ventilation occur between the 18th and 21st days.

TEMPERATURE

The aim is to keep the temperature within the egg at 100°F (38°C). As the embryo develops it gives off heat, so the environmental temperature inside the incubator is lowered gradually over the course of incubation so that the temperature of the egg remains constant.

Recommended temperatures for the incubators are:

for natural-draught incubators (capacity usually 100–150 eggs)—

1st week	*2nd week*	*3rd week*
102·5°F (39·5°C)	102°F (39°C)	100°F (38°C)

for forced-draught incubators (capacity several thousand eggs)—

100°F (38°C)	99·5°F (37·5°C)	98°F (36·5°C)

The bulb of the thermometer should be placed immediately above the eggs. The room in which the incubator is housed should be maintained at a constant temperature, preferably at 60°F (15·5°C).

HUMIDITY

If the environment is too dry too much water evaporates from developing eggs; if too wet insufficient water evaporates. The relative humidity in the incubator should be consistantly maintained, as closely as possible, at 60 per cent through incubation, and at hatching time increased to 80 per cent.

TURNING EGGS

Eggs must be turned at least twice each day to prevent the adhesion of the embryo to the shell membrane. By the 18th day the chick is sufficiently far developed for turning to be unnecessary and the process is discontinued at this time.

CANDLING FOR DEAD EMBRYOS AND INFERTILE EGGS

At the 7th or 8th day of incubation all eggs are candled. Infertile eggs appear clear (they are called 'clears'); fertile eggs show a darkspot from which blood vessels radiate (showing as a 'spider's web'), and dead embryos show a dark ring (the 'blood-ring'). The ring is due to the presence of blood which has separated out to the edge of the vascular system.

At the 14–15th day the eggs are again candled, when live embryos almost fill the egg, except for the air space, and show as a dark mass, while dead embryos show a less distinct air space and the dark area is cloudy. A dead embryo floats round as the egg is rotated.

HATCHING

Newly hatched chicks are removed to a brooder as soon as they have dried because they require more oxygen than is usually available in the atmosphere inside an incubator. Eggs in a hatching incubator at 18 days, when about 10 per cent of eggs have pipped, require the relative humidity to be increased to 80 per cent and maintained at that level until all chicks have hatched and dried off. Hatchability of fertile eggs should be from 80 to 85 per cent.

BROODERS

A brooder is a draught-free box or pen with a heat source. It should be so arranged that the chicks can never become chilled, but also so that the chicks can move away from the heat source if they become too hot. At first the brooder should provide a temperature of 90–95°F (32–35°C) at a point midway between the heat source and the floor or edge of the canopy. Study of the behaviour of the chicks is the best guide to adequacy of the heating system. Cold chicks huddle together in the warmest place. Overheated chicks will spread out as far from the heat source as possible, and may droop their wings and gasp. Comfortable chicks settle half-way between the heater and the edge of the pen or canopy. The temperature may be lowered by 5–7°F (3–4°C) each week to about 70°F (21°C) at the 4th or 5th week. For the first few days the chicks should be confined to a small area near the source of heat, as they may not find their way back to the warmth if they stray. The area of the pen may be increased as the chicks grow older and stronger. Allow 50 sq. in. of floor space per chick up to the age of 6 weeks and more thereafter. One 250-watt infra-red lamp hung 16 in. above the floor is sufficient for brooding 100 chicks.

LITTER

Commonly used litter materials are granulated peat moss, sawdust, chopped straw, or sand. If the litter becomes damp or if the chicks look as if they have been sweating it is a sure sign that the brooder is inadequately ventilated.

FEEDING

Dry mash or pellets may be fed. Water must be available at all times. Both food and water should be available to the chicks from the time they are put in the brooder. Fine oyster-shell or limestone grit may be provided in a separate hopper, but it is not strictly necessary for chicks given a well-balanced ration. Six linear feet of hopper space is recommended for each 100 chicks to feed at up to three weeks of age, increasing to 20 ft at sixteen weeks of age.

DISINFECTION OF INCUBATORS

Incubators should be disinfected between each hatching. Remove the thermometers and sterilize them separately. Wash the interior of the machine and leave the surfaces wet. Seal all appertures and ventilators. Have the machine at about 100°F (38°C). For a 150-egg incubator, place a bowl containing 15 g. potassium permanganate crystals on its floor and pour over this 30 ml. 40 per cent formaldehyde solution. Immediately close and seal the door, and leave for at least 3 hours before unsealing. Air machines thoroughly before using.

DEFINITIONS OF TERMS

The lists below have been prepared under the guidance of Dr M. Sabourdy and published by the International Committee on Laboratory Animals (ICLA Bulletin, No. 12 (annex), March 1963). It is recognized that the definitions given here will undoubtedly be modified in the light of public comment, and any such amendments will be published from time to time in the *ICLA Bulletins*. However, these lists are a noteworthy first attempt to provide an internationally acceptable set of definitions, and as such are welcomed by all persons working with laboratory animals.

The *ICLA Bulletin*, No. 12 (annex) says: 'The recommended definitions are the outcome of much investigation, in which full weight has been given to the opinions of a large number of competent experts. The Executive Committee of ICLA urges that very serious consideration be given to their widespread adoption. The Committee recognizes, however, that in spite of all the care that has been taken to formulate satisfactory definitions, there are some that will inevitably prove controversial. The definitions are, consequently, divided into three groups, arbitrarily designated *non-controversial*, *slightly controversial*, and *strongly controversial.* Comments on this list will be welcome, and should be addressed to Dr M. Sabourdy, Centre de Sélection des Animaux de Laboratoire, 5, rue Gustave Vatonne, Gif-sur-Yvette, (S.-et-O.), France.'

LIST I. NON-CONTROVERSIAL

TERM	RECOMMENDED DEFINITION
Closed colony	A colony not recruiting members from outside itself.
Primary type colony	A colony of any laboratory animals defined genetically and of known nutritional and disease status. Its function is to provide breeding stock for subcultivation elsewhere. Also known as Foundation Stock.
Primary type colony centre	A centre where one or more primary type colonies are being maintained. Also known as a Foundation Stocks Centre.
Secondary type colony	Represents a direct derivation of the individuals within the group from an earlier foundation stock.
Foundation stock	Breeding pairs of any inbred strain descended from recent common ancestors, maintained with elimination of sublines.
Pedigreed expansion stock	Breeding pairs of any inbred strain derived from the foundation stock, propagated for a few generations to increase the number of animals descended from the common ancestors of the foundation stock.
Family	A breeding group generally descended from a single pair of parents.
Substrain	As defined in 'Standardized Nomenclature for Inbred Strains of Mice', *Cancer Research*, Vol. 12, No. 8, 602–13 (1952): Any strain separated after *eight to fifteen* generations of brother × sister inbreeding and maintained thereafter in the same laboratory without intercrossing for a further *fifteen to twenty* generations shall be regarded as substrains. It shall also be considered that substrains have been constituted (*a*) if pairs from the parent strain (or substrain) are transferred to another investigator, or (*b*) if detectable genetic differences become established.' Ibid., Vol. 20, No. 2, 145–69 (1960): '. . . after *eight to nineteen* . . .', '. . . further *twelve or more* . . .'
Inbred strain	As defined in 'Standardized Nomenclature for Inbred Strains of Mice', *Cancer Research*, Vol. 20, No. 2, 145–69 (1960): 'A strain shall be regarded as inbred when it has been mated brother × sister (hereafter called b × s) for twenty or more consecutive generations. Parent × offspring mating may be substituted for b × s matings, provided that in the case of consecutive parent × offspring matings the mating in each case is to the younger of the two parents.'

TERM	RECOMMENDED DEFINITION
Subline	A division of a line.
Outbred	Individuals resulting from outbreeding.
Inbred	Resulting from continued matings between closely related animals.
Back cross	The cross of an F_1 hybrid to either of its parents; in mice, to animals of either parental strains.
Homozygous	Having identical alleles at a given locus.
Heterozygous	Having different alleles at a given locus.
Standards	Defined and accepted characteristics, tests, and regulations established by authority as a rule for the measure of value or quality.
Performance tested	Refers to an animal population (usually a strain or one of its subdivisions) shown to possess a continued ability to exhibit certain responses or characteristics for which the population is primarily maintained.
Commercial grade	Refers to animals of undetermined quality that do not necessarily conform to accepted minimum standards.
Accredited supplier	A supplier who raises stocks of laboratory animals which conform to defined and accepted minimum standards of production.
Laboratory animal technician	A person qualified by experience and training to provide specialized care and handling of laboratory animals.

LIST II. SLIGHTLY CONTROVERSIAL

TERM	DEFINITION
Species	(*a*) All the animals of the same kind that can (actually or potentially) mate together and produce fertile offspring. (*b*) A group of organisms with distinguishing characteristics reproductively isolated from other groups of organisms. (*c*) A group of actually (or potentially) interbreeding organisms reproductively isolated from other such groups.
Colony	(*a*) An animal population maintained under some degree of control for the purpose of reproduction. (*b*) All the animals of a species being maintained for reproduction in one laboratory or centre.
Strain	(*a*) A group of animals of known ancestry maintained by a deliberate mating system; generally with some distinguishing characteristics. (*b*) A stock of known ancestry maintained by a deliberate mating system; generally with some distinguishing characteristics. (*c*) A group of animals of known ancestry; generally with some distinguishing characteristics.

TERM	DEFINITION
Random breeding	(*a*) Matings made entirely at random within a population herd or breeding group without regard to either genotypic or phenotypic resemblance in the mated animals. (*b*) Mating of animals by chance, without regard to relationship. (*c*) Mating system in which the average relationship between mated individuals is the same as the average relationship between contemporary animals, i.e. the matings are random with respect to relationship.
Random bred	(*a*) Random breds are individuals resulting from random breeding. (*b*) Produced by random mating; generally implying also a fairly large number of parents in each generation, i.e. more than about ten pairs. (*c*) Refers to a population resulting from matings which are independent of relationship.
Outbreeding	(*a*) Mating system in which the relationship between mated pairs is less than the average relationship of contemporary individuals, i.e. the deliberate avoidance of inbreeding, even to the extent of introducing animals from outside. (*b*) Mating of animals less closely related than the average for the stock, especially when repeated for several generations.
Hybrid	(*a*) The immediate product of (*a*) an inter-specific cross, or (*b*) a cross between two inbred strains. (*b*) Individual resulting from a cross of parents having different inheritance. The hybrids may have various degrees of heterozygosity. (*c*) Progeny of mating between animals of unlike genetic constitution; in mice between animals from two inbred strains.
Selection	(*a*) Causing or permitting some kinds of individuals to produce more offspring than other kinds do. (*b*) A choice of individuals from which the breeder desires to obtain progeny. (*c*) Delete reference to 'the breeder' unless 'artificial selection' is to be defined. Selection, in laboratory strains, may be due to unconscious breeding practices, to heterosis, and other causes. (*d*) Matings of animals of designated types to increase or decrease the frequency of specific characteristics.
Isogenic	(*a*) Having identical genotypes. (*b*) Individuals, two or more different lines or families that have identical or, at least, most genes identical.
Enzootic	(*a*) A disease within an animal group which remains over a considerable period of time within the group. It has the same connotation as endemic diseases in man. (*b*) The occurrence in a colony of an illness or a group of illnesses of a similar nature not in excess of normal incidence and derived from a common or propagated source.
Epizootic	(*a*) A disease which affects many animals at one time having the same connotation as epidemic in man. (*b*) The occurrence in a colony of animals of an illness or a group of illnesses of similar nature, clearly in excess of normal incidence and derived from a common or a propagated source.

LIST III. STRONGLY CONTROVERSIAL

TERM	DEFINITION
Population	(*a*) Any group of animals of the same species whose characteristics as a group are being studied or described. (*b*) An interbreeding group of individuals of the same species, living at the same time on the same territory, and at least partially separated from other such populations. (*c*) Any group of individuals under consideration, either real or imaginary.
Stock	(*a*) A collection of animals being grown or maintained for breeding or for experimental use. (*b*) Represents a breeding group having some distinguishing characteristics of interest as geographical origin or genotypic content.
Line	(*a*) Part of a family of animals separated from other parts by one or more generations of independent ancestry. (*b*) Part of a strain, not necessarily an inbred strain, separated from other parts by one or more generations of independent ancestry. If part of an inbred strain, the separation is by an arbitrary fairly small number of generations. (*c*) Part of a strain separated from other parts by one or more generations of independent ancestry.
Mating system	(*a*) Any plan for propagating animals. (*b*) The manner in which pairs are chosen for mating for the production of offspring, the choice being governed by the relationship between the members of the pairs. (*c*) That method of animal breeding designed to produce a desired genetic result. (*d*) In my opinion this definition as given is completely erroneous. Mating systems need have nothing to do with planning; a naturally outbreeding or inbreeding group of animals or plants has its own mating system. In this generally accepted sense 'mating system' should probably not be included in the list at all. If you call it something like 'breeding programme', at least it is clearly distinct. But as you have already both 'selection' and items referring to inbreeding, etc., why an additional definition which presumably would have to cover both? (M.S.).

THE FOLLOWING ARE UNCLASSIFIED

TERM	DEFINITION
Production stock	Breeding pairs or trios of any inbred strain derived from a pedigreed expansion stock for further multiplication of animals descended from common ancestors of the foundation stock.
Genetically heterogeneous stock	A stock that has not been specifically bred for genetic uniformity.
Inbred derived	Progeny of mating between animals of a given inbred strain with recent common ancestors, but not necessarily related as brother and sister.

TERM	DEFINITION
Phenotype	The appearance or properties of an organism.
Genotype	The genetic composition of an organism.
Coisogenic	Having identical genotypes except for a designated difference.
Congenic	Having similar genotypes.
Certified grade	Refers to animals endorsed by competent authority as conforming to certain defined and accepted minimum standards.
Laboratory animal attendant	A person engaged in the care of laboratory animals whose primary duty is the maintenance of environmental sanitation and stability.
Specific pathogen free	Animals proved to be free of the causative agents or agents of one or more specific named diseases, but not necessarily free of the others not named.
Germ free	Animals that are free of all demonstrable organisms, resulting from use of a closed-system sterile technique.
Gnotobiotic	(*a*) An organism whose microbiota, if any, is known. (*b*) Pertaining to a gnotobiote or gnotobiotics.
Gnotobiotics	The science of rearing laboratory animals, the microfauna and microflora of which are specifically known in their entirety.
Gnotobiote	A specially reared laboratory animal, the microfauna and microflora of which are specifically known in their entirety.

REFERENCES

1. SCOTT, P. P., CARVALHO DA SILVA A. and LLOYD-JACOB, M. A., *UFAW Handbook*. Chapter 45, p. 479.
2. SANDFORD, J. C., *The Domestic Rabbit* (Crosby Lockwood).
3. ROWLAND, J. W., *Journal of Hygiene*, Vol. 47, No. 3 (1949).
4. BRUCE, H. M. and PARKES, A. S., *ibid.*, Vol. 46, No. 4 (1948).
5. SHORT, D. J., *UFAW Handbook*, Chapter 35 (1957).
6. ORSINI, MARGARET W., *Proceedings of Animal Care Panel* (1961).
7. YERGANIAN, G., *Private communication* (August 1962).
8. YERGANIAN, G., *Journal of National Cancer Institution*, Vol. 20, No. 4 (April 1958).
9. SHORT, D. J. and PARKES, A. S., *Journal of Hygiene*, Vol. 47, No. 2 (1949).
10. VAN WAGENEN, G., *Care and Breeding of Laboratory Animals* (Farris, 1950).
11. GRINHAM, W. E., *Journal of Animal Technicians Association*, Vol. 2, No. 4.
12. BRUCE, H. M., 'The importance of the environment on the establishment of pregnancy in the mouse', *Animal Behaviour*, **10**, 3–4, July, October (1962).
13. ROSS, M., 'Practical Application of the Whitton Effect', *Journal of Animal Technicians Association*, Vol. 12, No. 1 (1961).
14. CARTER, T. C., 'Breeding Mouse Stocks for Maximum Productivity', *Journal of Animal Technicians Association*, Vol. 12, No. 1 (1951).

15. LANE-PETTER *et al.*, 'Measuring Productivity in Breeding Small Animals', *Nature*, 1959, Vol. 183, p. 339 (1959).
16. LAMOTTE, J. H. L. and SHORT, D. J., *Journal of the Institute of Animal Technicians*, Vol. 17, No. 3, p. 85 (1966).
17. PAYNE, P. R., SEAMER, J. and SHORT, D. J., *Vet. Record*, Vol. 79, No. 2, p. 35, 9th July (1966)
18. DICKINSON, C. D. and SCOTT, P. P., *British Journal of Nutrition*, Vol. 10, p. 311 (1956).

CHAPTER TWENTY-THREE

Mammalian Physiology

Physiology is the study of the functions of the organs and tissues of the body. It is convenient to classify the functions of the body into systems. For instance, the cardiovascular (or circulatory) system comprises the heart and the blood vessels together with the blood. The function of this system is very similar in all mammals, and the information gained from the study of the smaller mammals will help us understand the function of our own circulatory system. In this chapter five aspects of the physiology of mammals are considered:

1. Structure and functions of the blood.
2. The cardiovascular system.
3. Structure and functions of the kidney.
4. The structure and function of the respiratory system.
5. The digestive system.

1. STRUCTURE AND FUNCTIONS OF THE BLOOD

Blood is the material which the circulatory system continually moves about the body in a system of tubes called blood vessels. Most of the functions of blood are related to satisfying the nutritional needs of the tissues through which the blood passes. It carries its own machinery for plugging any holes that might occur in a blood vessel, and in addition is concerned in the protection of the body against infection.

The functions of blood can be briefly summarized as:

1. Respiratory

The blood carries to all the remote cells of the body the oxygen which is absorbed in the lungs. From the outlying tissues it removes the carbon dioxide which the cells accumulate and carries this to the lungs for removal from the blood.

2. Nutrition

All the food that the tissues need is conveyed to them in the blood. In addition, blood removes ingested materials from the intestine and transports these materials as necessary, either to stores, to the liver for further processing, or to tissues that require the material.

3. Hormonal transport

Many of the body's functions are regulated by substances, called hormones, which are released into the blood by one organ to affect the working of another distant organ.

4. Transfer of heat

Heat is generated in different parts of the body at different times. For instance, in running muscles get hot. Blood, in travelling through muscles that are working, removes the heat and distributes it more evenly. Blood normally circulates freely in the skin, where it is cooled. By varying the amount of blood going to the skin, and thus the amount of cooling, the body can maintain an even temperature.

5. Protective

Blood contains cells and chemicals which can overcome bacterial and virus invasion. It also has a clotting mechanism to form a protective plug when a blood vessel is damaged.

CONSTITUENTS OF THE BLOOD

Blood consists of fluid and solid components. The fluid is a complex mixture of chemicals called plasma, and suspended in this plasma are vast numbers of blood cells which are the solid component.

The plasma constituents are:

Water	90–92 per cent
Proteins	7 per cent
Organic salts	1 per cent
Inorganic salts	1 per cent

Gases in solution and chemicals such as hormones . . . less than 1 per cent.

The plasma proteins may be classified in three main groups:

Serum albumin	4·0 per cent of plasma
Serum globulin	2·7 per cent of plasma
Fibrinogen	0·3 per cent of plasma

Fibrinogen is the protein actively concerned in the coagulation of blood. The substances which give us immunity to certain diseases are in the globulin fraction.

The blood cells

The blood cells (or corpuscles) are sometimes referred to as 'the formed elements of blood'. There are two main types of blood cells—the red cells and the white cells.

(a) The red corpuscles

A fully developed red corpuscle is called an erythrocyte. They are very small disc-shaped cells, and in mammals have no cell nucleus (Fig. 1). A cubic millimetre of human blood contains between 4·5 and 6 million erythrocytes.

The characteristic red colour of blood is due to the pigment haemoglobin, which is present in erythrocytes.

Erythrocytes have a life of about 120 days, and are continually being formed and destroyed. The cells are made in the bone marrow and the process of erythrocyte formation is called erythropoiesis—old erythrocytes are broken down by the spleen.

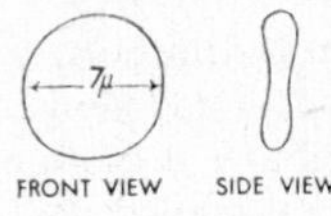

FIG. 1

The pigment haemoglobin which is contained in the erythrocyte is concerned largely in the transport of oxygen in chemical combination. By this means blood is able to absorb more oxygen than if the oxygen was in simple solution.

(b) The white corpuscles (leucocytes)

There are several different types of white cells in the blood. They are usually larger than erythrocytes and possess a nucleus but no haemoglobin. There are two main groups. There are the granulocytes—which are recognized by the presence of easily stained granules in their cytoplasm—and there are the agranulocytes—which as their name suggests—do not have any cytoplasmic granules.

There are far fewer leucocytes than erythrocytes, normally between 5,000 and 10,000 per cu. mm. of human blood.

Agranulocytes

There are two main types of agranulocytes:

(*a*) LYMPHOCYTES. Identified under the microscope by a nucleus that occupies most of the cell (Fig. 2). These cells are non-mobile and originate in lymph

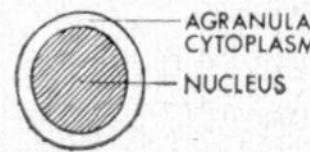

FIG. 2

nodes and in the spleen. Usually subdivided into two classes—large lymphocytes and small lymphocytes. The lymphocytes may be concerned in immunological reactions such as the production of antibodies and γ-globulins.

(*b*) MONOCYTES. Recognized microscopically by the kidney-shaped nucleus (Fig. 3). About three times the size of an erythrocyte, formed in red bone

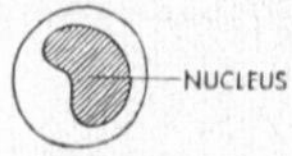

FIG. 3

marrow and have the ability to absorb bacteria and other foreign bodies in tissues (phagocytosis).

Granulocytes (polymorphonuclear cells)

They have peculiar multi-lobed nuclei, and their cytoplasm has granules which can be stained. There are several types of these cells with various staining properties. They are mobile cells—moving like amoeba by the use of pseudopodia, and they can absorb bacteria and other particles of matter. They are also able to leave the blood vessels and move into tissue spaces.

Three main types of granulocytes—which are identified by the staining properties of their cytoplasmic granules:

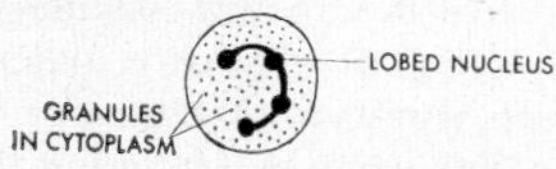

FIG. 4

(*a*) NEUTROPHILS. Granules stain with both acidic and basic stains, purplish with Leishmann's stains. These cells are active phagocytes.

(*b*) EOSINOPHILS. Granules stain with acid dyes (e.g. eosin), and thus appear red with most staining procedures. Function not understood, but the numbers of these cells circulating increase in allergic states or when an animal or man is exposed to stress (such as examinations).

(*c*) BASOPHILS. Granules stain with basic dyes, and hence are usually blue. They appear in chronic inflammatory states. They may also be concerned with the production of a naturally occurring material called heparin which prevents blood from clotting.

COAGULATION OF BLOOD

When a blood vessel is cut bleeding occurs. This bleeding is normally stopped by a combination of two factors—the local constriction of the blood vessel and the formation of a clot at the site of injury. The clot is composed of a

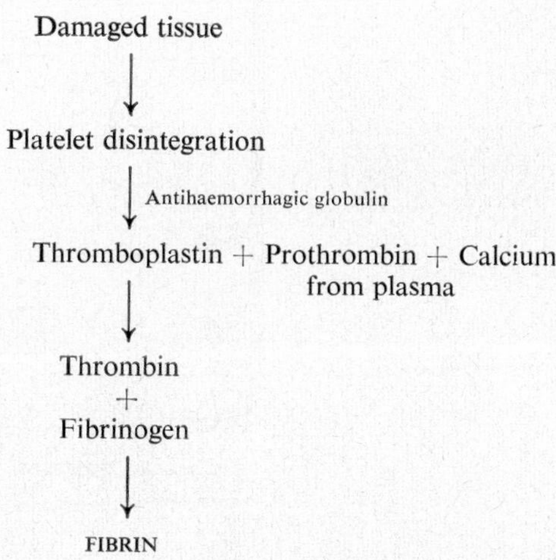

protein called fibrin which is precipitated from its soluble precursor in the blood called fibrinogen.

In the blood are numbers of minute elements called platelets or thrombocytes. Much smaller than red cells, there are 200–400 per cu mm. It is their disintegration at the site of injury which initiates the clotting process, for they produce thromboplastin (also called thrombokinase) when they disintegrate, and this initiates the process, summarized on p. 381, which ultimately causes the release of fibrin.

Blood clots whenever it comes in contact with a foreign surface. Thus, when it is necessary to prevent this an anti-coagulant is used. An anti-coagulant may be a chemically complex expensive material, such as heparin, or more simple materials, such as potassium citrate or oxalate, may be added to the blood. Citrate and oxalate work by rendering the Calcium in the blood unavailable, and thus interrupt the formation of thrombin.

2. THE CARDIOVASCULAR SYSTEM

The cardiovascular system consists of the heart and the blood vessels. There are three types of blood vessels: arteries, capillaries, and veins. *Arteries* are thick-walled blood vessels which lie deep in the tissues of the body and carry the blood which leaves the heart at high pressure. The large arteries divide into a network of small arteries which ultimately lead to the *capillaries*. Capillaries

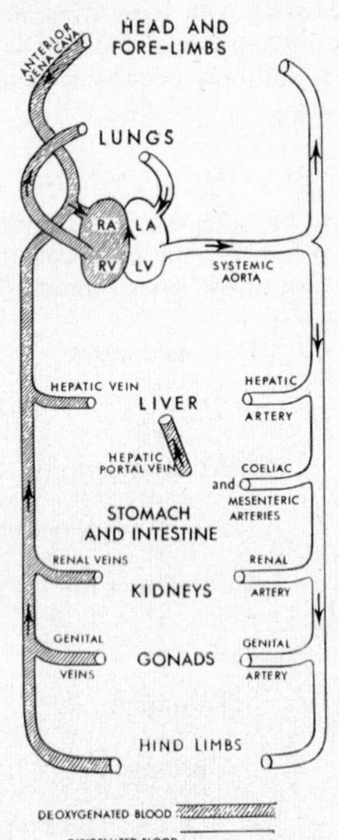

FIG. 5

are vessels with a very thin wall, usually only one cell thick, and here materials such as oxygen, carbon dioxide, etc., can readily enter or leave the blood vessels. Blood capillaries are in close association with the thin alveolar walls, and gaseous exchange takes place between the blood and the alveolar air. Blood returns to the heart in thin-walled blood vessels which lie quite superficially, called *veins*. The blood in the veins is at very low pressure, and veins are readily compressed. There is a system of one-way valves in the veins so that blood cannot flow backwards.

THE HEART

The blood is pumped around the body by the contractions of the heart. In the mammal the heart consists of two pumping systems working together. Each pumping system consists of two chambers, an auricle and a ventricle.

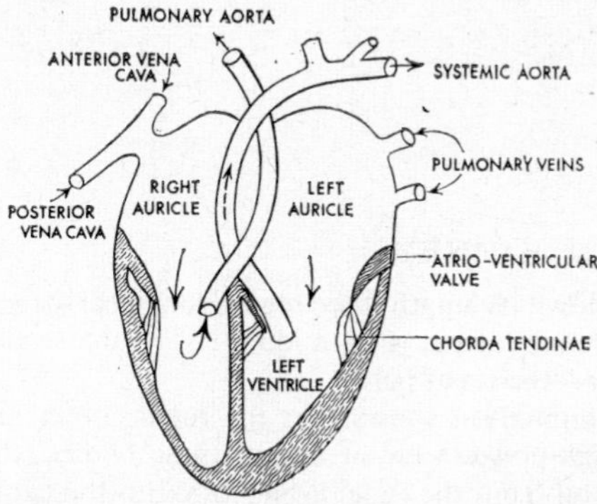

FIG. 6

The auricles (sometimes called atria) are thin-walled sacs which receive blood from the large veins and by their contractions force the blood into the thick-walled chambers called ventricles. The ventricles can develop great force during their contraction. The left ventricle develops enough force to pump blood round all the organs of the body except the lungs. The blood leaving the right ventricle is pumped through the lungs and then enters the left auricle.

It is important that the contractions of auricles and ventricles are synchronized, and this is accomplished by a special system in the heart. There is a conducting system which is arranged so that a contraction of the auricles causes activity in the system (Bundle of His), which is conducted to the ventricles through the Purkinje fibres. In this way the ventricles are caused to contract after the auricles and expel the blood which has been forced into them by the auricular contraction. The presence of valves between the auricles and ventricles and at the entry of the auricles prevents blood from regurgitating.

3. STRUCTURE AND FUNCTION OF THE KIDNEYS

The mammal normally has a pair of kidneys situated high in the abdomen, on the posterior wall and not far from the diaphragm. Each kidney is supplied with a large artery (the renal artery), and the blood is drained by a large vein (the renal vein). It has been calculated that one-sixth of the blood leaving the heart enters the kidneys.

A tube called a ureter leaves each kidney and enters the posterior wall of the bladder.

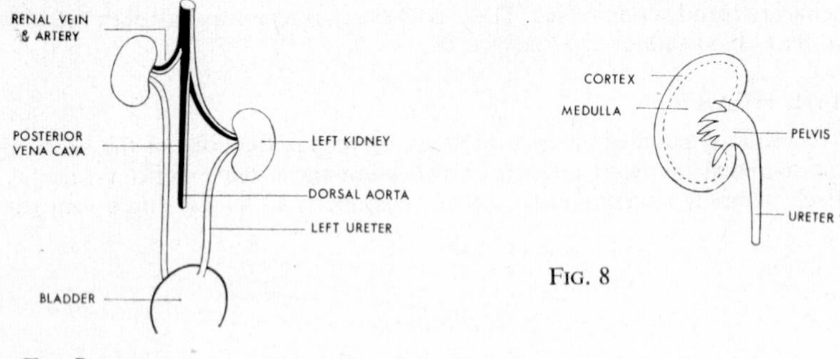

FIG. 7

FIG. 8

STRUCTURE OF KIDNEY

If a kidney is cut down its length three regions are clearly recognizable: (*a*) an outer layer—the renal cortex; (*b*) a middle layer—the renal medulla; (*c*) an inner spongy layer—the renal pelvis.

Microscopic examination shows that the renal cortex has many circular bodies in it. A high-power view of one of these bodies (the glomerulus) is shown below. Blood from the renal artery is led to the capillary network of the glomeruli, and here a high proportion of the substances in solution in the plasma are filtered off and enter the tubules. These tubules then dive towards the renal pelvis through the renal medulla. The medulla consists almost entirely of tubules surrounded by blood vessels. The blood vessels surrounding the tubules are formed from the veins leaving the glomeruli, and in the tubule they reclaim a lot of the useful materials from the filtered plasma. The materials that are not reclaimed are carried to the renal pelvis, where all the tubules converge into the ureter. These substances in solution are then carried to the bladder, from which they are periodically voided as urine.

FUNCTIONS OF KIDNEY

By the process of continuous filtration and reabsorption the kidney, together with the lungs, is able to control the composition of the blood. The metabolism of proteins leads to the continual production of materials such as urea, uric acid, and creatinine. If the concentration in blood of these substances is allowed to rise, then they may poison the organism—and the kidney normally removes these materials as they are formed, thus keeping their blood concentration at a safe low level.

The kidneys also regulate salt and water metabolism, getting rid of any excess that has been taken in, and retaining material when insufficient salt or water is taken. About $1\frac{1}{2}$ litres of urine are produced by the adult human each day from about 170 litres of glomerular filtrate formed in Bowman's capsule.

Water excretion is under the control of a hormone released from the posterior pituitary gland. This hormone, called the anti-diuretic hormone, is

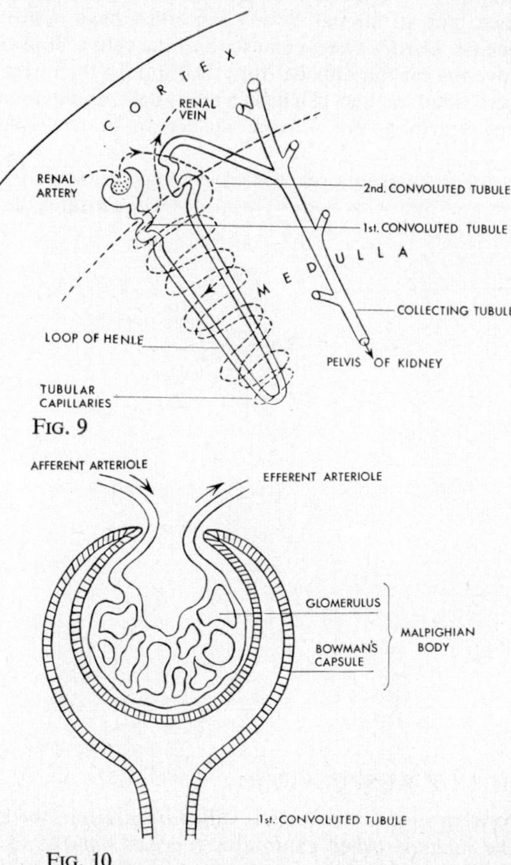

FIG. 9

FIG. 10

released into the blood whenever there is a need to conserve water. Thus, a higher proportion of the filtrate in the tubules is reabsorbed. The release of anti-diuretic hormone is governed by cells in the brain which are sensitive to the osmotic pressure of the blood.

4. THE STRUCTURE AND FUNCTION OF THE RESPIRATORY SYSTEM

The term respiration is used to cover two processes: (*a*) the process whereby air is breathed into and out of the lungs (external respiration); (*b*) the chemical processes taking place within cells whereby oxygen combines with foodstuffs to provide energy (internal respiration). Both processes are closely linked, external respiration being necessary for the gaseous requirements of internal respiration. For internal respiration to proceed, a system must exist to supply the cell with oxygen and to remove the carbon dioxide, which is formed during respiration. In unicellular organisms the exchange takes place at the cell surface, but an animal of any size must have a special system to ensure that the circulatory system can send to the cells a fluid rich in oxygen, and can remove the carbon dioxide from the fluid. In the insects organs have been developed called trachae, fish have a gill system, and mammals use lungs. All these organs arrange for a large surface to be available for gaseous exchange.

The possession of internal organs such as lungs means that a mammal has to pump air in and out of its body. This process constitutes the mechanics of respiration.

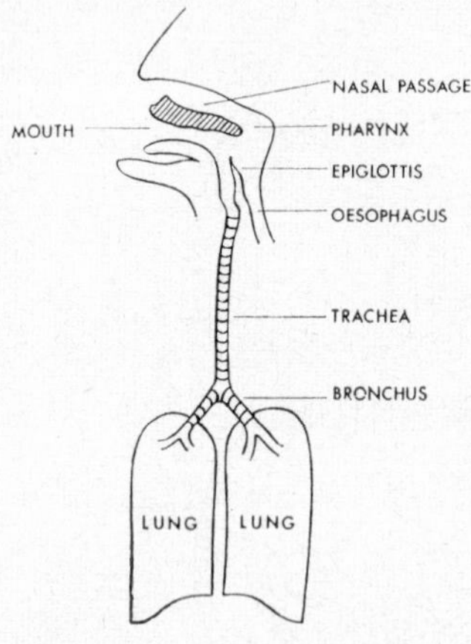

FIG. 11

MECHANICS OF RESPIRATION

The act of drawing air into the lungs is called inspiration, the act of blowing air out of the lungs is called expiration. Certain muscles of the body are responsible for the chest movements which lead to inspiration and expiration. The muscles are: (*a*) the diaphragm—a sheet of muscle which separates the abdomen from the thorax, and (*b*) the muscles of the rib cage, which can move the ribs.

Inspiration is caused by the muscles expanding the chest and thus reducing the pressure in the chest. The opening into the chest leads into the lungs, and thus air entering the body from outside to equalize the pressure enters the lungs. For expiration to occur, the chest cavity is reduced, thus compressing the air in the chest and forcing air out of the lungs.

The tubes leading to the lungs are called the respiratory passages.

The trachea passes from the pharynx into the thorex, where it divides to form two bronchi, which enter the lungs.

The actual tubes in the lungs are the *bronchi*. These are tubes with walls stiffened with cartilage. The bronchi divide into smaller tubes, the *bronchioles*, which lead to blind sacs in which the actual gas exchange takes place. These blind sacs are called *alveoli*.

The amount of air entering and leaving the lungs is closely related to how hard the body is working. Hard work requires the consumption of more oxygen, so the amount of air breathed per minute is substantially increased. A number of terms are commonly used to describe the quantities of air involved in respiration.

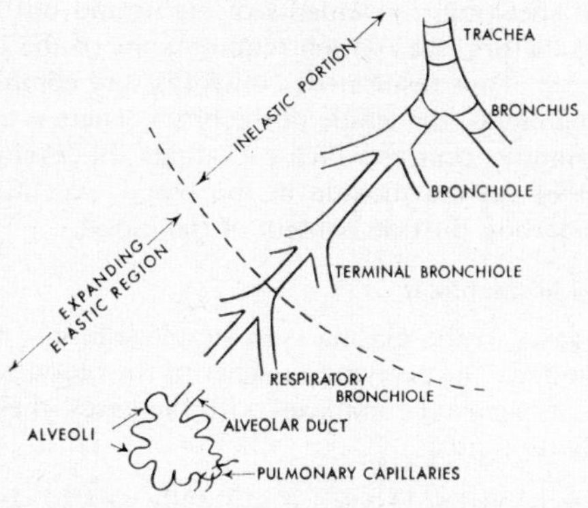

FIG. 12

Total capacity

This is the theoretical amount of air which maximally distended lungs can hold. We divide this into two fractions.

(*a*) RESIDUAL AIR. This is the amount of air which is left even when we breathe out hard—we cannot totally collapse our lungs—so some air is always left in the lungs.

(*b*) VITAL CAPACITY. This is the amount of air that we can expel from our lungs after a maximal inspiration (a really deep breath).

When breathing quietly we only breathe out a much smaller amount of air, and this volume is called '*tidal air*'. The amount over and above the tidal air which we can breathe out in a forced expiration is called '*complemental air*'.

Composition of inspired, expired, and Alveolar air

The composition of the inspired air is naturally the composition of the air around us. Since the lungs absorb oxygen from the inspired air and get rid of carbon dioxide, expired air is poor in oxygen and rich in carbon dioxide. Since the inspired and expired air is not directly exchanged with the air in the alveoli, the concentration of the gases in alveolar air is different again.

	Per cent oxygen	Per cent carbon dioxide	Per cent nitrogen
Inspired air	21·0	0·04	79·0
Expired air	16·5	4·0	79·5
Alveolar air	14·0	5·5	80·0

The depth and frequency of respiration is controlled by the brain. Obviously, say when speaking, movements of air in and out of the lungs are dependent on two factors, the oxygen requirements of the body and the requirements of speech. Thus respiratory control is very complex, involving, in man and higher mammals, the whole of the brain. There is evidence, though, of lower, more primitive centres which can adjust the level of respiration to the concentration of carbon dioxide in the blood. An increase in activity quickly raises the carbon dioxide content of the blood.

Transport of gases in the blood

A proportion of gases in the plasma is in simple solution, but this carrying capacity is very limited. The carrying capacity of the blood is vastly increased by the formation of chemical complexes with the gases, these chemical reactions being readily reversible.

(*a*) TRANSPORT OF OXYGEN. Oxygen is normally carried combined with the red pigment of the erythrocytes, haemoglobin. The reaction may be summarized.

$$O_2 + Hb \rightleftharpoons HbO_2$$

$$\text{Oxygen} + \text{haemoglobin} \rightleftharpoons \text{Oxyhaemoglobin}$$

This is a reversible reaction.

The oxyhaemoglobin is bright red, and haemoglobin is a much darker colour. This accounts for the differences in colour between arterial and venous blood.

(*b*) TRANSPORT OF CARBON DIOXIDE. Carbon dioxide is more soluble than oxygen, and hence a larger proportion of the gas is carried in solution. In addition, it forms a complex with haemoglobin after the haemoglobin has given up its oxygen. This complex is called carbamino-haemoglobin. Most of the carbon dioxide is carried in the plasma as sodium bicarbonate.

The actual formation of bicarbonate takes place in the erythrocytes. The formation of bicarbonate in the erythrocytes is accompanied by a loss of chloride from the erythrocytes—the 'chloride shift'.

5. THE DIGESTIVE SYSTEM

The digestive system renders food into a form that can be absorbed into the blood stream. The detail of the digestive system varies in different groups of mammals, the principal groups being:

(*a*) Herbivores—those animals which live on vegetable matter alone, e.g. rabbit.

(*b*) Carnivores—those animals which normally have a very high proportion of meat in their diet, e.g. dog and cat.

(*c*) Omnivores—those animals which eat meat or vegetable matter with equal facility, e.g. man.

The ruminants constitute a major sub-group of the herbivores and are characterized by the possession of a large multi-chambered stomach in which a population of bacteria actively aid the digestive process.

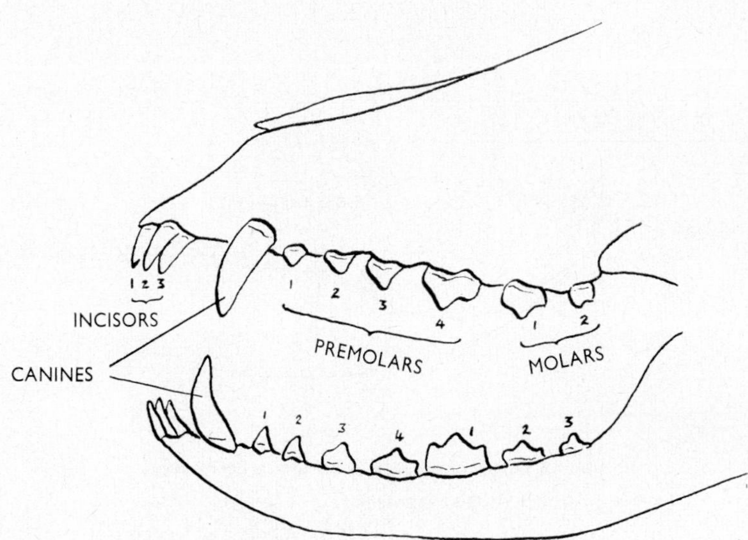

FIG. 13. Diagram to show dentition in carnivores.
Dog jaw; side view
Dental formula: $i\frac{3}{3}\,c\frac{1}{1}\,pm\frac{4}{4}\,m\frac{2}{3} = 42$

Food enters the digestive system when it enters the mouth. Here food is chopped and ground by the teeth, and saliva, secreted into the mouth from the salivary glands, is added to form a slipper bolus. Saliva contains an enzyme which can break starch down to sugars.

There are substantial differences in the arrangement of teeth between species and between groups of mammals. Primates have four kinds of teeth: incisors, canines, pre-molars, and molars. The incisors and canines chop the food, and it is crushed by the pre-molars and the molars. In rodents the canine teeth may continue to grow throughout life, but in many species the permanent teeth do not grow appreciably during life.

During the act of swallowing the food is actively propelled into the oesophagus, a muscular tube which in turn propels the food into the stomach.

The stomach has an external muscular coat which by organized contraction churns the stomach contents, and an inner glandular coat which secretes various digestive juices on to the stomach contents. These juices include a number of enzymes, including pepsin, rennin, lipase, and hydrochloric acid, which digest protein and fat and curdle milk.

After a period of time in the stomach (the length of time depends on the nature of the food) the partially digested contents are emptied into the small intestine. The first section of the small intestine is called the duodenum and,

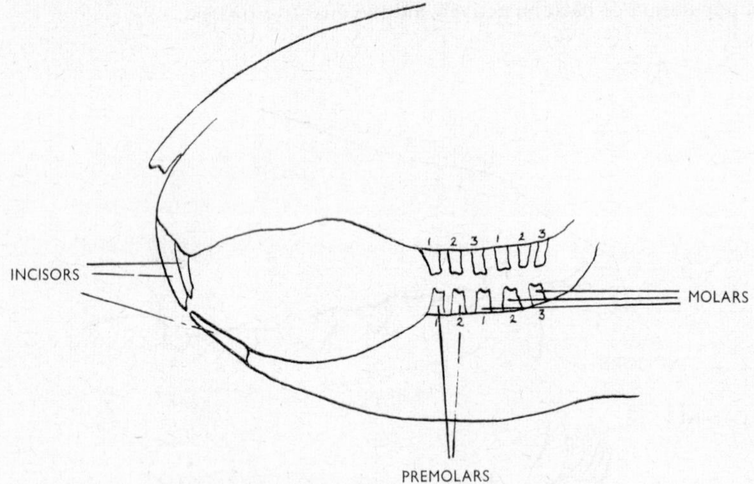

FIG. 14 Diagram to show dentition in herbivores.
Rabbit jaw; side view
Dental formula: $i\frac{2}{1}\,c\frac{0}{0}\,pm\frac{3}{2}\,m\frac{3}{3} = 28$

like the stomach, consists of an outer muscular coat and an inner glandular coat. The glands secrete a variety of enzymes, including erepsin, amylase, maltase, lactase, sucrase, lipase, and nucleases, which further digest food constituents. The pancreas, which lies in a loop of the small intestine, secretes digestive enzymes into the duodenum when food arrives there from the stomach. These enzymes include trypsin, chymotrypsin, amylase, and lipase, which aid in the breakdown of fats, sugars, and proteins. Bile also enters the duodenum from the bile duct. This fluid contains a variety of complex salts which aid in fat digestion.

Partially digested matter is continually churned by intestinal movements and is also propelled onward by a travelling wave of contraction called peristalsis. Ultimately, the residual contents of the unabsorbed and undigested matter pass to the large intestine, where a high percentage of the remaining water is absorbed. The partially solidified remnant then passes to the rectum and is periodically voided as faeces. The characteristic colour of faeces is due to the presence of modified bile pigment.

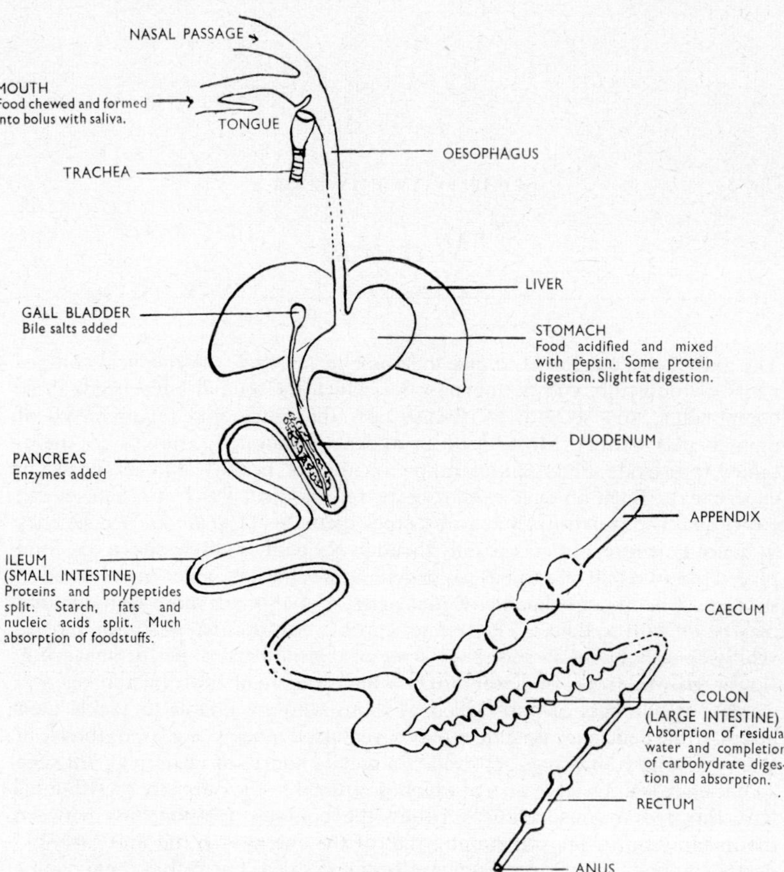

FIG. 15 Diagram to show the alimentary canal (Rabbit) and a summary of the process of digestion.

FURTHER READING

The following books provide introductions to the subjects of anatomy and physiology.

ROWETT, H. G. Q., *Dissection Guides*. No. 3: The rat, with notes on the mouse. No. 4: The rabbit (1950–53).
Basic anatomy and physiology (1959), John Murray, London (1950–53).

Teach Yourself Anatomy. The English Universities Press Ltd., London.

Teach Yourself Physiology. The English Universities Press Ltd., London.

WALKER, K., *Human physiology*, Pelican Book No. A102, Penguin Books, Harmondsworth.

CHAPTER TWENTY-FOUR

Nutrition

The aim of this chapter is to give in simple terms, first, the chemical components of foodstuffs and the purpose for which the animal body needs these components, and second, to discuss how the nutritional requirements of animals may be met. Most breeding stocks of laboratory animals are maintained to provide a continuous supply of normal, healthy animals for use in experiments. Such animals cannot be reared on diets which are deficient in either quality or quantity. Since most stock diets are fed *ad libitum*, a deficiency in quantity is rare and occurs only through accident or carelessness, e.g. food placed out of reach of animals, or providing wet mashes which are allowed to sour and become unpalatable. Deficiencies in quality are more common and may be difficult to detect. The state of chronic, sub-optimal nutrition may reveal itself only by an all-round lowering of the standard of performance, e.g. slower growth rates, or fewer litters which are small both in number and weight, or numbers of pups killed by dams who are unable to suckle their litters. If the deficiency is acute it may reveal itself quickly, e.g. an outbreak of scurvy among guinea pigs deprived of a dietary source of vitamin C, but such a deficiency is unusual in an age which is attuned to the publicity the vitamins have enjoyed in this century. Today the complex relationships between the metabolism of the protein and that of the energy-rich (fat and carbohydrate) components of the dietary are being revealed, but before these can be discussed the components themselves must be considered.

All the material needed for the growth and maintenance of the animal body must be taken in by mouth. The nutrients required are proteins, carbohydrates, fats, salts, vitamins, and water. These are the digestible and absorbable constituents of foodstuffs. The fibre (roughage) of foodstuffs is non-digestible, but some animals (especially the ruminants) are able to utilize it to various degrees owing to the action of the micro-organisms present in their digestive tracts.

Food contains chemical energy which is converted, in the body, to work and heat. Protein and carbohydrate yield energy equivalent to 4 Calories per gram; fat yields 9 Calories per gram. (A Calorie is the amount of heat required to raise the temperature of one litre of pure water from 16–17°C.)

Foodstuffs have to be broken down into simple substances before they can be utilized by the body. First, the digestive enzymes act upon foodstuffs, splitting them into less-complex substances which can be absorbed through the gut wall. The breaking-down process is continued by other enzymes in the body tissues. This breakdown is called *catabolism*; the reverse process (e.g. the building up of body proteins) is called *anabolism*; the term *metabolism* includes both anabolism and catabolism. Some enzymes can only function properly when in combination with co-enzymes. Many vitamins act as co-enzymes.

PROTEINS

These are composed of some twenty different *amino acids* (nitrogen-containing organic compounds). Protein is broken down to its constituent amino acids in the body. Some of the mixture of amino acids is used to form new body proteins, and the rest is broken down with the liberation of energy. The waste product from protein metabolism is *urea*, which is excreted through the kidney.

Amino acids may be divided into two groups—the *essential amino acids*, which must be provided from the diet, and the *non-essential amino acids*, which can be made in the animal body. The amino acids essential for most animals are lysine, methionine, leucine, isoleucine, valine, tryptophan, histidine, phenylalanine, and threonine.

Plants can synthesize amino acids using simple nitrogen-containing salts, such as ammonium sulphate, but animals have only a limited ability to synthesize amino acids from nitrogenous organic compounds.

Animal sources of protein (meat, fish, eggs, milk) contain all the essential amino acids in approximately the right proportions for optimal growth. Plant proteins (e.g. protein from cereals, beans, nuts) may contain all the essential amino acids, but not in the best proportions for animal growth, or may contain only some of the essential amino acids. A mixture of plant protein may, however, contain all the essential amino acids in the right proportions for animal growth and be equivalent to animal sources of protein.

Proteins constitute about one-sixth of the body mass. Animals therefore require a liberal supply of protein throughout life for the growth and replacement (maintenance) of body tissues.

CARBOHYDRATES

As their name implies, these are composed of the elements carbon, hydrogen, and oxygen. They are the chief source of energy for the animal body. Carbohydrates are catabolized (broken down) to simple sugars and then to carbon dioxide and water with the liberation of energy. Sugars and starches (which are complex forms of sugars) are carbohydrates.

Plants, but not animals, can synthesize carbohydrates from carbon dioxide and water. Glucose is the commonest simple carbohydrate which animals can utilize.

Cellulose (fibre or roughage), a complex carbohydrate, is not itself digestible, but some animals (e.g. rabbits and ruminants) are able to utilize cellulose, which is first broken down into digestible components by micro-organisms normally present in the gut.

Animals store carbohydrate as *glycogen* ('animal starch') in the liver and muscle. Glycogen is the most readily available energy store of the animal body. Carbohydrate can also be converted to fat, and stored as such, within the animal body.

FATS

Plants and animals contain substances, insoluble in water but soluble in chloroform, ether, and benzene, which are properly called LIPIDS. This group includes neutral FATS and *other similar compounds.* Quantitatively and economically the fats are the most important members of the group, but some of the other compounds have important physiological functions.

Fats are composed of carbon, hydrogen, and oxygen and are metabolized in the body to carbon dioxide and water. Like carbohydrates, they are a

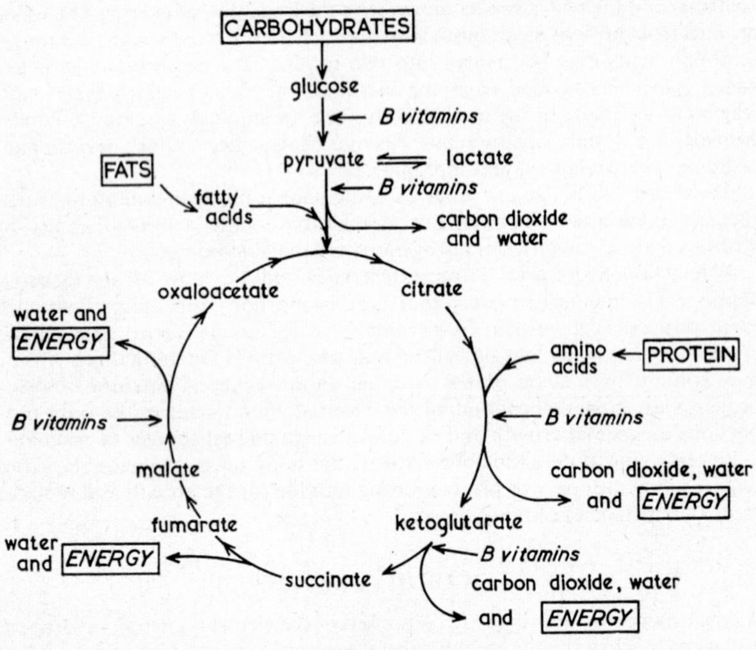

FIG. 1

Diagram to indicate how protein, carbohydrate, and fat may be oxidized to yield energy through a common metabolic path known as the citric acid cycle.

source of energy, and weight for weight, fats yield more than twice as much energy as carbohydrates. Although carbohydrate may replace fat as a source of energy, it is desirable that diets should contain some fat to promote the absorption of fat-soluble vitamins. Furthermore, *choline* and the *essential fatty acids*, which are known to be essential food factors, occur in foodstuffs as components of lipids. Oils obtained from plants are much better sources of the essential fatty acids than are animal fats. In practice, it is impossible to devise a diet composed of natural foodstuffs which would also be devoid of fat.

MINERAL SALTS

Mineral salts are widely distributed in nature and occur in all foodstuffs. They are required by animals for the growth, replacement, and proper functioning of all body tissues.

Calcium, *phosphorus*, and a small amount of *magnesium* are the chief constituents of bones and teeth. Calcium and phosphorus cannot be efficiently utilized except in the presence of vitamin D. During the period of rapid growth considerable amounts of calcium and phosphorus (and vitamin D) are needed, but the requirement falls off as the adult state is reached. During pregnancy and lactation there is a renewed demand for calcium and phosphorus for the formation of bone in the foetus and because these salts are secreted in milk. Calcium also occurs in blood plasma, and plays an essential role in the clotting of blood.

Sodium, *potassium*, *and chlorine* are present in the soft tissues and fluids, where their function is to maintain the osmotic pressure and the acid–base equilibrium. These salts also play an important part in water metabolism. Sodium, potassium, and chlorine are not stored in the body, any excess being excreted in the urine. These salts must be available from the diet throughout life.

Iron is present in the body in a small, but important, amount. It is a constituent of the respiratory pigment haemoglobin, present in red blood cells. Once iron has been absorbed by the body it is held tenaciously and is not readily excreted, even in milk. The requirement for iron in adult life is therefore low, except during reproduction. Most young are born with considerable reserves of iron. The notable exception are piglets, which frequently suffer from anaemia (perhaps because sows have been bred to throw numerically large litters), and it is good practice to administer iron to young pigs. Anaemia due to iron deficiency can occur in any animal at any time of life should the supply of iron fall below the requirement for haemoglobin formation, e.g. following haemorrhage.

It should be remembered that there are other types of anaemia due to specific diseases of the blood-forming organs and to protein and vitamin deficiencies.

Iodine is present in the body in very small amounts and is found chiefly in the thyroid gland. The thyroid secretes an iodine-containing hormone called thyroxine, which controls the metabolic rate of the body. A deficiency of iodine leads to enlargement of the thyroid gland—a condition known as goitre. Iodine must be supplied from the diet throughout life. Goitre is uncommon among laboratory animals.

Other elements are known to be essential, in small amounts, to the animal body (e.g. copper for blood formation). These are known as the '*trace elements*'. Many other elements are normally present in the body, perhaps performing some useful function or perhaps there by accident, having been ingested with the food.

Although mineral salts occur in all foodstuffs, it is customary to ensure that adequate amounts are present in a diet by adding up to 1 per cent of chalk and common salt (to supply calcium, sodium, and chlorine) and a mixture of trace elements. All cereals are rich sources of phosphorus. Other good sources of elements in a natural diet are milk (calcium), fish (iodine), liver and muscle (iron), and vegetables.

VITAMINS

Vitamins act like catalysts in many metabolic reactions; that is, the vitamins take part in these reactions without themselves being used up. Because vitamins behave like catalysts they are required only in minute amounts. Some vitamins are quickly excreted from the body and therefore need to be replaced frequently from the food. Other vitamins (notably the fat-soluble vitamins) can be stored within the body for several months.

FAT-SOLUBLE VITAMINS

VITAMIN A

Sources

Vitamin A does not occur, as such, in plants, but its precursor, *carotene*, does occur in plants. Carotene can be converted to vitamin A within the alimentary tract. Carotene is found in green, leafy materials and, generally speaking, the greener the leaf, the more carotene it contains. Grains, roots, and tubers, with the exception of carrots, are poor sources. Halibut- and cod-liver oils are rich sources of vitamin A.

Storage and stability

Vitamin A is quickly absorbed by the animal body and is stored in the liver. Animals can store sufficient of this vitamin to meet their needs for two or three months.

Outside the body, both vitamin A and carotene are easily destroyed by oxidation when exposed to air and light.

Function and signs of deficiency

Vitamin A is necessary for the physiological process of seeing, and a deficiency results in slow dark adaptation and night blindness.

Further, a deficiency of vitamin A leads to a general degenerative change in the epithelial tissues of the body, thus lowering the resistance of these tissues to attack by infective organisms. The tissues of the eye, skin, and reproductive organs are the most obviously affected, and blindness and sterility may ensue.

The malformation of growing bone in vitamin-A deficiency may result in pressure on nerves by bone. Blindness and deafness have been reported from this cause.

VITAMIN D

Sources

Of all the vitamins, vitamin D has the most limited distribution in natural foodstuffs.

The only rich source is cod-liver oil. Unprocessed milk fat contains a variable amount. Seeds and growing forage crops are poor sources.

Action of sunlight

The ultra-violet radiations of the sun can produce vitamin D from precursors which are present in the skin and sebaceous glands of animals. The activated substance so produced can be absorbed through the skin. However, the effect of sunlight may be relied upon only to a limited extent and only for animals having access to open runs.

When hays are *sun-cured* vitamin D is formed by the action of radiant energy upon *ergosterol*, the precursor of vitamin D which plants contain. Sun-cured, soft, green, leafy hays have a higher vitamin-D content than pale, stemmy hays.

Storage and stability

Vitamin D is stored by the body, but to a lesser extent than is vitamin A. Outside the body, vitamin D is less readily oxidized than vitamin A, but considerable losses can occur. Rapid loss of vitamin D occurs if it is mixed with mineral salts, especially calcium carbonate (chalk).

Function and signs of deficiency

Vitamin D is required, together with calcium and phosphorus, for the normal calcification of bone. *Rickets* is a disease associated with abnormal bone formation, and it can arise from a deficiency of calcium or phosphorus in the absence of sufficient vitamin D.

Whatever the cause, a deficiency in bone nutrition during the period of rapid growth arrests the process of bone formation, and abnormalities occur in bone structure. Such bone as is formed is weak and mis-shapen. Tension from muscles pulls the bones out of shape, and the weight of the body bends the leg bones, which may fracture (greenstick).

The requirement for vitamin D (and calcium and phosphorus) is greatest when growth is rapid and during pregnancy and lactation.

VITAMIN E

Sources

Wheat-germ oil is the richest source. Whole cereal grains are good, the vitamin being found particularly in the germ. Green, leafy materials (including hay) are good sources.

Storage and stability

Vitamin E is stored throughout the animal body for long periods. A female rat, born of a normal dam but reared on a vitamin-E-deficient diet may well have been born with a reserve of vitamin E sufficient to carry her through her first pregnancy.

Vitamin E is resistant to heat, but is easily oxidized. It stores well in ready-mixed foods, except in the presence of rancid fat, when the vitamin is rapidly oxidized. It should be noted that rancidity will have occurred long before it can be detected by its characteristic smell.

Function and signs of deficiency

The precise function of vitamin E has yet to be elucidated, but it is known that a deficiency of this vitamin affects many species in various ways. This vitamin is essential for normal reproduction in the rat (affecting both the male and the female), the female mouse, and the male hamster. Pregnant rats deficient in vitamin E may resorb the foetuses, but if pups are successfully born they may suffer from paralysis before reaching weaning age. Cattle, sheep, pigs, rabbits, monkeys, and guinea pigs deprived of vitamin E exhibit muscular weakness and paralysis.

The wide distribution of vitamin E in foodstuffs and the considerable ability of animals to store it makes a deficiency of this vitamin an uncommon occurrence.

VITAMIN K

Sources

Green, leafy materials, fresh or dried, are rich sources. The micro-organisms present in the intestines of all animals synthesize sufficient vitamin K to supply the animals' normal requirements without supplement from a dietary source. Birds, which have short digestive tracts, are the only creatures requiring this vitamin in their food.

Function and signs of deficiency

Vitamin K influences the formation, in the liver, of prothrombin. In vitamin-K deficiency the prothrombin level in the blood falls and the clotting time is increased. Haemorrhages occur in the skin and muscles.

Vitamin-K deficiency can be induced in laboratory animals by feeding purified diets and by suppressing the activity of the intestinal micro-organisms.

WATER-SOLUBLE VITAMINS

VITAMIN C (ASCORBIC ACID)

Only man, monkeys, and guinea pigs require a dietary source of vitamin C. Other species have the ability to synthesize within their own bodies sufficient vitamin C to meet their requirements.

Sources

Fresh, green, leafy vegetables, citrus fruits, tomatoes, and potatoes are the important sources of this vitamin.

Storage and stability

The animal body has little ability to store this vitamin, which must be supplied continuously in the diet.

Outside the body, vitamin C is quickly and easily oxidized. The vitamin C content of stored foods falls rapidly, and there is considerable destruction of this vitamin during cooking.

Function and signs of deficiency

Vitamin C is essential for the formation of the intercellular material in soft tissues and bone. If the intercellular material breaks down the bones become weak and may fracture; the gums swell and bleed and the teeth loosen; the walls of small blood vessels rupture, allowing the escape of blood into the surrounding tissues, causing haemorrhages throughout the body. Early signs of a deficiency in guinea pigs are loss of weight, staring coat, and a mincing gait, as if the feet were too painful to walk upon.

Scurvy is the name given to the deficiency disease associated with a lack of vitamin C.

THE VITAMIN-B COMPLEX

The known vitamins of the B-complex are aneurin (vitamin B_1, thiamine), riboflavin (vitamin B_2, vitamin G) nicotinic acid, pantothenic acid, pyridoxine (Vitamin B_6), *p*-aminobenzoic acid, biotin, folic acid, cyanocobalamin (vitamin B_{12}), and inositol. All are concerned with the metabolism (the breakdown and rebuilding) of the digested and absorbed components of foodstuffs. If food components cannot be metabolized energy cannot be released for use by the animal, and new tissue cannot be laid down nor existing tissues maintained. It is rare to encounter a deficiency of only one B-vitamin. The clinical picture is usually of a multiple deficiency—loss of weight; loss of appetite; apathy; skin lesions; anaemia; loss and/or greying of hair.

All species require B-vitamins for the proper functioning of their metabolic processes. All species are able to satisfy, to a greater or lesser extent, their requirement for B-vitamins from the synthesis of these vitamins by their intestinal micro-organisms. The ruminants are able to supply practically all the B-vitamins they require from this source. Other species, however, require a supplement of some B-vitamins from a dietary source.

Sources of B-vitamins

Liver; yeast; the germ and outer seed coats of cereals; green, leafy materials; milk.

None of the B-vitamins is stored by the body to any appreciable extent; like vitamin C, they have to be supplied continuously from the diet to all species standing in need of them. *Specific diseases* associated with deficiencies of particular B-vitamins are polyneuritis in birds (beri-beri in man) due to a deficiency of vitamin B_1, and black tongue in dogs (pellagra in man) due to a deficiency of nicotinic acid.

MEETING THE NUTRITIONAL REQUIREMENTS OF ANIMALS

In designing a diet to be composed of natural foodstuffs the first consideration is the amount and quality of protein the diet is to contain. Protein is an expensive item; therefore the amount of protein in a diet should not be greatly in excess of the consumer's requirements, and plant proteins, which are cheaper than animal proteins, should be used to the best advantage.

Enough is now known about the amino-acid content of foodstuffs and the amino-acid requirements of animals for it to be possible to equate, with some accuracy, the supply of these nutrients with the demand for them.

The quality of cereal proteins is usually limited by their lysine content. Lysine is one of the essential amino acids in which animal proteins are rich. It is common practice to use a mixture of cereals to supply the bulk of the protein portion of a diet, and to add a small amount of animal protein to guard against a possible deficiency of lysine. Animal protein is also a source of vitamin B_{12}, which does not occur in plant foodstuffs.

The quantity of protein in any one cereal depends on the variety of the cereal (e.g. soft, English wheat contains about 10 per cent of protein; hard, Manitoba wheat about 15 per cent). Recent work indicates that the soil on which a crop grows and the weather conditions during growth exert a great influence on the yield per acre, but little or no influence on the chemical composition of the crop.

It will be found necessary to have 80–90 per cent of cereals in a diet to obtain sufficient protein from this source. This amount of cereal provides not only protein but also carbohydrate in quantity, fat, fibre, minerals, and some vitamins. Fat of plant origin is a good source of essential fatty acids. In effect, the cereal provides all the required carbohydrate and fat, and most of the protein and vitamins. Thus, the remaining considerations are: (i) to enhance the quality of the protein by the addition of some animal protein, and (ii) to study the vitamin and mineral content of the diet and to make any necessary additions, e.g. calcium would need to be added to balance the high phosphorus content of cereals.

Since the major problem in designing a diet is to supply protein of the right value, i.e. of adequate quality in sufficient quantity, it is worth while considering this problem in a little more detail.

The need for protein is greatest during the period of rapid growth and reproduction, when new tissues are being laid down, and during lactation, when protein is secreted in the milk for the sustenance of the suckling. The qualitative requirements of different species for amino acids show a remarkable similarity, though the amounts required differ both from species to species and within any one species for different purposes, e.g. tissue maintenance and reproduction. The biochemical reactions involved in the metabolism of amino acids are only beginning to be understood.

Two diets having the same protein content as determined *chemically* do not necessarily have the same nutritive value when fed to animals, because the protein *quality* may be different. Diets having a high protein quality are those which supply all the necessary amino acids in approximately the right proportions and amounts for maintenance, or growth, or reproduction. Such a diet would give maximum results for minimum protein intake. Another diet may have its amino acids so ill balanced as to be inadequate to support growth (or maintenance, or reproduction) no matter how much of it is eaten by the animal. Amino acids which are not utilized for their specific function of protein nutrition are not entirely wasted, but are catabolized and provide heat and energy. The provision of energy is the function of the carbohydrates and fats; to use protein (the most expensive component of the diet) for this purpose is wasteful and uneconomic.

Ever since it has been known that the quality of proteins can differ widely much effort has been devoted to evaluating these differences. Methods of measuring protein quality have been developed which take into account *digestibility* (defined as the percentage of the ingested protein which is absorbed) and the *biological value* (the percentage of the absorbed protein which is retained in the body). These measurements are made by biological assay in which diets under tests are fed to young, growing rats. The product of the figures obtained for digestibility and biological value, divided by 100, is termed *Net Protein Utilization* $\left(\frac{D \times BV}{100} = \text{NPU}\right)$, and is a measure of protein *quality*.

However, the efficiency of protein utilization also depends on the total energy value of the diet, that is, the amount of energy which could be obtained from the diet if *all* the protein, as well as the carbohydrate and fat, were burned for this purpose.

The animal body does not distinguish between protein, carbohydrate, and fat as possible sources of energy in order to reserve all the available amino acids for tissue growth and replacement, and some protein is always burned to supply heat and energy.

The value for NPU falls as the protein concentration of a diet is increased above a certain level. When the protein concentration is high the concentration of carbohydrate and fat is correspondingly low, and therefore the body must burn protein in addition to carbohydrate and fat to satisfy its energy requirements.

If food intake is *restricted* so that an animal cannot completely satisfy its hunger, then some of the protein has to be burned to provide energy, and the efficiency of protein utilization (NPU) is again impaired. In this case the body has been forced to burn protein to satisfy its immediate need for energy at the expense of its need to renew or build body tissues. The feeding of additional carbohydrate and fat to supply energy would *spare the protein* for its proper function, and protein utilization (NPU) would be improved.

Obviously, then, when assessing the protein value of the diet as a whole account must be taken not only of the quality and quantity of protein present but also of the total calorie (energy) content of the diet, for this must always be sufficient to meet the energy requirement of the animal being fed.

Classically, the *quantity* of protein in a diet is measured by determining the nitrogen present in the diet and converting the nitrogen figure obtained to protein, using the assumption that proteins contain 16 per cent of nitrogen (i.e. protein = nitrogen × 6·25). But the quantity of protein in a diet can also be expressed in terms of energy, i.e. as the percentage of the total calories of the diet which could be supplied by the protein portion if it were used for energy purposes only. Energy is measured in Calories (Cals), and each gram of protein could provide 4 Cals. Then,

$$\text{Protein Cals \%} = \frac{\text{Nitrogen of diet} \times 6{\cdot}25 \times 4}{\text{Total Cals per 100 gm of diet}} \times \frac{100}{1} = \frac{2{,}500\ \text{N}}{\text{Total Cals}}$$

The product of the values obtained for protein quality (NPU) and protein quantity (protein Cals per cent), divided by 100, is termed *Net Dietary-*

protein Calories per cent (NDp Cals per cent) and is a measure of both quality and quantity of protein. Where direct biological assay methods are not available for the determination of NPU, protein values can be 'scored' from data to be found in food composition tables. The method of scoring is illustrated in Appendix I to this chapter, where Diet 41B is used as an example.

It is known that for reasonable growth in rats a diet in which the NDp Cals per cent = 8 is necessary. Such a value cannot be reached by diets having less than 9 per cent protein, however good its quality, nor with a protein having a quality of less than NPU = 50, whatever its concentration in the diet. It is of interest to note that it would be difficult to devise a diet consisting mainly of wheat which had an NDp Cals per cent of less than 8. Stock diets commonly contain about 15 per cent of protein from mixed sources; these well-tried diets are known to support growth and reproduction.

CUBED DIETS

Ready-made diets for laboratory animals have been introduced and widely used during the last twenty years. Such diets should be bought in small quantities (not more than about one month's supply) and used up completely before the next batch is opened. Storage of a diet increases the opportunity for infestation by parasites and moulds, and for rancidity to develop. Furthermore, some considerable time may elapse between the date of manufacture of a diet and the date of its delivery to a purchaser.

In using cubed diets great reliance is placed, initially, on the persons who formulated the diets, and thereafter on the compounders of the diets. It should be remembered that these recipes were designed to provide moderately priced diets which are adequate, rather than optimal, for growth and reproduction. Undoubtedly the growth rate of rats may be increased by supplementing the standard cubed diets. Whether or not it is desirable to do so depends upon personal opinion and preference, economic necessity, and experimental conditions.

The manufacture of cubed diets for laboratory animals represents an insignificantly small part of the milling industry of Great Britain and is consequently unlikely to command the undivided attention of millers. None the less, millers are conscientious persons who manufacture the diets according to the recipes, using ingredients of specified standards. No variations are made to the recipes without consultation with the buyer. It is, on rare occasions, a question of going without diet or accepting diet made up to a modified recipe. In this event it is preferable to have the diet made with a better (and more expensive), rather than a poorer-quality ingredient, since to use a poorer-quality ingredient might so reduce the overall quality of the diet as to render it inadequate for reproduction and/or growth. It is not easy to standardize processed foodstuffs. Fishmeal, which is widely used as a source of protein in animal diets, is one of the most difficult of foodstuffs to process without damage to its protein. It says much for the manufacturers of animal foodstuffs that it is rare for a laboratory to be faced with severe or enduring trouble arising from a dietary deficiency.

APPENDIX I

CALCULATION OF THE PROTEIN VALUE OF A DIET

The quality of a protein may be calculated by 'scoring' the amino-acid pattern of the protein against a known, adequate amino-acid reference pattern (see *FAO Nutritional Studies*, Nos. 15 and 16, 1957). The amino acids of the unknown protein are scored as a percentage of the amino acids of the reference protein pattern, and thus a score of 100 or more indicates that the protein is equivalent to the reference protein pattern. In theory each amino acid should be scored in turn to find which of the essential amino acids is 'limiting' the value of the protein as a whole. In practice, it has been found sufficient to consider only three amino acids: (i) the 'sulphur-containing amino acids' (SAA), *methionine and cystine* (the latter can partly replace methionine) for diets containing animal protein; (ii) *lysine* for diets in which cereals are the chief source of protein; and (iii) *tryptophan* for diets in which maize is the chief source of protein. The Food and Agriculture Organization of the United Nations gives recommendations for the amounts of the essential amino acids which should be present in each gram of protein nitrogen. These factors are:

Sulphur amino acids (SAA)	0·27 gm/gm N
Lysine	0·27 gm/gm N
Tryptophan	0·09 gm/gm N

The following table gives the amounts of amino acids, in grams of amino acids per gram of nitrogen, in some common foodstuffs. These figures were determined by biological assay in the Medical Research Council's Human Nutrition Research Unit.

FOOD	SULPHUR AMINO ACIDS (SAA) (gm/gm N)	LYSINE (gm/gm N)
Rice	0·197	0·155
Maize	0·156	0·127
Potato	0·160	0·130
Oats	0·202	0·168
Wheat	0·195	0·132
Meat	0·216	0·540
Fish	0·243	0·540
Milk	0·230	0·500
Legumes (peas and beans)	0·122	0·450
Groundnut	0·122	0·225
Soya	0·190	0·400

The amino-acid composition of proteins can be found by reference to published food tables, and then the amount of any chosen amino acid per gram of protein nitrogen may be calculated. The appropriate FAO factor is applied to this figure to obtain a 'score' for the protein value of the diet. The quality and quantity of protein in the diet may then be expressed as NDp Cals per cent.

The two following worked examples show: (i) the step-by-step calculation, and (ii) the data set out in tabular form—an arrangement which facilitates the calculation of a diet having several ingredients.

Examples

Diet: Whole wheat 90 per cent.
Skim milk powder (SMP) 10 per cent.

Chosen basis for scoring: sulphur amino acids (SAA).
Relevant data from food tables:

Wheat—8·9 per cent protein: 333 Cals per 100 gm; 0·20 gm SAA/gm N
Skim milk powder—34·5 per cent protein: 326 Cals per 100 gm; 0·23 gm SAA/gm N
Protein is assumed to contain 16 per cent of nitrogen.

Calculation of total nitrogen of diet

90 gm of wheat contains $\frac{8{\cdot}9 \times 90}{100}$ gm of protein

or $\frac{8{\cdot}9 \times 90}{100} \times \frac{16}{100}$ gm of nitrogen

$$= \frac{8{\cdot}9 \times 90}{100 \times 6{\cdot}25} = 1{\cdot}28 \text{ gm N}$$

Similarly, 10 gm SMP contains

$$\frac{34{\cdot}5 \times 10}{625} = 0{\cdot}55 \text{ gm N}$$

Then, total nitrogen in 100 gm diet is

$$1{\cdot}28 + 0{\cdot}55 = 1{\cdot}83 \text{ gm}$$

Calculation of sulphur amino acids in diet

90 gm of wheat contains 1·3 gm N, of which 0·20 gm/gm is SAA (sulphur amino acids).

Then 90 gm of wheat contains

$$1{\cdot}3 \times 0{\cdot}20 = 0{\cdot}26 \text{ gm SAA}$$

and 10 gm of SMP contains

$$0{\cdot}55 \times 0{\cdot}23 = 0{\cdot}13 \text{ gm SAA}$$

Therefore total SAA in 100 gm diet is 0·39 gm OR $\frac{0{\cdot}39}{1{\cdot}83} = 0{\cdot}21$ gm SAA/gm N.

Scoring

But FAO recommend 0·27 gm SAA/gm N
therefore score is

$$\frac{0{\cdot}21 \times 100}{0{\cdot}27} = 79 \text{ per cent}$$

This scoring gives numerical expression for the quality of the protein in the diet. The quantity of protein in the diet must now be expressed in terms which can be related to this expression of the protein quality.

Calculation of protein calories per cent of the diet

90 gm of wheat contains

$$\frac{90 \times 333}{100} = 299{\cdot}7 \text{ Cals}$$

10 gm of SMP contains

$$\frac{10 \times 326}{100} = 32{\cdot}6 \text{ Cals}$$

Therefore 100 gm of diet contains 332·3 Cals.

The total nitrogen in 100 gm of diet, calculated above, is 1·83 gm, which is equivalent to 1·83 × 6·25 gm of protein.

Proteins yield 4 Cals per gram.

Then the protein from 100 gm of diet would yield 1·83 × 6·25 × 4 Cals.

This amount as a percentage of the total Calories is

$$\frac{1{\cdot}83 \times 6{\cdot}25 \times 4 \times 100}{332{\cdot}3} = 13{\cdot}8 \text{ per cent}$$

It is now necessary to read off the value for NDp Cals per cent of the above diet from a nomograph which has been constructed to allow for the effect on NPU of the concentration of protein in the diet. On this nomograph it will be seen that the horizontal line for a protein score of 79 per cent intersects the vertical line for a

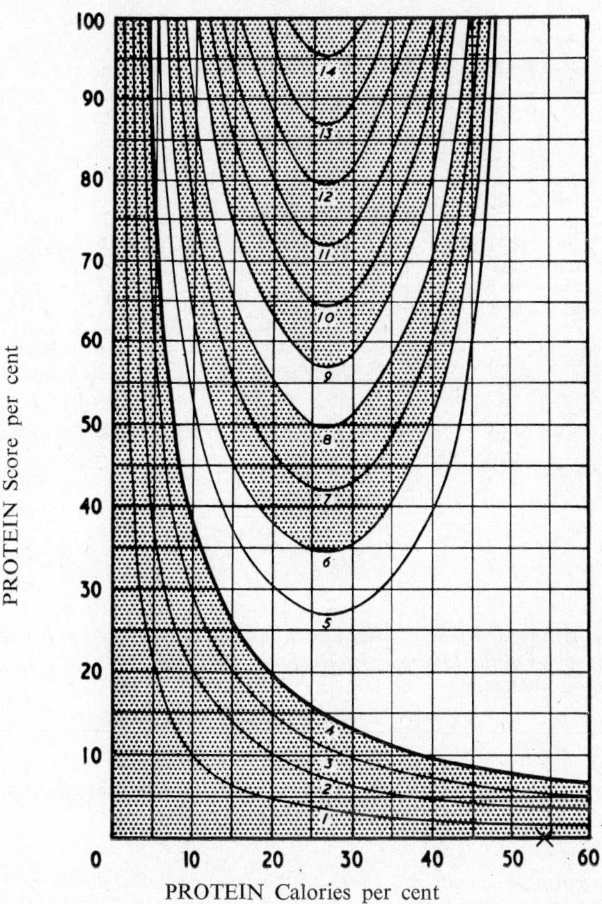

FIG. 2

Nomograph for the prediction of protein values of diets. The curves are lines of equal NDp Cals per cent.

protein Cals per cent of 13·8 close to the parabola marked 9. The above diet therefore has a calculated NDp Cals per cent of 9·2.

The foregoing calculations may be simplified by charting the data for any given diet as shown below, where diet 41B is used as an example.

Inspection of the nomograph reveals that: (i) when the protein Cals per cent is increased beyond 27 for a protein of any score, then the NDp Cals per cent begins to fall, because the protein is being metabolized for heat and energy; (ii) if the protein Cals per cent is less than 9 it is never possible to achieve an NDp Cals per cent of 8—the value thought to be required for a reasonable growth rate in rats; and, similarly, (iii) if the protein score is less than 50 it is never possible to achieve an NDp Cals per cent of 8.

DIET 41B

CHOSEN BASIS

SULPHUR AMINO ACIDS (METHIONINE AND CYSTINE)

Weight	Food	% P	% Cals	AA/gN[3]	Ng	Cals	AAg
a		b	c	d	$\frac{ab}{625}$	$\frac{ac}{100}$	$d\frac{ab}{625}$
47	Wheat flour[1] English, 100% extr.	8·9	333	0·20	0·67	156·0	0·134
40	Oatmeal[1]	12·1	404	0·20	0·77	162·0	0·154
8	White fishmeal[2]	61·0	365	0·22	0·78	29·2	0·172
1	Dried yeast[2]	41·5	360	0·10	0·07	3·6	0·007
3	Skim milk powder[1]	34·5	326	0·23	0·17	9·8	0·039
1	Salt (NaCl)	Nil	Nil	Nil			
100	Mixture	—	—	—	2·46	360·6	0·506
					e	f	g

$$\text{Protein Cals } \% = \frac{e \times 2500}{f} = 17{\cdot}1$$

$$\text{Score} = \frac{g}{e} \times \frac{100}{\text{FAO}} = 76\%$$

$$\text{NDp Cals } \% = 9{\cdot}7$$

SOURCES

1. *The Composition of Foods*, by McCANCE, R. A., and WIDDOWSON, E. M.
2. *Bulletin No. 48 of the Ministry of Agriculture, Fish, and Food.*
3 See table in the text.

APPENDIX II

NOTES ON CHEMICAL COMPOSITION OF FOODSTUFFS USED IN ANIMAL DIETS

Cereals

Wheat has a protein content of 8–15 per cent, depending on the variety. Soft, English wheats contain less protein than the hard wheats grown in Canada. The average figure for mixed wheats is 11·5 per cent of protein, 62 per cent of carbohydrate, 2 per cent of fat, 13 per cent of water, and fibre and mineral salts. The fibre content is low, about 2 per cent.

Oats and oatmeal contain about 12 per cent of protein. The fibre content of Sussex ground oats is about 11 per cent, but oatmeal (from which the hull has been removed) contains less fibre. Oats contain more fat than other cereals; ground whole oats 5–6 per cent of fat, and oatmeal 7–8 per cent.

Maize contains about 10 per cent of protein, but is notoriously deficient in tryptophan and lysine, two of the essential amino acids. *Yellow* maize, alone among the cereals, contains carotene (the precursor of vitamin A) equivalent to about one International Unit of vitamin A per gram. Maize contains about 4 per cent of fat as corn oil and is a good source of essential fatty acids.

Barley and barley meal contain about 10 per cent of protein and 5 per cent of fibre.

All cereals yield about 3·5 Cals per gram, and the whole grains are good sources of B-vitamins. Vitamin E is present in cereals, notably in wheat, but vitamins A, D, and C do not occur in cereals. Cereals contain considerable amounts of phosphorus and other minerals.

Cereal by-products. Bran, middlings, and offals are by-products of the milling industry obtained during the manufacture of flours and meals when the outer layers of the grains are removed. *Fine middlings* from wheat has a composition similar to that of whole wheat, but with a slightly higher fibre content. *Fine bran* (wheat) has a fibre content of 9–10 per cent. *Wheat germ* is rich in protein (30 per cent) and oil (9 per cent), and also contains considerable amounts of B-vitamins and Vitamin E. Wheat germ is liable to become rancid during storage, because of its high fat content. Vitamin E is destroyed in the presence of rancid fat.

Hays and grass meals

The protein content of hays varies with the ripeness of the crop. Hays cut before flowering and seeding have higher protein and lower fibre content than late-cut hays. The ranges are 7–15 per cent of protein and 20–30 per cent of fibre. Timothy, sainfoin, and seed hays have a higher fibre content than good-quality meadow hay. *Sun-cured* hays contain small and variable amounts of vitamin D.

Lucerne and grass meals also vary in composition with the ripeness of the crop. Lucerne in bud contains 22 per cent of protein and 18 per cent of fibre, but in early flower the values are 16 per cent and 25 per cent, respectively. The figures for grass meal are similar. These meals are good sources of carotene.

Greenfoods

Cabbage and similar vegetables are fed chiefly to supply vitamin C, though they are also good sources of carotene. Greenfoods contain about 90 per cent of water and about 2 per cent of protein. Good, fresh cabbage contains about 1 mg/gm of vitamin C. The carotene content is associated with the colour of the leaf, and the greener the leaf, the higher its carotene content. Cabbage leaves may contain carotene equivalent to between 50 and 2,500 IU of vitamin A per 100 gm. Kale contains more vitamin C and carotene than cabbage.

Food supplements in common use

Raw ox liver is a good source of vitamins. It contains about 75 per cent of water, 16 per cent of protein, and 14 mg/100 gm of iron.

Cod-liver oil is an excellent source of vitamin A (750 IU/gm) and vitamin D (90 IU/gm). It also contains iodine.

Whole milk has about 87 per cent of water, but the solids consist of all the nutrients, in approximately the right proportions, known to be required for animal growth. The vitamin A and D content of summer milk is higher than that of winter milk. The vitamin C content falls rapidly during storage, from 1·5 to 0·5 mg/100 gm in one day. Riboflavin is destroyed rapidly if the milk is exposed to sunlight.

Dried yeast is a good source of B-vitamins.

Protein-rich foodstuffs

Skim milk powder contains about 35 per cent of protein of good quality, 1 per cent of fat, 50 per cent of carbohydrate (as lactose or 'sugar of milk'), 5 per cent of water, and minerals. *Whole milk powder* (full-cream milk powder) contains about 27 per cent of protein of good quality, 30 per cent of fat, 40 per cent of carbohydrate, 1 per cent of water and mineral salts. Milk powders also contain B-vitamins and some vitamin C. Vitamins A and D are present in whole milk powder, but not in skim milk powder. Whole milk powder is sometimes *fortified* by the addition of vitamins A and D.

White fishmeal is a very variable product, its protein *quality* depending on the amount of heat-damage which had occurred during manufacture. Though, when analysed chemically, a fishmeal may be shown to have a high protein content, some of the protein may have been so badly damaged during processing that it can no longer be fully utilized by the animal body. The biological value of fishmeal can range from 20 to 80 (the maximum biological value attainable is 100). White fishmeal is manufactured from non-oily fish. In addition to protein it contains mineral salts, and is an important source of iodine and calcium (from the fish bones).

Oil-seed cakes and meals. These may be from linseed, groundnut, cottonseed, palm kernel, rapeseed, soya-bean, and many other oil seeds. They are by-products of the margarine, cooking fat, soap, and cosmetic industries. The removal of the oil leaves a product relatively rich in protein (30–50 per cent). The amount of residual oil depends on the extraction process, the lowest oil content being found in meals which have been solvent extracted. Meals which have been extracted with chlorinated hydrocarbons are toxic to animals, but other solvents are harmless. Oil-seed cakes and meals are specified as 'decorticated' or 'undecorticated' (indicating whether or not the seed coat or shell has been removed before processing) and as 'solvent extracted' if they have been subjected to such treatment.

Before cottonseed is used as animal feed it should be ascertained that the sample is free from gossypol—a toxic substance which sometimes occurs in cottonseed. Groundnut meal has recently suffered unfavourable criticism because some samples were found to be toxic, though the toxic effect does not manifest itself until the groundnut is fed at the 20 per cent level. Manufacturers of oil-seed cakes and meals are aware of the possibility of the presence of these toxins, and batches of oil seeds are tested and treated, if necessary, before they are put on the market.

The *quality* of the protein from oil-seed cakes and meals is not good, and these substances cannot be relied on as the sole source of protein in a diet. Nevertheless, oil-seed meals obtained from suitable processing remain a valuable source of protein for animals.

REFERENCES

(i) GENERAL READING

MAYNARD, L. A. and LOOSLI, J. K., *Animal Nutrition.* McGraw-Hill, 4th Edition (1956).

TYLER, C., *Animal Nutrition.* Chapman and Hall Ltd. (1950).

(ii) FOOD COMPOSITION TABLES AND DIET RECIPES

McCANCE, R. A. and WIDDOWSON, E. M., *The Composition of Foods.* Medical Research Council Special Report Series, No. 297 (3rd revised edition of Special Report No. 235). HMSO (1960).

CHATFIELD, C., *Food Composition Tables for International Use.* Food and Agriculture Organization of the United Nations, Nutritional Studies Nos. 3 and 11.

EVANS, R. E., *Rations for Livestock*, Bulletin No. 48 of the Min. Agric., Fish, and Food. HMSO (1960).

ALBRITTON, E. C., *Standard Values in Nutrition and Metabolism*. W. B. Saunders and Co. (1955).

(iii) THE EVALUATION OF THE NUTRITIVE VALUES OF DIETS

MILLER, D. S. and PAYNE, P. R., 'Problems in the Prediction of Protein Values in Diets: (i) 'The Influence of Protein Concentration', *British Journal of Nutrition*, **15**, 11 (1961). (ii) 'The Use of Food Composition Tables', *Journal of Nutrition*, **74**, 413 (1961). (iii) 'Caloric Restriction', *Journal of Nutrition*, **75**, 225 (1961).

TABLE 1

SUGGESTED DIETS AND SUPPLEMENTS FOR LABORATORY ANIMALS

All amounts are in gms or mls per animal/day unless otherwise stated

SPECIES	BASIC DIET	AVERAGE AMOUNT EATEN DAILY	SUPPLEMENTS	AVERAGE DAILY INTAKE OF WATER
Mice	Any of the recognized cubed diets, e.g. 41B	5	Usually none necessary, though some inbred strains may have to be supplemented	6
Rats	,, ,, ,,	15	None necessary	24
Cotton Rats	,, ,, ,,	15	5 cabbage	24
Hamsters	,, ,, ,,	10	5 cabbage	8
Guinea pigs	SG1 pellets + a source of vitamin C + hay	30	VITAMIN C ESSENTIAL, usually supplied by 50 cabbage	85
Rabbits	SG1 pellets	100	None necessary except hay for lactating does	330
Ferrets	50 raw meat + up to ¼ pt. whole milk	—	—	Give water in addition to whole milk
Monkeys	As for mice + a source of Vitamin C	4% of body weight	VITAMIN C ESSENTIAL, usually supplied by 70 cabbage	Varies with size
Dogs	40 proprietary tinned dog meat (NOT meat and cereal) + 30 skim milk powder + 41B powder to satisfy appetite (up to 300) + 10 mls cod-liver oil. Mix and damp slightly with water		Double meat and milk ration for pregnant and lactating bitches. Warm bread and milk for pups from 3 to 10 weeks of age (in addition to basic died)	Varies with size
Cats	No complete proprietary diet yet available Suggested diet: Cooked horsemeat 50% Cooked peeled potatoes 43·25% Biscuit meal 6·5% Mineral and vitamin mixture 0·25%	Adults, 4% of body weight Growing, 6–8% of body weight	1·0 gm. special, feeding bone meal per cat per day ...	100–150
Fowls	Chicks: chick pellets (25% protein fortified) Growing: pellets 16% protein Adults: complete pellet	110–225	None	Varies with age

See relevant chapters of this book for rations for other species and *UFAW Handbook* for general information

CHAPTER TWENTY-FIVE

The Production and Use of Pathogen-free Animals

INTRODUCTION

Breeders of laboratory animals are aware that the presence of natural disease in breeding colonies causes economic loss. Experimentalists, on the other hand, are aware that if their animals are already infected at the beginning of an experiment, even if the disease is unapparent or latent, then it is unlikely that the experiment will yield satisfactory results. Intercurrent deaths may make it necessary for experiments to be repeated. The effects of subclinical disease may be less obvious, but experimental results may be so influenced as to be completely misleading or, at best, more difficult to interpret. The frustration suffered by research workers on this account over the years has led to a demand for animals variously described as either 'healthy', 'infection-free', or 'disease-free'.

In breeding colonies the prompt removal of clinically affected and in-contact animals has reduced overt disease, but even the highest standards of husbandry and hygiene have failed either to control the commoner diseases or to eliminate parasitic infestations. The problems created by latent and unapparent infections remain as a threat to producer and investigator alike. Similarly, control measures based on mass chemotherapy or on the use of broad-spectrum antibiotics have generally failed to produce lasting success. In the literature there are accounts of small colonies of animals being developed free of a single specified pathogen after a tedious programme of testing, with the slaughter of all 'positive' animals. Such colonies were extremely useful for the research programmes of their originator, but they had few attributes for general use. Nelson and Gowen (1931) described such an attempt to establish a rat colony free from respiratory disease. The colony was built up by selection and isolation. A brother–sister mating system was followed. The parents were killed and examined after their first litter had been weaned. If the parents showed no abnormalities, then the weanlings were retained for further breeding. After eight generations the colony was expanded by random breeding. Even by this elaborate procedure, chronic respiratory disease was not eliminated as older animals continued to develop progressive inflammatory pulmonary lesions. However, otitis media (PPLO infection) was eliminated, and attempts to isolate *Streptobacillus moniliformis*, a common pathogen of the upper respiratory tract, were unsuccessful. Similarly, an attempt by King (1939) to build up a colony of healthy laboratory rats by using wild grey Norway rats as foster-mothers resulted in the elimination of otitis media but not of endemic pneumonia. By 1950 it had been established by Reyniers and his co-workers (see Reyniers, 1957) that stock obtained by careful aseptic hysterectomy techniques and introduced into, and hand reared in, the sterile environment of a germ-free isolator had, in one operation, been freed not only of all bacterial pathogens and parasites but also of many virus infections. Nelson (1951) made a second and successful attempt to establish a rat colony free from pulmonary disease using as his foundation stock three rats which had been obtained by Caesarean section and hand reared on the bench in Reynier's laboratory. At least one commercial rat production colony free from chronic murine pneumonia (Cumming and Elias, 1957) and a similar institutional breeding colony (Innes *et al.*, 1957) were established using animals from Nelson's colony as their foundation stock. In both instances the stocks were shown to be free of internal and external parasites. Henthorne and Veenstra (1957) were able to rear Caesarean-derived mice on lactating pathogen-free rats and in due course Caesarean-derived hamsters, multimammate rats and deermice were raised by the pathogen-free mice. Cannabalism of the young was prevented by the use of highly scented cedar shavings as bedding.

Subsequent work has amply confirmed the fact that the placenta acts as a most efficient barrier against bacterial and parasitical agents, but it is not so effective against viruses. First Foster (1957) and later Davey (1959) showed that it was possible to set up 'healthy' or 'specified-pathogen-free' colonies of rats by rearing a nucleus of Caesarean-derived animals to maturity on sterilized or pasteurized food behind a 'barrier' designed to exclude their natural pathogens. Some form of 'barrier' is now generally recognized as being essential to the prolonged operation of any type of SPF production colony. The original group was then expanded to the size required by natural breeding. The animals in these colonies acquired a basic non-pathogenic flora, probably through the medium of the animal technicians, but Davey (1957) fed two strains of lacto-bacilli to his rats. In this way the intestinal tract was rendered structurally and functionally similar to that of conventionally raised animals. Reyniers (1957) has suggested that 'clean' animals can best be obtained by starting with germ-free animals and bringing them up to a desired level of contamination by deliberately introducing pure cultures of acceptable organisms into their environment. This system may be used with success in rats and mice, but so far it has proved impossible to transfer guinea-pigs from the germ-free to the conventional environment. We have been able to hand rear Caesarean-derived guinea-pigs and rabbits in both conventional and SPF environments, simply allowing them to acquire a bacterial flora from the environment. However, Caesarean-derived guinea-pigs being hand reared as foundation stock in a new SPF unit which had been most carefully cleaned and disinfected before the stock was introduced appeared to thrive better after a strain of Lactobacillus isolated from a healthy conventionally reared guinea-pig was introduced into their environment.

DEFINITION OF THE TERM SPF (SPECIFIED-PATHOGEN-FREE)

This designation, despite its many drawbacks, is generally accepted as being an appropriate description for animals specially reared for laboratory use and which are not infected with those pathogens and parasites commonly found in conventionally raised animals of the same species. At the third I.C.L.A. Symposium (Sabourdy, 1963) the nomenclature sub-committee presented the following definition of the term SPF:

'Animals that are free of specified micro-organisms and parasites but not necessarily free of others not named.'

This definition is still the subject of much debate, and it might be more acceptable if it commenced 'Animals produced in colonies regularly examined and tested for the specified micro-organisms . . .' and included some reference to internal and external parasites and pathogenic protozoa.

It cannot be too often stressed that SPF animals are functionally and structurally identical with their conventional counterparts, but their flora and fauna are to some extent controlled. They are not germ-free, but they should be pathogen-free and parasite-free. They are not produced in a sterile environment, and therefore, although ideally everything passing into their environment should be sterilized, there is no objection to allowing materials which have been properly pasteurized to pass in. There is, as yet, no generally agreed definition of 'pasteurization' in this context, but as a working guide, materials throughout their mass should be raised to a temperature of 70°C for 30 min. or to 90°C for 3 min. Such procedures would certainly destroy vegetative bacteria, parasites and their eggs, and also encysted protozoa.

PRODUCTION OF SPF ANIMALS

Most of the published work concerning SPF colonies and their operation has been concerned with rats and mice, but it is equally important that SPF guinea-pigs and rabbits and possibly other species should be available to research workers because overt disease and inapparent infections in these species have caused much frustration (see Paterson, 1956; Paterson, 1962; Sacquet, 1962). The underlying principles of SPF technology apply to all species, but naturally there are considerable differences in the application of these principles to the different species of laboratory animals, just as there are in the operation of their conventional counterparts.

The technical points which require consideration may be allocated to one or other of the following general topics, viz.:

1. The building itself.
2. The environment in the animal rooms.
3. The barrier system and its maintenance.
4. Preliminary sanitization of the interior of the building.
5. Introduction of stock.
6. Colony management.
7. Transport of animals.

1. The building

Where it is possible to start operations in a new building, this should be done because the adaptation and alteration of existing buildings for SPF operation is seldom entirely satisfactory. However, if production is to be conducted on a moderate scale for a limited time satisfactory modifications can often be made.

The general principles to be followed in the design and construction of buildings for the production of SPF animals are similar to those which have been found to be satisfactory for the production of healthy conventional animals, but the following points must be given very close attention:

1. The building must be rodent and vermin-proof. It must be well insulated and constructed of durable, fireproof materials and with a minimum of external openings. Windows must be sealed to the wall structure, and should be kept to the absolute minimum. A possible solution is to place a narrow run of hollow glass bricks at high level in corridors and animal rooms. This would mean that artificial light would be required for normal working, but the natural light would be ample for personnel to carry out essential tasks if there were a power failure during usual working hours. If it is considered necessary for aesthetic reasons to have windows for personnel to see out of, then these should be double glazed, sealed to the fabric of the building and placed either in rest rooms or at the end of blind corridors.
2. Non-glare artificial lighting (15 lumens at 30 in. above floor level) must be provided in all animal rooms.
3. Interior walls and ceilings should be vapour sealed and have a smooth and impervious finish. The floor likewise should be damp-proofed, smooth, and impervious.
4. Changing rooms with showers must be provided for personnel to decontaminate themselves as they enter the building.
5. A double-ended autoclave, steamer, or ethylene oxide chamber is required to sterilize or pasteurize ingoing materials and for passing out waste materials. Alternatively, waste materials may be sucked or blown out of the unit by a suitably designed pneumatic system.
6. A chemical dunk-tank and a tunnel fitted with air-tight doors and equipped with ultra-violet light are useful adjuncts for passing objects which cannot be subjected to heat either into or out of the building.

2. The environment

A stable environment is desirable, and this can be best achieved by using a single air-conditioning unit which will sterilize, heat or cool, and humidify the incoming air as required, and circulate it throughout the building. Whenever climatic conditions allow, circulation of 100 per cent fresh air is desirable to keep down animal smells, but the development of activated carbon filters may lead to the greater use of recirculated air in buildings situated in the colder parts of the world. To provide maximum flexibility, each individual room should be provided with its own temperature and humidity control preset to ensure optimal environmental conditions for the species it houses.

All mechanical equipment relating to environmental control should be placed outside 'the barrier' so that it may be maintained without service personnel having to pass through 'the barrier' and enter the building. Alarms to indicate the occurrence of high and low temperature at selected points, power failure, air-flow failure and steam-pressure failure, must be provided and set to provide visual and sound warnings at a point which is manned 24 hours each day. As an insurance against power failure, a standby diesel-powered electric generator may be installed.

3. The barrier system

The principle behind this concept is that it is possible, in large measure, to control the entry of pathogens into the animal area. Firstly, by arranging the air pressures in the building so that the animal rooms (0·15 in. water gauge above atmosphere) are slightly positive to the adjacent corridors (0·1 in. water gauge). The latter, in turn, are held positive to the changing rooms (0·05 in. water gauge). Thus the whole building, except the entrance air lock, is above atmospheric pressure. Pressure variations are achieved by balancing the input and extract ventilation rates to the desired level in each section. The differential pressures must be monitored frequently by reference to inclined manometers. Any fall in pressure must be investigated and the fault corrected. The possibility of a break in the physical barrier, e.g. a broken window, a cracked ceiling, or a badly fitting door, should not be overlooked. Secondly, precautions must be taken to ensure as far as is practicable that pathogens and parasites are not conveyed into the building by vermin and insects, personnel, bedding, food, or water, or by other materials such as transport boxes, record cards, and animal caging. Care must also be taken to ensure that animals and waste materials orginating in the building must be brought out without endangering the security of the barrier.

AIR STERILIZATION. Air entering the ventilation system must first be cleaned by passing through a screen to remove gross particles and insects. It is then passed through a roughing filter to remove particles down to 10 microns. After passing the fan, the air is sterilized as it passes through a high-efficiency filter. Towards the end of each duct, close to the point where air is delivered into a room or passageway, it is advisable to install a high-efficiency filter, referred to later as the terminal filter.

The extract system must also be fitted with filters. First, in the animal room itself a removable 'rough filter' pad must be fitted to each extract grille to prevent the gross particles of dust and animal hair originating in the animal house from entering the ducts. Immediately behind the rough filter, a high-efficiency filter should be fitted. To guard against the possibility of a 'blow-back' of pathogens or insects from the atmosphere, a high-efficiency filter is placed in the main extract duct before the fan. This is a real danger during periods of power failure or other mechanical breakdown when positive pressure in the animal quarters will be lost, but the danger may be minimized if the delivery and extract fans are linked so that if one fails, the other stops automatically. (See Chap. 2, part two, for further details.)

VERMIN AND INSECTS. The use of step-over vermin barriers 24 in. high in external doorways is recommended. The upper edge should have a 4-in. outward-turned lip. Barriers may be permanent, and if so, they should be sealed to the floor and door-frame: if they are removable to allow the passage of trolleys or barrows into the building they must be replaced immediately after the movement has been completed. Special attention must be directed to every point where service ducts or pipes pass through the walls, ceilings, or floors. These must all be plugged and sealed to prevent wild mice, rats, and even insects from entering and establishing themselves in the building. A special danger arises if the outer wall is of the hollow vented type. The exclusion of crawling and flying insects is not simple, and according to the magnitude of the problem, the outer air lock in the entry vestibule should be provided with continuous insecticidal vapour generators or self-priming aerosol dispensers which are operated automatically each time the outer door is opened.

PERSONNEL. Personnel entering the building are potentially a serious danger to the security of the barrier. They should enter by a system of three air-locks which may be fitted with electrically interlocked doors to ensure that the door ahead is not opened until the door behind is secured. All clothing except under-garments are left in the second lock. In the third lock the face, hands, and feet are carefully washed before donning working clothes consisting of shirt, blouse or surgical gown, trousers, caps, and shoes. These should be provided each day, either freshly laundered or sterilized. In an emergency the same clothes may be worn for not more than three consecutive days, provided the clothes remain inside the unit and are not brought out further than the innermost lock. Where stricter precautions are deemed necessary, e.g. in the case of valuable nucleus colonies, a whole-body shower system may be interposed between air-locks 2 and 3. Showering may be omitted when leaving the unit.

BEDDING. The most satisfactory system for passing bedding into the unit is to have a double-ended autoclave or ethylene oxide chamber as part of the barrier system in which bedding may be sterilized as it crosses the barrier. In an autoclave wood shavings packed loosely in rot-proofed hessian sacks may be sterilized at 30 lb/sq. in. (134°C) for 30 min. Foster (1962) has reported that bales of wood shavings may be sterilized at 15 lb/sq. in. (121°C) for 15 min providing the cycle begins and ends with the drawing of a 27-in. vacuum. In preliminary observations with a similar process and drawing 27-in. vacuums, we found that sacks of loosely packed wood shavings were sterilized in 3 min. at 30 lb/sq. in. (134°C). Sawdust was more difficult to sterilize, and a longer period of treatment (not less than 6 min.) was required. In the high-vacuum process materials are dried during the pulling of the final vacuum and by flushing the autoclave with sterile (filtered) air. Ethylene oxide gas is an efficient sterilizing agent, and its efficiency is improved if a high-vacuum technique is used (Foster, 1962). The process is slow, but running costs are low. Highly absorbent materials, such as peat moss, or dried corn cobs, are more satisfactorily sterilized by ethylene oxide than by steam.

FOOD. Every process available for the sterilization or pasteurization of food causes some loss of nutrient value, but colonies of rats and mice have been maintained on diets treated by the following methods:

(*a*) Baking to 600°C (Foster, 1962), *B. subtilis* was the only organism recorded from treated diet.

(*b*) Autoclaving at 5 lb/sq. in. for 30 min in a conventional autoclave (Davey, 1962). No bacteriological results available.

(*c*) Preliminary high vacuum (27 in.), followed by autoclaving at 107·2°C for a period of 15 min. No undue destruction of vitamins was noted (Foster, Black, and Pfau, 1964).

(*d*) Exposure to 2·5 Megarads gamma-irradiation using a Cobalt-60 source (authors' unpublished observations). By this treatment, the diet is completely sterilized, and work which is still in progress indicates that although destruction of vitamins does occur, it is not excessive and can be allowed for by incorporating an excess of the more labile substances in the formula. An established SPF colony of rats whose diet is sterilized by irradiation (2·5 Megarads) has continued to flourish for six years. Colonies of SPF mice, guinea-pigs, and rabbits have been successfully raised and expanded on irradiated diets. The general use of irradiation for sterilizing laboratory animal diets is feasible, but the cost of irradiation and transport charges make it a relatively expensive process. This might be substantially reduced if demand rose appreciably.

WATER. Normally water from the public main may be used but as an additional precaution, water may be either chlorinated (12 ppm) or acidulated to pH 2·5 with hydrochloric acid (McPherson 1963) on the premises. If open reservoirs serve the premises, there may be a danger of contamination with worm eggs or coccidial oocysts and, in these instances, it may be necessary to pass all water entering the building through 'Berkfeld' type filters to hold back parasitic eggs and protozoal cysts, neither of which are destroyed by mild chlorination or acidulation. The use of ultra-violet light units has been shown to be a practical alternative to chlorination. Lastly, there is the possibility of rendering water safe by heat distillation processes, but this latter method may involve considerable recurring expense.

4. Preliminary sanitization of the building

This important task should not be undertaken until the building is ready for occupation. It matters little whether the building was previously used to house conventional animals and has been adapted for SPF production or whether it is a new building. The following system of cleaning, disinfection, and fumigation has been found effective and is recommended. The ventilation system of many buildings is not designed or built to retain penetrating fumigants such as formaldehyde and ammonia, and there is the possibility that these gases may leak from the ducting and permeate parts of the building outside the areas being fumigated. Therefore, before these agents are employed, expert advice should be sought from the ventilating engineer. The underlying principle is first to clean the inside of the building physically and then to sterilize the surfaces as far as possible. It is important that each step is carried out in the order given. Fumigation may stop short of the entrance porch but cleaning and disinfection must be carried through to the outer door.

1. The heating and ventilation systems are run for at least seven days without the terminal filters in place. During this period much of the dust and also possible contaminants lodged in the duct-work of the delivery ventilation system will be blown into the building, while that which has lodged in the extract system will be blown to the exterior.

2. (*a*) The terminal filters are then placed in position and the system is run for a further 72 hours, during which time the functioning of the thermostats and alarm systems are checked. Finally, the air balances are adjusted and measured manometrically.

(*b*) While the above tests are in progress the ceilings, walls, floors, and doors of the entire unit and all fittings and equipment—benches, chairs, desks, cupboards, racking, and cages, etc.—are scrubbed or swabbed with hot water containing 4 per cent washing soda (sodium carbonate crystals), particular attention being paid to areas in which grease or dirt may be trapped. When this cleaning is completed the whole is washed down with very hot clean water.

3. (*a*) With the ventilation system shut down, the short lengths of trunking between the room grilles and the terminal filters may be decontaminated by fogging or spraying the space with 10 per cent formalin at the rate of 1 ml/cu ft. If coccidia are a possible hazard, then the space must be similarly treated with 10 per cent ammonia not less than 24 hours after the formalin treatment.

(*b*) The main duct work may be decontaminated with either formaldehyde or ammonia if suitable tube points for introducing these agents with steam are provided at intervals along the ducts. The volume of the duct to be decontaminated is calculated and, as detailed below, the appropriate amount of formalin or '880' ammonia solution diluted in tap water, is boiled to near dryness in a portable, closed, vessel connected by wide-bore ($\frac{1}{2}$ in.) polythene tubing to the points or the duct work. Twenty-four hours after each fumigation is completed the ventilation is switched on for a short time and the system cleared of fumigant.

(*c*) All ventilation grilles in the rooms and passageways are then sealed, and the walls, ceilings, floors, fittings, and equipment are sprayed with a 1 per cent solution of Tego (103G) and left for 24 hours.

4. The lid on the 'dirty' side of the still empty dunk tank and the outer door of the ultra-violet tunnel are sealed; the corresponding 'clean' side closures are fixed in an open position. The building is then ready for fumigation, and this is carried out in the following manner, each room, corridor, passageway, or air-lock being considered as a separate unit.

(*a*) The cubic volume of each unit is calculated. A vessel of appropriate size with and integral 1-kW or 2-kW electric heating element is placed on the floor. For each cubic foot (0·0285 cubic metre) of space, 1·0 ml of a 10 per cent solution of commercial formalin (40 per cent formaldehyde) in tap water is placed in the vessel and evaporated to near dryness. The metric equivalent is 35 ml of the 10 per cent formalin solution to each cubic metre of volume. It is convenient if the electric current to each vessel is controlled by a separate time-switch set so that the current will be switched off when the formalin solution has been evaporated down to the level of the top of the heating element. To overcome the unlikelihood of time-switch failure, the electric element should be of the type fitted with a 'kick-out' device which operates when the vessel has boiled to near dryness.

(*b*) The rooms or units are then sealed, particular attention being paid to the doors. Adhesive tape 2 in. (5·08 cm) wide is useful for making good imperfect closures.

(*c*) The current to the heaters is switched on by remote control, and evaporation of the formaldehyde and water proceeds, usually for 2–4 hours, until switched off by the time-switches.

(*d*) After a period of 24 hours the building is entered for the purpose of removing the caps on the inlet and outlet ventilators. The person making the entry undergoes the full routine of personal decontamination and, in addition, must wear a respirator designed to afford protection against formaldehyde gas. The ventilation system is then run for 24 hours to remove the formaldehyde.

(*e*) The ventilation is then switched off and a second 'clean' entry is made. The caps are replaced on the inlet and outlet ventilators. The vessels used for the formaldehyde fumigation are first thoroughly washed with water and then recharged with a 10 per cent dilution of '880' ammonia in tap water, 1·0 ml of the dilute solution being allowed for each cubic foot of space. The units are again sealed and the ammonia solution boiled in the same manner as the formalin solution. The purpose of this second fumigation is to destroy coccidial oocysts which are not affected by formaldehyde.

(*f*) Twenty-four hours later a further 'clean' entry is made and wearing a respirator proof against ammonia fumes, the operative removes the caps from the ventilators, and seals the lid of the dunk-tank and the door of the ultra-violet tunnel on the clean side. The ammonia is cleared from the unit by switching on the ventilation system (this may take 24 hours or more).

(*g*) The last step is to fill the dunk-tank with a germicidal solution. In our experience a solution containing Tego (103G) and formalin (40 per cent commercial), each to a final concentration of 1·0 per cent in distilled water, has given satisfactory service. The prepared solution is run into the dunk-tank from the dirty side until it is about to overflow. The dirty air trapped on the clean side is allowed to pass to the dirty side through a metal tube passing through the panel dividing the upper part of the dunk tank. (*Note:* there must be a method of sealing the dirty end of the tube when it is not in use.) Twenty-four hours after the tank has been filled, the level of the fluid should be adjusted to the desired level by running the surplus to waste on the clean side.

(*h*) When the dunk-tank fluid has to be discarded and the tank refilled, the clean-side lid must be closed and sealed with tape. After the tank has been drained to the dirty-side, it should be cleaned out by swabbing with sterile cloths soaked in fresh formalin/Tego solution. The tank should then be refilled immediately as detailed in the preceding paragraph.

The 1 per cent Tego/1 per cent commercial formalin solution is very stable and does not attack stainless steel or aluminium alloys, the usual materials of which dunk-tanks are constructed. It retains its activity for many months. Vegetative bacteria on clean surfaces are killed within a few minutes, and bacterial spores in 1–2 hours. Tego solutions leave an active residual film on treated surfaces, and when packages are accepted on the clean side they may be left to drip-dry before being opened. Losses due to spillage, removal by dunked articles, or by evaporation must be made good by topping up the tank with freshly prepared fluid of the same formulation.

It must be stressed that dunk-tanks must not be expected to sterilize dirty or grossly contaminated surfaces. Their purpose is to allow materials which cannot be transferred by any other way to be passed rapidly into the unit. It follows, therefore, that the materials themselves must have been properly sterilized previously by some physical or chemical means and thereafter protected from contamination by at least two layers of effective non-permeable wrapping material. Immediately before the transfer, the outermost wrapping is removed, great care being taken not to allow its exterior surface to touch and contaminate any part of the inner wrapping. When packages are accepted on the clean side it is advisable to lay them aside for at least 2 hours before opening them.

5. Introduction of stock

Foundation stocks of rats or mice may be weaned stock from a germ-free unit. Normally such animals survive well when introduced into an SPF environment, but sometimes part of, or even the whole, batch may die from bacterial enteritis caused by *E. coli* or *Cl. welchii.* Alternatively, stock may be obtained from a reliable existing SPF colony. These procedures are not yet a practical method for setting up SPF colonies of the other commonly used species—rabbits, guinea-pigs, cats, and dogs. Until such time as reliable SPF stocks of these animals are available, Caesarean-derived young will have to be hand-reared within the barrier. In these latter cases it is likely that one of the institution's own strains would be chosen, but in any event, once the colony is established, further Caesarean-derived young of the original foundation strain or of additional strains may be introduced to be reared by established foster mothers.

(*a*) GERM-FREE STOCK. If the germ-free unit is close by, the easiest way of transferring the animals is to place the germ-free animals in a suitable container of appropriate size and then, after sealing its air-filtration ports, to pass it out of the germ-free unit and into the SPF unit via the dunk-tank as speedily as possible. Care must be taken to ensure the container chosen contains sufficient air to enable all the animals to survive without any distress being caused. To avoid contamination of the outer surfaces of the container if it has to be conveyed between buildings or, for that matter, more than a few yards inside a building, it should be placed in a plastic bucket and covered with 10 per cent formalin or other quick-acting disinfectant, e.g. 10 per cent 'Chloros' (ICI) (hypochlorous acid solution).

If the donor unit is at some distance from the SPF unit the transport container must be fitted with some form of high-efficiency bacterial air-filter to provide fresh air for the animals. On arrival at the SPF unit all external surfaces must be effectively cleaned and disinfected, and then if it is possible to seal the air filters it may be passed through the dunk-tank to be opened up on the clean side. Otherwise the container may be taken to the shower area to be opened at 'the barrier' (see below under (*b*) for details of this operation).

(*b*) SPF STOCK. If the numbers of animals are small and suitable containers are available the methods outlined above for germ-free animals may be used. However, it is possible that containers will be too large or may be otherwise unsuitable for passing through the dunk-tank. In such cases the outside of each container should be cleaned with a suitable detergent solution and then treated with an efficient disinfectant. It may then be taken to the inner barrier of the shower room, where the lid is removed and a 'clean' technician standing on the 'clean' side extracts the animals from the container, taking great care not to contaminate himself or the animals by touching the outside of the containers. This manoeuvre is greatly facilitated and the risks of contamination reduced if the animals are packed in removable wire-mesh baskets fitted with handles for ease of lifting.

(*c*) STOCK OBTAINED BY ASEPTIC HYSTERECTOMY (elsewhere referred to as Caesarean-derived, or CD, stock).

(i) *General considerations.* Basically the same procedure may be used for rats, mice, and rabbits, but a somewhat different technique is, in our opinion, needed for guinea-pigs. A pregnant female at full term is prepared for aseptic laparotomy, and when all is ready she is killed by cervical fracture. The entire uterus is removed and transferred via the germicidal dunk-tank into the SPF unit, where the young are removed from the uterus and revived. With guinea-pigs, however, using this technique, few foetuses can be revived. Better results are obtained if a sterile hysterectomy is performed and the young are removed from the uterus in a sterile operating hood, where they are resuscitated before being passed with aseptic precautions through the dunk-tank into the SPF unit.

(ii) *Diagnosis of pregnancy.* In rats, mice, and guinea-pigs this is easily determined by simple inspection, but in the rabbit, abdominal palpation may be needed to confirm the presence of forming foetuses. This is most easily done in the third week of pregnancy between the 14th and 21st days.

(iii) *Timed mating and planned hysterectomies.* A good survival rate cannot be expected if the foetuses are premature. If acceptable survival rates are to be obtained, care must be taken to ensure that the donor mothers are at, or close to, the point of parturition. The best survival rates in mice and rats achieved by the authors have occurred when timed matings have been carefully made and parturition has been delayed by up to 24 hours by the use of progesterone (Cook, 1965). In rats and mice this allowed the hysterectomies to be carried out at pre-arranged times, a desirable procedure when carrying out a large-scale programme or where the numbers of donor females and/or foster mothers are limited. Mice of various strains whose normal gestation period is 19 days were placed with fertile males for 24 hours. On the 17th, 18th, and 19th days of pregnancy each mouse was given 0·2 mg progesterone sub-cutaneously (0·1 ml of 'Protormone' (10 mg/ml) Burroughs Wellcome Ltd., diluted 1 in 5 in sterile ethyl oleate). The hysterectomy was carried out at 9 a.m. on the 20th day, i.e. 19 days and 21 hours after the females were first placed in contact with the male. Female rats of the Chester Beatty inbred strain, whose normal gestation period is 22 days, were inoculated with 0·2 ml neat 'Protormone', i.e. 2·0 mg progesterone on the 21st and 22nd days of pregnancy, and the hysterectomy was performed at 2 p.m. on the 23rd day (23 days and 2 hours after the females were first placed in contact with the males).

Parturition in the rabbit (normally 31 days) may also be delayed by injecting progesterone on the 30th and 31st day of pregnancy, the hysterectomy being performed on the morning of the 32nd day. The dose is 2·0 mg/kg bodyweight. Using 'Protormone' (10 mg/ml), Burroughs Wellcome Ltd., if the donor mother's weight in kg is divided by 5, the answer equals the number of ml to be injected. Thus, if the rabbit weighs 5·5 kg the dose in ml is 5·5 ÷ 5, i.e. 1·1 ml, which contains 11·0 mg progesterone.

The gestation period of the guinea-pig varies between 59 and 72 days, the average being 63 days. The careful timing of mating and planning of hysterectomy times is not, therefore, a profitable exercise. Good survival rates are obtained when the donor mothers are at full term, and this is readily de-determined by observing the degree of separation of the pelvic bones. In the non-pregnant animal and until 7 days before parturition, the right and left pubic bones overlap slightly. As parturition approaches, the bones part company and a small but definite gap is discernible on the 5th day before parturition. This gap increases steadily, and when it reaches 2·5 cm (approximately the width of the average human thumb) parturition is imminent and the animal may be used for the Caesarean operation.

(iv) *Aseptic hysterectomy*. The technique is essentially the same for mice, rats, and rabbits, but it has to be modified for guinea-pigs. For species such as the dog and cat and for farm animals it would have to be considerably modified (see Betts, Lamont, and Littlewort (1960), Sweat and Dunn (1965), and Edward, Mills, and Calhoon (1967)), as in these species true Caesarean section—hysterotomy—may be carried out.

As the time for the operation approaches, the trunk of the donor animal is clipped and depilated and painted with a persistent germicide, such as dilute tincture of iodine (*Liq. iodi. mitis. B. vet. C.*, 1965). If the animal is not to be used immediately she is placed in a sterile box to await operation. We do not favour the use of anaesthetics and prefer to kill the donor mother by breaking the neck (cervical fracture). This method when properly carried out is humane and does not depress foetal viability. Immediately after it has been killed the animal is completely immersed in a freshly prepared 10 per cent solution of 'Chloros' in distilled water held at 40°C. The wet body is then placed on its back on a sterile tray and the surface of the abdomen is thoroughly dried with sterile lint swabs. Drying is speeded up and undue chilling of the body is avoided if the work area is heated to 40°C by an overhead infra-red lamp. The body is then removed to a second sterile operating tray or board, where it is secured in a position suitable for operating by lengths of adhesive tape passed over the thorax and the superior part of the hind legs. A sterile self-adhesive membrane (Steridrape, Minnesota Mining and Manufacturing Co. Ltd.) of appropriate size is placed over the entire body and made to adhere firmly to the prepared site on the abdomen. The operation then proceeds using good aseptic technique. All operatives scrub up and wear sterile gowns, hats, masks, and gloves; all instruments and swabs are sterilized. The abdomen is opened using scalpel and dressing forceps; the uterine horns are freed from their attachments, and as the uterus is lifted out on to the surface of the membrane it is clamped with a single pair of artery forceps close to the cervix. The uterus is severed at the neck immediately posterior to the forceps and placed in a sterile container filled with warm (40°C) 10 per cent Chloros for 1 min. When this period has elapsed it is removed from the germicidal solution by an operative wearing a sterile arm-length rubber gauntlet and passed rapidly through the dunk-tank into the SPF unit. Within the unit the foetuses are expelled from the uterus, freed from their membranes, and gently massaged and dried with soft sterile cloths. The umbilical cord may be severed manually or by cautery. To avoid undue chilling of the foetuses, all manipulations are best carried out in a work area which is heated locally by infra-red lamps to a temperature of approximately 37°C.

Rabbit foetuses revive very rapidly and become well oxygenated and vigorous within a few minutes, but rats and mice are rather slower and may require additional massage to stimulate them to breathe regularly. The survival rate of mice and rats is improved if the young are placed in an incubator (37°C) for 30 min before being fostered.

The survival rate of guinea-pigs obtained by the technique outlined above is low, possibly because their advanced state of development makes them more sensitive to the physiological effects of high carbon dioxide or low oxygen levels in the circulating blood. Good results may be obtained using a sterile operating-hood system, and the following procedures were used to establish an SPF colony of 220 females and 55 males.

The operating isolator, which was autoclavable and constructed of aluminium and glass, was fitted with standard glove ports on each side to enable two operatives to work. It had a dunk-tank at one end, and a warm sterile air supply was provided. The operating area was formed by a circular hole in the base of the isolator. This was covered by an autoclavable self-adhesive membrane. The apparatus containing all necessary operating instruments and transfer jars was prepared and sterilized by autoclaving as for germ-free operations.

The torso of the selected donor mother was prepared by clipping, depilation, and painting with iodine, as already described. The animal was killed by cervical fracture. After immersion in germicide and drying the body was strapped to a metal plate and placed on a pneumatically operated lifting device controlled by a foot pump, immediately below the operating area. The body was raised until the abdomen firmly adhered to the membrane. One operator carried out the hysterectomy, handing the uterus to the second operator, who released the foetuses from their membranes and resuscitated them.

Removal of the uterus from the abdomen caused the adhesive membrane to fall below the level of the floor of the isolator, but there was no loss of adhesion between the abdominal wall and the membrane. A second sterile membrane, stored within the operating isolator, was placed over the operating area and firmly secured to the floor of the isolator. The air pressure in the pneumatic lift was then released and the plate with the remains of the donor mother was allowed to fall away gently. As much as possible of the first membrane was carefully cut away from underneath and discarded with the body. The isolator was then ready for the next hysterectomy, it being unneccessary to resterilize. The youngsters meantime had been dried and revived. They were then quickly transferred in large-mouth bottles (4-lb glass fruit-preserving jars) from the operating isolator to the adjacent SPF unit via their respective dunk-tanks. In one period of 14 days eighty operations were performed, and of some three hundred young obtained, 95 per cent were resuscitated and transferred into the SPF unit.

(v) *Fostering procedures.* The fostering of rabbits, rats, or mice is not a difficult task and, providing the foster-mothers have good breeding records, very few of the fostered young will be rejected. Foster-mothers should have littered down during the preceding 24 hours. The chosen foster-mother is removed from her cage or box and her natural litter discarded. The youngsters to be fostered are placed in the vacated nest and the entire space is lightly sprayed with a scented aerosol. The foster-mother is similarly sprayed and returned to the nest. This spraying procedure materially increased the percentage of successes in mice, rats, and rabbits. Guinea-pigs may be fostered simply by placing in ones or twos in cages containing recently born young. It is not unknown for rats and mice to produce a solitary youngster many hours after the birth of the rest of her litter. This can cause confusion in inbred/line-bred strains if the fostered young and the foster-mother are of the same colour. When it is essential that the fostered young are positively identified it is advisable to snip off the tip of the tail of each foetus as it is removed from the uterus. This appears to stimulate breathing rather than cause any harm and, at weaning, serves as an aid to positive identification. Litters of guinea-pigs may be identified one from the other, by marking with dyes. Rabbits are very difficult to colour satisfactorily when very young, and if there is any possibility of confusion they should be fostered by females of a different breed and colour.

(vi) *Hand-rearing of guinea-pigs and rabbits.* It is important that new-born young do not suffer chilling. Guinea-pigs are highly developed anatomically when born and are able to move about freely; rabbits, mice, and rats, on the other hand, are devoid of fur and helpless. It follows, therefore, that totally different environmental conditions must be provided and that the techniques of rearing will be different.

(a) *Guinea-pigs.* These may be reared without hand feeding. Newborn young with freedom to move about in a warm environment and given a nutritionally adequate and palatable liquid diet in shallow dishes will learn to lap within a few hours of birth. We housed our guinea-pigs in solid-floored metal boxes measuring 16 in. long, 12 in. wide, and 8 in. high. Each box was individually heated from above by a 250-watt energy-regulated ceramic-rod infra-red heater fitted with a polished metal reflector so placed to produce a temperature of between 30° and 32°C over approximately one-third of the floor area, the remainder being at the ambient temperature of the room, 20–21°C. This variation in temperature permitted each guinea-pig to choose its own area of comfort and made it possible to place the dishes containing the fluid diet in the cooler area, where evaporation losses were not great.

Guinea-pig milk contains approximately 8·7 per cent protein, 4·8 per cent carbohydrate, 5·0 per cent fat, and 0·9 per cent ash. A spray-dried milk substitute diet was prepared for us by Messrs. Pritchitts, Deptford, London, from bovine milk supplemented with protein and fat. This powder supplemented with vitamins and minerals was packed in 500-gm lots in glass jars and sterilized by gamma-irradiation (2·5 Megarads).

The fluid diet was prepared by homogenizing 25 gm of the milk substitute powder in 90 ml water, plus 10 ml of an aqueous solution containing 0·225 gm potassium acetate and 0·2375 gm of magnesium acetate. The guinea-pigs were fed twice daily. The liquid diet was freshly prepared on each occasion, 100 ml being usually sufficient for four animals. The fluid diet was continued for four weeks, but from the seventh day onwards a pelleted diet (Diet RGP—Paterson, 1967) and hay, both sterilized by gamma-irradiation (2·5 Megarads) were available *ad lib.* These were consumed in variable amounts, and to ensure that an adequate intake of ascorbic acid was maintained, each guinea-pig was dosed daily *per os* with 10 mg ascorbic acid in 0·1 ml distilled water. When sufficient Diet RGP was being consumed (usually about the 20th day) this oral dosing was discontinued. Diet RGP should contain not less than 1,000 mg ascorbic acid in each kilogram. The morning milk feed was reduced on the 28th day and withdrawn on the 29th day. The afternoon feed was gradually reduced from the 29th to the 35th day, when it, too, was withdrawn. Of 300 youngsters obtained by aseptic Caesarean operation, a total of 264 were successfully raised, an overall survival rate of 88 per cent. Using the same powdered milk substitute diet and similar methods, hysterectomy-derived guinea-pigs have been reared at the Laboratory Animals Centre (Bleby, 1966) and the Rowett Institute, Aberdeen (Bullen, 1966).

(b) *Rabbits.* Newly born young are dependent on mothers' milk, and because they are hairless the maternal nest is vital to their survival. Therefore, if hysterectomy-derived rabbits are to be hand-reared they must be bottle fed regularly and provided with a warm environment. A satisfactory substitute for rabbits' milk is difficult to formulate because, compared with cows' milk, rabbit milk contains roughly four times as much protein, fat, calcium, and phosphorous. A further problem arises because, as Williams-Smith (1966) and Roderiguez (1966) have shown, a lipase produced in the stomach of the baby rabbit reacts uniquely with the fat present in its mother's milk to produce a series of products, mainly octoic and decoic acids, which are highly bactericidal. It would seem likely that these compounds are responsible for the virtual absence of bacteria in the gut of suckling rabbits. These anti-bacterial substances are not produced if the young rabbit is fed on bovine milk and many die from enteric infections.

We have successfully reared Caesarean-derived rabbits on a diet based on cows' colostrum. Fresh colostrum was first thoroughly homogenized and then freeze-dried. The resultant powder was fortified with minerals, vitamins, and antibiotics (see table) and packed in approximately 400-gm amounts in suitable glass jars of the type used for fruit preserving. Finally, the diet is sterilized by gamma-irradiation (2.5 Megarads).

DIET FOR HAND-REARING RABBITS

Colostral powder *	380 gm
Albevite †	4 gm
DL-Methionine	1·4 gm
L-Tryptophan	0·5 gm
Choline	0·6 gm
i-Inositol	1·0 gm
Ascorbic acid	0·25 gm
Ferric ammonium citrate	0·24 gm
Thiamine	20 mg
Cyanocobalamin	2 mg
Biotin	10 mg
Folic acid	5 mg
Penbritin powder ‡	10 gm
Framomycin §	2 sachets
Calcium glycerophosphate	28 gm

* Prepared from cows' colostrum collected within 24 hours of parturition and freeze-dried after homogenization.

† 'Albevite'—a water-miscible vitamin and mineral powder. The Crookes Laboratories Ltd., Basingstoke, Hants, England.

‡ 'Penbritin Veterinary Powder' contains 100 mgm of ampicillin per 3·4 gm. Beecham Research Laboratories, Brentford, Middlesex, England.

§ 'Framomycin' packs of 100 sachets each contains 250 mgm of framycetin sulphate. The Crookes Laboratories Ltd., Basingstoke, Hants., England.

For use, 40 gm of the fortified colostral powder, 10 ml arachis oil, and 100 ml of sterile water at 60°C are homogenized mechanically. The Silverson homogenizer we used is fitted with a fine-mesh stainless-steel sieve to ensure that the milk-substitute feed is free from particles which might block the hole in the teat of the feeding bottle. Because rabbits feed their young only once daily, we have adopted the same procedure. On the first day of life 5 ml were fed, and this was gradually increased until approximately 40 ml was being taken on the 14th day of life. During the preliminary trials with this diet osteogenesis imperfecta (brittle bones) was common, and supplementary calcium and phosphorus was given in the form of finely powdered calcium orthophosphate ($CaHPO_4$) simply by withdrawing the teat several times during each feed and coating it with the phosphate and quickly giving it back to the youngster to suck. This method of administration was cumbersome and inaccurate, but roughly 0·1 gm was fed on the first day, increasing gradually to 0·25 gm on the 14th day. The inclusion of calcium glycerophosphate in the diet has overcome this problem.

The fluid diet is continued until the 35th day, after which it is reduced steadily over the next seven days as the animals are weaned on to a solid diet of RAF pellets and hay, both sterilized by gamma-irradiation (2·5 Megarads).

6. Colony management

The daily routine of animal care in an SPF colony does not vary greatly from that in a well-run conventional colony, but very great care must be taken to ensure that the integrity of the barrier is not broken. A sensible arrangement is to draw up sets of rules governing the various operations and to post them at each vital point in the barrier. The rules themselves must be simple, precise, and clearly understandable. They should cover the movement of staff and their personal hygiene; the procedures to be adopted when materials or animals are passed into or out of the building, and will therefore detail such matters as:

(*a*) the operation of the sterilizer chambers;
(*b*) the maintenance of dunk-tanks;
(*c*) the removal of refuse from the building;
(*d*) the routine for sterilizing clothing;
(*e*) the monitoring of air pressure differences.

It is possible to produce animals under a 'total barrier system' as described for rats and mice by Foster (1962). Under this plan, once the building has been equipped and stocked, there is no movement of equipment through the barrier. The watering system is automatic, and the cages are cleaned *in situ* by a vacuum system which conveys waste materials from the animal cages direct to the outside of the building. A 'split barrier system' may be said to be operated when the cages, water bottles, and other equipment are removed from the animal rooms to a central wash area. If this area is within the barrier the sanitized equipment is simply returned to its original location, but if the wash area is outside the barrier, then all equipment must be re-sterilized as it is returned into the SPF area.

7. Transport of animals

If the animals are required for use as truly SPF animals, and this will certainly be the case if they are to be used as foundation stock for a new colony, then it is essential to use transport boxes fitted with high-efficiency free-air filters. Glass-fibre media is at present the material of choice for such filters, but other materials may prove to be alternatives, e.g. polyester fibre or autoclavable paper. The prepared transport boxes must be sterilized as they are passed into the SPF unit. The animals are placed in their container with sufficient food for their journey and the container is sealed with tape. The provision of water for animals in a sealed container presents many problems, and in our experience, for journeys of up to 18 hours it is better to dispense with water altogether and concentrate on ensuring that the journey is cut to the minimum by making certain that everyone concerned with the carriage of the animals and the recipient are fully aware of the travel arrangements. If the package or packages have to cross national frontiers or, in some countries, state boundaries it is essential that prior arrangements are made with the Customs Authorities that the containers will be passed through without being opened and that the formalities of Customs clearance will be conducted with the minimum of delay.

If, however, the animals are to be used under 'conventional' conditions at their destination less strict precautions may be permissible during transport, providing they are unlikely to be contaminated by contact with other animals of the same species during their journey.

SPF ANIMALS UNDER EXPERIMENT

For pharmacological, toxicological, and similar experiments not involving infectious agents precautions may be limited to pasteurization or sterilization of food and bedding, and maintaining a slight positive air pressure in the animal rooms. Incoming air must be subjected to filtration through high-efficiency filters, and a very high standard of cage and animal room hygiene must be observed. Staff must discard outer street clothes, thoroughly scrub up, and don sterile protective overalls before entering the animal area.

Where infectious agents are employed, the same level of precautions must be employed, but a negative air pressure must be maintained in the experimental animals' immediate environment. This would be the whole room if the animals were kept in open cages, but if the cages were installed in negatively pressurized laminar-flow cabinets the room could be at normal air pressure. The air exhaust system of such animal laboratories must be fully protected at the delivery point by high-efficiency filters to prevent the infectious agent or agents being discharged into the atmosphere. The level of precautions to be taken by the staff for self-protection will depend upon the nature of the agent under investigation and should be laid down by the safety (medical) officer of the unit.

PRODUCTION AND USE OF SPF FARM ANIMALS

There are no obvious technical difficulties to prevent the production of SPF pigs, lambs, kids, foals, or calves. The principles would be those already described, but the hysterectomy techniques might vary. For example, it might be economically worthwhile to attempt to salvage the carcases of the donor mothers for meat, and this would limit the choice of anaesthetic. Generally, the gravid uterus would be obtained by aseptic surgery and passed into a sterile operating hood, where the young would be removed from the uterus and revived. Later they would be passed, with suitable precautions, through the barrier into the SPF quarters to be reared. This method is particularly suitable for swine (Betts *et al.*, 1960; Young *et al.*, 1955). For other species it might be more practical and convenient to fix a sterile plastic operating tent directly on to the sterilized skin of the anaesthetized donor mother and to perform a hysterotomy. The viable young would be transferred into a sterile transport box or bag for transport to the SPF rearing area.

SUMMARY AND CONCLUSION

As biological tools, SPF animals are physiologically and structurally identical with conventional animals. They differ only in the fact that, as produced, they are free from parasites and their natural infections, and if proper precautions are taken by experimentalists they will remain in this state until the experiment, trial, or observation is completed.

There are several possible pitfalls in the production and use of SPF animals. For example, the ventilation system may fail to maintain a positive pressure, or the air filtration system may be faulty; the water may not be sterile, or the food and bedding may be imperfectly sterilized. The greatest hazard, undoubtedly, lies with the staff, who, despite careful personal decontamination, may carry the agents of disease—viruses, bacteria, protozoal cysts, worm eggs—into the unit under finger-nails, in the hair, or even in the nose and throat. Lastly, there is the possibility of accidental damage to the fabric of the building causing a break in the 'barrier system'.

However, if the ideal cannot yet be achieved and maintained, genuine SPF animals when properly used should not, like their conventional counterparts, be riddled with disease and act as a source of frustration to animal technicians and experimentalists.

REFERENCES

BETTS, A. O., LAMONT, P. H., and LITTLEWORT, M. C. G., 'The Production by Hysterectomy of Pathogen-free Colostrum-deprived Pigs and the Foundation of a Minimal Disease Herd', *Vet. Rec.*, **72**, 461 (1960).

BLEBY, J., Personal communication (1966).

BULLEN, J. J., Personal communication (1966).

CANÃS-RODRIGUES, A. and WILLIAMS-SMITH, H., 'The Identification of the Antimicrobial Factors of the Stomach Contents of Sucking Rabbits', *Biochem. J.* **100**, 79–82 (1966).

COOK, R., 'A Practical Method for Ensuring Full-term Pregnant Rats and Mice for Caesarean Operation', *J. anim. Tech. Ass.*, **16**, 34 (1965).

CUMMING, C. N. W. and ELIAS, C., 'The Establishment by a Commercial Company of a Colony of Rats Free from Certain Pathogens', *Proc. Anim. Care Panel*, **7**, 41–49 (1957).

DAVEY, D. G., 'Establishing and Maintaining a Colony of Specific-pathogen-free Mice, Rats and Guinea-pigs', *LAC Collected Papers*, **8**, 17–34 (1959).

FOSTER, H. L., 'Facilities for Commercial Production of Pathogen-free Rats', *Proc. Anim. Care Panel*, **8**, 91 (1957).

FOSTER, H. L., 'The Problems of Laboratory Animal Diseases', *Establishment and Operation of SPF Colonies*, Acad. Press, London, pp. 249–59 (1962).

FOSTER, H. L., BLACK, C. L., and PFAU, E. S., 'A Pasteurization Process for Pelleted Diets', *Lab. Anim. Care*, **14**, 373–81 (1964).

INNES, J. R. M., DONATI, E. J., ROSS, N. A., STOUFER, R. M., YEVICH, P. P., WILSON, C. E., FARBER, J. F., PANKEVICIUS, J. A., and DOWNING, T. O., 'Establishment of a Rat Colony Free from Chronic Murine Pneumonia', *Cornell Vet.*, **47**, 260 (1957).

HENTHORNE, R. D. and VEENSTRA, R. J., 'The Development and Maintenance of Disease-free Animal Colonies at the Walter Reed Army Institute of Research', *Proc. Anim. Care Panel*, **7**, 50–55 (1957).

KING, H. D., 'Labyrinthitis in the Rat and a Method for its Control', *Anat. Rec.*, **74**, 215 (1939).

MCPHERSON, C. W., 'Reduction of *Pseudomonas aeruginosa* and Coliform Organisms in Mouse Drinking Water Following Treatment with Hydrochloric Acid and Chlorine', *Lab. Anim. Care*, **13**, 737–44 (19763).

NELSON, J. B., 'Studies on Endemic Pneumonia of the Albino Rat. IV. Development of a Rat Colony Free from Respiratory Infections', *J. exp. Med.*, 1951, **94**, 377–86 (1951).

NELSON, J. B. and GOWEN, J. W., 'The Establishment of an Albino Rat Colony Free from Middle Ear Disease', *J. exp. Med.*, **54**, 629 (1931).

PATERSON, J. S., 'Control of Disease in Rabbits', *L.A.C. Collected Papers*, **4**, 37 (1956).

PATERSON, J. S., 'The Problems of Laboratory Animal Disease', *Guinea-pig Disease*, Academic Press, London, p. 169 (1962).

PATERSON, J. S., 'The Guinea-pig.' *UFAW Handbook* (3rd Edition), E. & S. Livingstone, Edinburgh & London (1967).

REYNIERS, J. A., 'The Control of Contamination in Colonies of Laboratory Animals by the Use of Germ-free Techniques', *Proc. Anim. Care Panel*, **7**, 9–28 (1957).

SABOURDY, M. A., 'The Need for Specification in Laboratory Animals: Suggested Terms and Definitions', *Food Cosmet. Toxicol.*, **3**, 19 (1965).

SACQUET, E., *The Problems of Laboratory Animal Disease*, Academic Press, London, p. 57 (1962).

WILLIAMS-SMITH, H., 'The Antimicrobial Activity of the Stomach Contents of Sucking Rabbits', *Jour. Path. Bact.*, **91**, 1–9 (1966).

YOUNG, G. A., UNDERDAHL, N. R., and HINZ, R. W., 'Procurement of Baby Pigs by Hysterectomy', *Amer. J. vet. Res.*, **16**, 121–31 (1955).

CHAPTER TWENTY-SIX

Germ-free Animals

During the processes of evolution, higher and lower organisms have developed in close association with each other, and the normal (or 'conventional') animal lives in intimate contact with myriads of micro-organisms in its alimentary tract, its superficial tissues and its immediate environment. Most of our knowledge of the biological processes has been gained by studying animals carrying their natural microbial burden, and we cannot be sure that any particular process is brought about solely by the host animal or whether it is partly or entirely due to the action of the accompanying microbes. To answer such questions it is essential to be able to maintain the host devoid of its usual microbial burden; we can then determine how far its reactions are modified by the introduction of known micro-organisms.

The idea of rearing animals without bacteria has fascinated scientists for many years. Louis Pasteur, as long ago as 1885, was probably the first to speculate on the possibility. Although several attempts were made between then and the early part of this century, knowledge of bacteriology, nutrition, and engineering was far from adequate, and no one succeeded in maintaining animals healthy and germ-free for very long. The pioneers of modern germ-free techniques were Professor J. A. Reyniers and his colleagues of the Lobund Laboratory, University of Notre Dame, U.S.A., and Professor B. E. Gustafsson at the University of Lund, Sweden. Descriptions of their early attempts to develop the technique can be found in the publications recommended for further reading at the end of this chapter.

A germ-free animal is, by definition, an animal dissociated from any other detectable form of life. The term '*axenic*' (literally, 'without strangers') is also occasionally used. If a single strain of micro-organism is introduced the animal is said to be *monoassociated*; introduction of several kinds of organism results in a *polyassociated* system. A useful term, *gnotobiote* (from the Greek *gnosos*, knowledge, and *bios*, life), has been introduced to describe an animal living in association with known organisms. Thus a rat contaminated with a strain of *Lactobacillus* is a gnotobiote; if a strain of *Streptococcus* or other recognized species is introduced it is still gnotobiotic, but if the rat becomes associated with one or more unidentified organisms it may no longer be referred to as a gnotobiote. In other words, a mono- or polyassociated animal is only gnotobiotic if the identity of the contaminants has been established. The normal animal, living under ordinary laboratory conditions, is usually described as 'conventional' or 'classic'. Apart from their use in fundamental research, germ-free animals may form the foundation stock for a 'specified-pathogen-free' colony. For this purpose they are reared in germ-free conditions and then transferred to a pathogen-free, but not sterile, environment, from which they rapidly acquire a mixed microflora. At this stage they are unlikely to be truly gnotobiotic, because, although they are free of specified pathogens, their population of non-pathogens is usually not fully identified.

HOUSING OF GERM-FREE ANIMALS

Design of isolators

The apparatus used to rear germ-free animals is generally called an isolator. It may be of various designs according to economic considerations, the number and size of the animals, and their experimental purpose. Basically it must consist of a container that can be sterilized and kept sealed against the entry of extraneous microbial contaminants. There must be some means of introducing the animals and of supplying them with food, water, and sterile air. It is essential to be able to observe the animals and to manipulate them inside the container. The latter requirement is achieved by fitting long-sleeved rubber gloves in the side of the apparatus; several pairs of gloves can be fitted in one isolator, but its dimensions are restricted to the distance that can be reached by the operators.

Until recently isolators were generally made of metal, preferably stainless steel, which is extremely durable but expensive. It has the advantage of withstanding treatment by steam under pressure, which is one of the most effective methods of sterilization. The Reyniers isolator (Fig. 1) is really a modified autoclave, cylindrical in shape and constructed of heavy stainless steel, which can be sterilized by blowing in steam under pressure. Because of the high internal pressure sustained during the sterilization procedure, the rubber gloves must be protected with steel plates to prevent rupture, and only one or two small observation windows of toughened glass can be incorporated into the body of the apparatus. For the introduction of heat-stable materials into the isolator a small autoclave is provided at one end. If the door between the small autoclave and the isolator is sealed objects may be admitted through the outer door of the autoclave and sterilized inside it by steam under pressure. When the process is complete the inner door is opened and the contents of the autoclave are passed into the isolator. For the introduction of materials that will not stand autoclaving a metal tube dipping into a 'dunk tank' of germicide can be attached (see p. 435). The small autoclave has a further use; with its inner door sealed, the outer opening can be attached to a second isolator, forming a tunnel that can be sterilized. If the doors at both ends of the tunnel are then opened animals can be transferred from one isolator to the other. This operation will be necessary when animals have grown and become overcrowded, or when some must be transferred to a different experimental treatment, or simply because it may be easier to move them to fresh quarters than to refurbish the old.

The Gustafsson isolator (Fig. 2) is also made of steel. The whole apparatus is placed in an autoclave for sterilization, in the course of which the steam pressure is the same on the inside and outside of the apparatus. It can therefore be constructed of fairly lightweight metal, and a plate-glass window occupying the whole of the upper surface can be incorporated without fear of breakage during sterilization. Thus it has the advantage of a lower cost and better visibility than the Reyniers unit, but an autoclave of considerable size is an essential adjunct. The larger Gustafsson isolator is fitted with a small autoclave

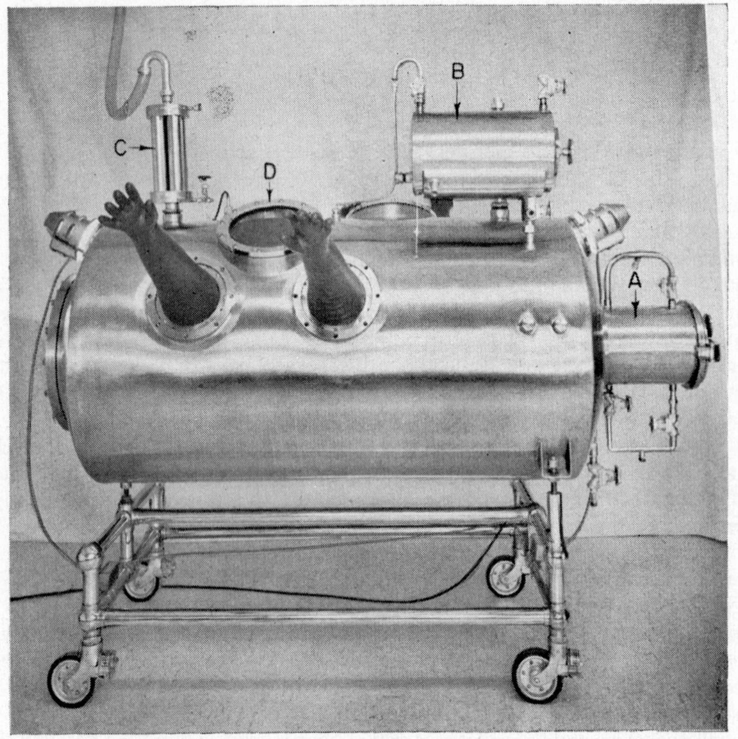

FIG. 1. Reyniers isolator. A, small autoclave; B, reservoir of drinking water; C, air filter; D, viewing port. (*Photograph by courtesy of Professor J. A. Reyniers, Tampa, Florida.*)

which is used as described for the Reyniers apparatus. In the smaller type illustrated in Fig. 2 there is no room for an autoclave, and all introductions into the isolator are made by way of a built-in germicidal trap, as described on p. 434.

Since P. C. Trexler first introduced the idea of using plastic film instead of steel, the cost of housing germ-free animals has been greatly reduced. Plastic isolators can be purchased ready made, or can be easily constructed in the laboratory from rigid sheet or flexible film. An example of the latter type is illustrated in Fig. 3. Apart from nylon film, which can be autoclaved (Lev, 1962), most of the plastics used in germ-free equipment will not stand high temperatures, and isolators made from them must be sterilized chemically. A germicidal solution, usually 2 per cent peracetic acid, is sprayed into the isolator and allowed to remain there for several hours. Sterile air is then passed through the apparatus until all traces of the germicidal vapour have been removed, which may take as long as two days. Efficient sterilization by this process depends on direct contact between the germicide and the organisms,

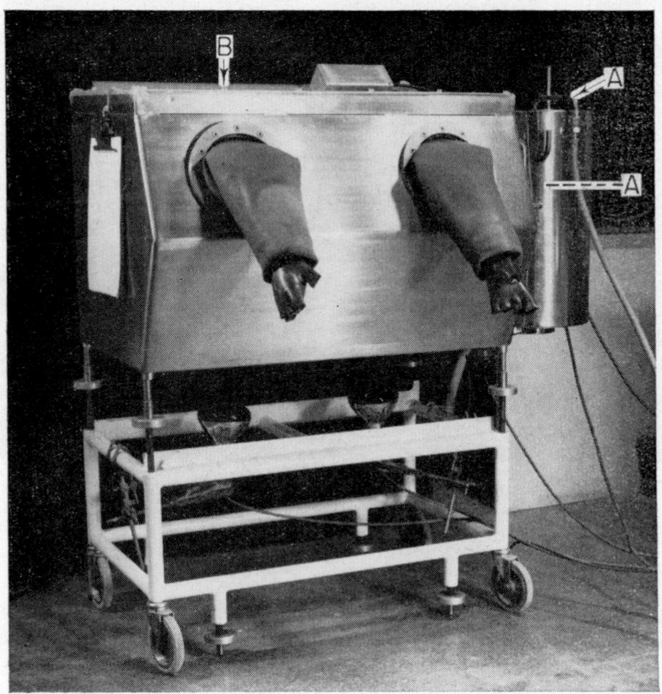

FIG. 2. Small Gustafsson isolator. A, air sterilizer; B, strip light above plate glass window. (*Photograph by N. J. Gruber.*)

hence it is essential that the disinfectant should penetrate thoroughly to all parts of the isolator. A wetting agent is usually incorporated into the spray to facilitate its spreading over surfaces, and if plastic film is to be sterilized by this means great care must be taken to avoid folds which might harbour bacteria out of reach of the sterilizing vapour.

Metal and plastic may be combined; thus a metal frame might be used to support a plastic film, and metal fittings incorporated so that a small autoclave or germicidal trap can be attached as in the steel isolators. In the absence of an autoclave a sterile lock is provided for introduction of objects into plastic isolators. It consists of a double-ended plastic container, one end opening into the isolator and the other to the exterior. It is used in the same way as the small autoclave on the Reyniers or Gustafsson apparatus, except that materials introduced through it are sterilized by treatment with peracetic acid.

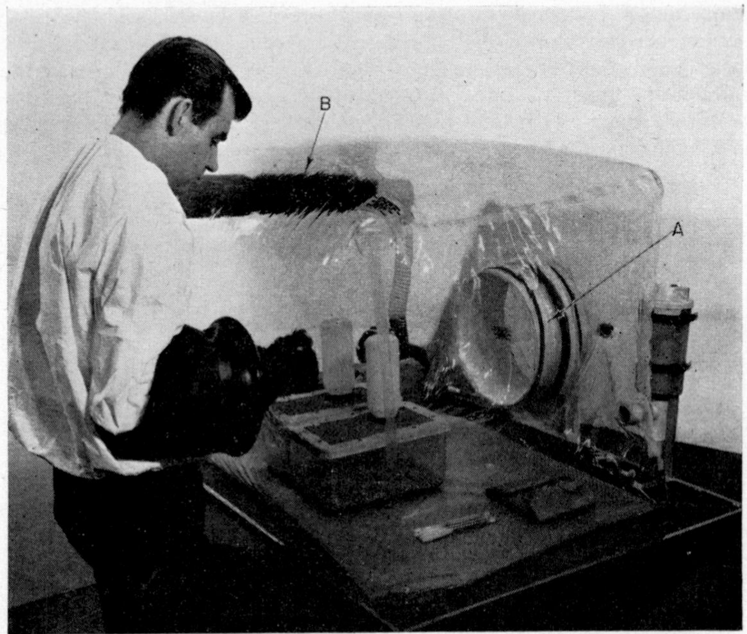

FIG. 3. Flexible film isolator. A, sterile lock; B, air filter. (*Photograph by courtesy of P. C. Trexler, Wilmington, Mass.*)

Use of a germicidal trap

A germicidal trap or 'dunk tank' for entry of heat-labile materials into an isolator is essentially a bath of disinfectant that forms a barrier between the germ-free interior and the contaminated environment of the laboratory. The trap may be permanently built-in to the equipment, as in the small Gustafsson apparatus. Alternatively, the isolator may be fitted with a wide tube with a removable closure. When the entry trap is needed the tube is immersed in a bath of disinfectant and opened below the surface of the liquid. After the entry has been made the tube is closed again and the container of disinfectant removed. The principle is illustrated in Fig. 4, from which it can be seen that the germicide forms a liquid seal through which objects may be passed into or out of the isolator. A hook and chain is useful for drawing things out of the trap into the interior. The liquid in the trap must be an efficient germicide and, since some of it may be carried into the interior in the course of the entry operation, it should be non-corrosive and of low toxicity. The quaternary ammonium compounds are satisfactory, but care must be taken to avoid foaming, which might lead to bubbles of contaminated air being carried over into the isolator. Iodine is an excellent disinfectant, but is volatile, so that frequent 'topping-up' is necessary to maintain the desired concentration.

Aseptically delivered foetuses must be taken into an isolator through a germicidal trap, and hatching eggs, bacterial cultures, and vitamin solutions are often introduced this way. Animals can be transferred from one isolator to another by passing them, in a suitable container, through the germicidal traps; they can be placed in a small glass jar or plastic bag, and if the transfer is made rapidly they do not suffer any discomfort. It is important to realize that the germicidal bath is only a barrier and cannot be relied upon to sterilize the materials passed through it. Apart from bacterial cultures, anything

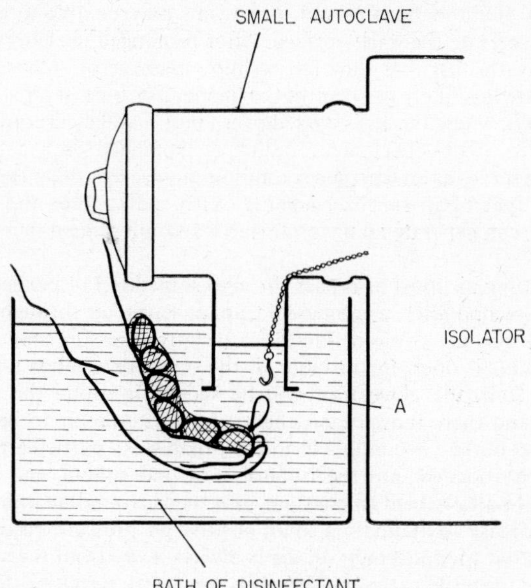

FIG. 4. Removable entry trap on large Gustafsson isolator. A, position of steel closure plate which is removed while trap is in use. Note that the level of liquid in the entry tube is lower than that in the rest of the bath of disinfectant, because pressure inside the isolator is slightly above atmospheric. (*Diagram by C. Machin.*)

introduced through the trap must have been previously sterilized, and the outside of the container as well as its contents must be completely free from contaminants. Packets of diet, for instance, should be doubly wrapped before sterilization; when passing them into the isolator the outer wrapping is torn open and only the inner, sterile packet is slipped into the germicidal bath. Glass ampoules may be soaked in chromic acid to sterilize their outer surface. Sterilized gloves or forceps should be used for manipulations through the trap. When possible, it is wise to leave objects soaking in the germicide for 5–20 minutes so that organisms which may have fallen on to them as they entered the trap are exposed to the action of the disinfectant.

Air supply to the isolator

All living organisms must be removed from the air before it enters the isolator. This can be achieved by passing it through a bacterial filter. Suitable filters can be bought ready-made, or can be constructed in the laboratory from materials such as glass or cotton-wool. Glass-wool filter down No. F.G.50 (American Air Filter Co. Inc., Louisville, Kentucky) has been widely used, and Lev (1964) reported that four layers of it retained particles as small as 0·5 μ.

Faulty air filtration, allowing contamination of the isolator, may result from any of several causes. Cracks can develop in the filter after repeated autoclaving. If filters become wet, organisms may be able to grow through them and penetrate the inner surface. After prolonged use filters can become so clogged with dust that they fail to allow passage of sufficient air. These difficulties are less likely to occur in the incinerator type of air sterilizer (Gustafsson, 1959), where the air is passed over finely divided carborundum heated electrically to 350°C. The carborundum is supported on a pad of asbestos wool, which serves as an emergency air filter in the event of an electrical breakdown. This type of air sterilizer is more costly than a filter and, even though well-lagged, can generate an uncomfortable amount of heat in a small laboratory.

The outgoing air must also pass through some kind of barrier, which may consist of a second filter, a germicidal trap, or repassage through the incinerator. The reasons are twofold. First, if a negative pressure should develop in the isolator (as it does, for instance, if the operator's hands are too rapidly withdrawn from the gloves) air will be sucked in along the path of least resistance, and there must be no chance for unsterile air to be drawn back through the outlet. Secondly, if micro-organisms, particularly pathogens, have been introduced into the isolator it is undesirable, and may even be dangerous, to allow them to pass out into the surrounding atmosphere.

It is customary to maintain a small positive air pressure inside a germ-free isolator so that the movement of air is always away from the sterile interior towards the unsterile exterior. Then if a small leak develops (for instance, a puncture in a rubber glove or a plastic film) unsterile air cannot easily enter, and it is quite often possible to repair the hole before the isolator becomes contaminated.

In large germ-free installations it is usual to provide dehumidified air to the isolators. This is not essential, but has the twofold advantage of minimizing the risk of wet filters and reducing the humidity inside the apparatus. The volume of air that can be passed through an isolator is limited, depending on the power of the air pump and the resistance of the inlet and outlet filters, but a high rate of ventilation with dry air helps to carry away much of the excess water vapour which accumulates as a result of the animals' respiration and of evaporation from the excreta and water vessels.

PRODUCTION AND REARING

Chickens and quail

In many ways the chicken is one of the simplest subjects to obtain in a germ-free state. Inside the egg the developing embryo is sterile provided the parent stock is healthy. After proper decontamination of the shells fertile eggs can be introduced into an isolator and left there to hatch.

Selection and disinfection of eggs

The usual procedure is to incubate clean eggs in a commercial incubator until the embryos are about eighteen days old. Eggs containing live embryos are then selected by candling. About six to a dozen eggs are packed into a net bag and washed for 2 minutes in a detergent solution; a gentle massaging action is used to ensure that the detergent reaches all parts of the shell. They are transferred to a 2 per cent solution of mercuric chloride, where they are held for a further 8 minutes. The purpose of the detergent is to remove surface dust and to ensure that the germicide spreads readily over the shell surface. The solutions must be at the same temperature as the incubator (about 38°C); cooling at this stage would not only chill the developing embryo but also, by contracting the air inside the shell, might cause organisms on its surface to be drawn in through the pores. After the disinfection procedure the eggs are passed into the isolator through a germicidal trap. In plastic isolators with a sterile lock the eggs may be decontaminated in the lock by spraying them with 2 per cent peracetic acid and detergent solution and leaving them exposed to the vapour for 30 minutes before passing them into the main compartment of the isolator.

Although the methods outlined above are not entirely satisfactory because they do not always result in complete decontamination of the shell, more drastic disinfection processes are liable to have a lethal effect on the developing embryos. Care in management of the hens and selection of the eggs can contribute much towards success in obtaining germ-free chicks. Only perfectly clean eggs, free from cracks, should be chosen, and any showing the slightest trace of mud or faecal stain must be rejected. The practice of marking eggs with pencil or a rubber stamp should be avoided. Eggs are likely to be cleaner if the hens are kept on wire-floored runs, and 'rollaway' nest boxes are advantageous. In this device the egg is laid on a gently sloping surface, and immediately rolls out of reach of soiling by the hen's feet or droppings. The most likely time for organisms to penetrate the shells is soon after laying. When laid the egg is warm and moist, and as it cools down microbes on its surface may be drawn in through the pores of the shell. This danger can be mitigated by dipping the eggs within an hour or two after lay in a germicide; a hypochlorite solution containing 0·03 per cent available chlorine is effective. The eggs should be stored in a cool dust-free atmosphere, and incubated in a clean incubator until they are put through the final disinfection process and transferred to the isolator.

Hatching

Inside the isolator environmental conditions must be closely controlled, and wet- and dry-bulb thermometers are essential articles of equipment. Temperature must be maintained at about 38°C. A few degrees increase is lethal to the embryos, and lower temperatures delay hatching and may cause malformed chicks. The relative humidity should be about 65 per cent. At low humidity the egg membrane becomes dry and hard and the young bird has difficulty in pecking its way into the air space; in high humidity water accumulates in the lungs and the chicks 'drown' before they can hatch. Good ventilation is important. The necessary rate of air flow will vary according to the design of the isolator, but it has been suggested that the CO_2 content should not be allowed to exceed 150 parts per 10,000 parts of air.

It is not easy to achieve close control of the physical environment in sterile conditions. Heat can be easily and cheaply applied to metal isolators by infra-red lamps, but plastic is likely to be scorched if the lamps are placed too close. Although some kind of thermostatic control is desirable during the hatching period, most of the common laboratory thermostats are damaged by the sterilization process. A thermistor will withstand autoclaving (the necessary relay, etc., can be left outside the isolator and connected by way of a cable passing through the germicidal trap), but such equipment is costly and gives a finer control than is strictly necessary. The simplest solution may be to stand the isolator in a small room in which the temperature can be controlled during the hatching period. Humidity can be raised inside the apparatus by exposing pans of water or wet sponges; with the incinerator type of air sterilizer humidity is very easily controlled by dripping water on to the heated carborundum, the size and rate of the drip being regulated to attain the desired humidity. Although the volume of air passing through the isolator may be adequate, the rate of flow is usually not sufficient to ensure that the air is well mixed, and dead pockets of air in the region of the eggs may cause poor hatchability. The risk could be overcome by fixing a fan inside the isolator, but the problem of either sterilizing the motor or fitting it externally is not easily overcome. Although the process of disinfection interferes to some extent with hatching, if optimal environmental conditions can be maintained within the isolator a hatch of about 75 per cent should be achieved.

Rearing

After hatching the temperature is maintained at 35°C for the first few days then gradually reduced to about 25°C or less over the next three weeks. There are two particular problems that should be watched for in the management of germ-free birds. The first concerns the accumulation of dust from diet and feather debris which may be deposited in the air-outlet filter and eventually block it. The trouble can be prevented by fitting a coarse filter than can readily be changed inside the isolator, in front of the air outlet. The second problem is the development of a high humidity. Chicks do not drink readily from a drip supply and are inclined to splash water from a drinking trough. This, combined with expired water vapour and evaporation from droppings, can raise humidity inside the isolator to saturation point. A high relative humidity does not seriously distress the birds, but condensed water vapour on the sides of the isolator interferes with visibility. The best defence is to increase ventilation as much as possible and to reduce exposed water surfaces; collection of droppings into plastic bags or metal containers removes a further source of water vapour.

On adequate diets chicks grow well in a germ-free environment, and there is evidence that their early growth is superior to that of hatchmates in a conventional environment. They have been extensively used in studies of nutrition and metabolism, and in cancer research. Bantams have been kept germ-free for long enough to reproduce a further generation, but in general chickens are uneconomic for long-term germ-free studies because of their large size and slow rate of maturity and reproduction.

Quail

The Japanese quail (*Coturnix*) is being developed as a useful alternative to the chicken because it is extremely small and rapidly maturing. Adult quail weigh only about 120 g, and reach full sexual maturity at 10 weeks of age. Thus a production flock of quail can be maintained in the laboratory in a relatively small space. If given adequate light they will remain in production for 11 months of the year. There is some difficulty in obtaining eggs clean enough for the production of germ-free quail, because the females have no regular nesting habits and eggs are laid casually in the run. They easily become soiled or cracked unless the run is sloped to allow them to roll away out of reach of the birds' feet. Optimal conditions for hatching have not yet been determined, but they seem similar to those for chicken eggs. The normal incubation time is 17 days, so for germ-free studies the eggs are best introduced into the isolator on the 14th or 15th day. The shells are very fragile, so the eggs must be handled with extreme care, especially during the disinfection procedure. Reyniers and Sacksteder (1960) recommend treatment with a detergent before incubation, and disinfection for 5 minutes with 2 per cent mercuric chloride solution immediately before passing into the isolator. Quail eggs require a very high humidity for hatching, and should be lightly sprayed with water several times during the hatching period. Reyniers and Sacksteder (1960) have bred quail inside germ-free isolators; incubation of the eggs was carried out in the small side autoclave, which was adapted for the purpose by surrounding it with an external heating coil.

Mammals

In the mammal, as in the bird, the developing foetuses in a healthy mother are sterile. To obtain germ-free mammals the foetuses must be aseptically removed from the mother just before term and transferred at once to a germ-free environment. The operation may be a hysterotomy (Caesarean section), in which an incision is made into the uterus *in situ* and the young are taken out one by one, or a hysterectomy, in which the entire uterus is removed from the mother before the foetuses are dissected from it. The preliminary stages are the same in both methods. Surgery is delayed until a few hours before spontaneous delivery is expected. The mother's abdomen is shaved or depilated and the skin treated with iodine solution. The animal is immobilized by a fatal or near-fatal blow at the base of the skull. Alternatively, the spinal cord may be cut under ether anaesthesia, but deep narcosis should be avoided, because it might adversely affect the survival of the young.

The hysterotomy procedure requires a special isolator or operating hood with an opening, sealed with plastic, in the floor. The interior of one kind of apparatus for delivery of germ-free mammals is illustrated in Fig. 5. The abdominal skin of the mother is washed with a good skin-sterilizing solution (e.g. iodine) and then pressed tightly into contact with the plastic floor. From inside the apparatus a longitudinal incision is made through the plastic floor and the abdominal wall. In small animals the exposed uterus is usually then removed from the mother into the isolator, where the foetuses are taken from it. In large animals the whole process takes longer, and it is advisable to perform a hysterotomy so that the foetuses remain for as long as possible in contact with the maternal blood supply. If the young are to be reared in the same apparatus in which they were delivered the incision in the floor must be sealed with fresh plastic before removal of the mother's body. Alternatively, the young may be transferred to a standard isolator for rearing.

In small animals hysterectomy is a more convenient method of delivery. No special apparatus is required, but a very strict aseptic routine must be observed during the procedure, which should be carried out in a dust-free

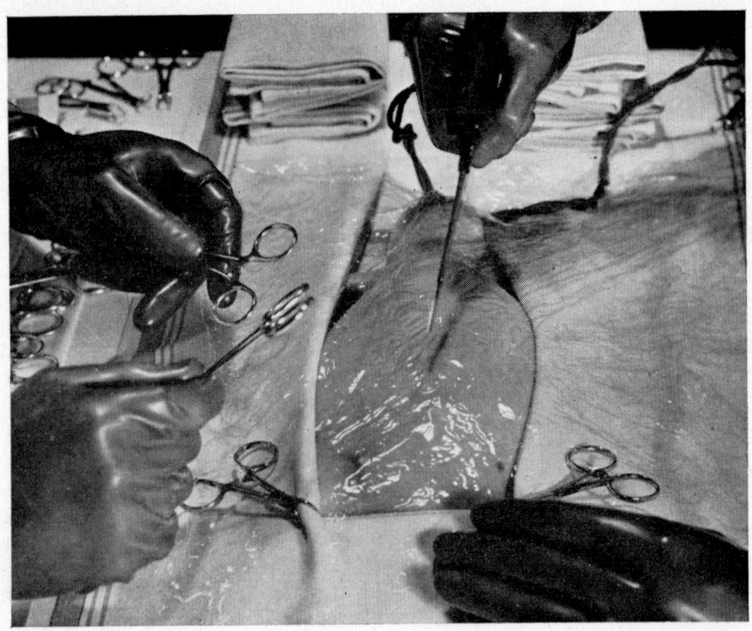

FIG. 5. Interior of special isolator for delivery of mammals. An incision is about to be made with a cautery through the plastic floor and the skin of the mother's abdomen. (*Photograph by courtesy of Dr J. A. Pleasants, University of Notre Dame, Indiana.*)

atmosphere or under a hood. The disinfected abdominal skin of the mother is covered with sterile gauze or plastic, and an incision is made through the covering and the abdominal wall. A ligature or clip is placed around the body of the uterus between the cervix and the lowest foetus. The uterus is then removed as rapidly as possible to the germicidal trap of the isolator and passed into the interior, where the foetuses are dissected out. If the operation is performed at some distance from the isolator the uterus may be immersed in a covered container of germicide for transport to the trap. Solutions in the germicidal trap and transport vessel should be warmed to about 37°C.

Rearing

The newly-delivered animals are usually confined for the first few days to a gauze-lined tray or box. They require a temperature of about 35°C. It may also become necessary to humidify the atmosphere; the baby animals present a dry, wrinkled appearance if the relative humidity falls too low.

Guinea pigs are relatively simple to rear, as they are well developed at birth and readily take solid diets. Most other mammals require hand-feeding, and the development of suitable milk formulae and techniques of artificial feeding occupied many years of painstaking research. A method of rearing rats and mice on an autoclaved liquid diet prepared from cow's milk has been described by Pleasants (1959). The young animals were force-fed at frequent intervals through a fine stomach tube made from rubber latex. Natural defaecation and urination were stimulated by massaging the anal region after each feed. At the present time there is little need to undertake the difficult task of hand-rearing rats and mice, because several breeding colonies have now been established and it is possible to purchase germ-free mice and rats from a number of commercial breeders. Lactating does from an established colony can be used as foster-mothers for the surgically delivered young of another strain.

Germ-free rabbits have been reared (Pleasants, 1959) and have reproduced, but so far their nutritional requirements are not fully understood and no large-scale production has yet been achieved. Hamsters, lambs, monkeys, cats, and dogs have all been produced in a germ-free state, but their successful maintenance is still a matter for investigation. Of the larger animals the germ-free pig is currently receiving considerable attention because of its potential use in the study of diseases that cause serious economic losses on the farm.

The fact that germ-free rats and mice have bred through many generations is proof of their ability to live successfully, but one departure from normal which may have serious consequences is an enlargement of the caecum, a condition which has been frequently observed in most germ-free rodents but not in chickens. The muscular tone of the caecal wall is reduced and the caecal sac is grossly distended with fluid; the caecal contents may weigh from six to ten times as much as in the conventional animal. In severe cases the caecum constitutes 20–25 per cent of the total body weight and may eventually rupture; in pregnant females it can also obstruct foetal growth. Removal of a germ-free animal to a conventional environment, or contamination with certain organisms, results in a reduction of caecal size within a few days. The cause of the phenomenon is not yet understood, but it may be due in part to the action of toxic substances that in the conventional animal are destroyed or inactivated by the intestinal microflora.

TRANSPORT OF GERM-FREE ANIMALS

Germ-free animals can be satisfactorily transported over long distances provided that an adequate air supply is maintained on the way. The transport of animals inside standard isolators calls for a special vehicle fitted with a pump to force air through the apparatus. Smaller transport units with a battery-operated pump have also been devised. Small animals, such as rats and mice, are commonly transported in a lightweight container ventilated by diffusion of air through a large area of filter material. Movement of the animals inside the container causes sufficient exchange of air through the filter. The method of transfer of animals to and from the transport unit will depend on the design of the unit and of the isolators in the sender's and recipient's laboratories, but it is clearly essential that a sterile connexion can be made between the unit and the receiving isolator—for instance, by means of a metal ring that can be attached to the small autoclave or sterile lock.

DIETS FOR GERM-FREE ANIMALS

At the present time little is known about the precise nutritional requirements of animals reared in a germ-free environment. In general, it appears that a diet that is nutritionally complete in conventional conditions will support good growth and reproduction in germ-free animals, although this statement may have to be modified in the light of further experience with species other than mice, rats, and chickens. There are indications, for instance, that germ-free guinea pigs and rabbits have a need for higher dietary levels of some minerals than their conventional counterparts. As all diets for germ-free animals must be sterilized, and as most sterilization procedures cause some degree of damage to the nutrient content of the diet, it is often necessary to add extra supplements of some nutrients, particularly vitamins, in order to allow for losses during sterilization. However, breakdown products formed in the diets during sterilization may be harmful to the animals, and there is also a risk that nutrients that cannot easily be replaced may be destroyed. It is unwise, therefore, to apply a more drastic procedure than is strictly necessary to achieve sterility.

Methods of sterilization of diets

Autoclaving. A temperature of 121°C for 20–30 minutes is usually recommended. In order to ensure that the correct conditions for sterilization are attained throughout, the diet should be packed in layers not more than $\frac{1}{2}$ in. deep, in containers that are permeable to steam (e.g. muslin bags or paper sacks). With pelletted diets it is helpful to draw a vacuum first, to remove air between and within the pellets and allow proper access of steam. A number of important nutrients are readily destroyed by autoclaving, particularly thiamine, vitamin B_6, vitamin A, and lysine. The extent of their destruction is influenced by other components of the diet; vitamins of the B complex are more stable if the water content of the diet is high; ascorbic acid may protect thiamine, and other antioxidants such as the tocopherols and hydroquinone help to stabilize vitamin A.

In the so-called 'high-vacuum autoclaving' process the diet is raised to 135°C (30 lb pressure of steam) for only 1 or 2 minutes; a very high vacuum (10 mm Hg) is drawn before and after the heat is applied. Preliminary results indicate that losses of nutrients during this procedure are less than in the more usual autoclave treatment.

Gamma-irradiation. In this process the diet is exposed to gamma-rays, usually from a ^{60}Co source. The irradiation equipment is expensive and demands extensive safety precautions in its operation, hence few plants are at present available, and the cost of treating diets by this process is high. Because of the intense penetrating power of the rays, diets can be packed in relatively large bulk and in metal containers without risk of the inner layers escaping treatment. Very little heat is generated during irradiation, and dry diets suffer little damage; increased breakdown of nutrients occurs with increases in water content of the diet. Vitamin K is particularly unstable to irradiation. Some losses of vitamins A, E, B_1, and B_6 can be expected, although experience so far indicates that they are small.

Fumigation by ethylene oxide. Diets can be sterilized by exposure to ethylene oxide gas, although this method has not been extensively used in work with germ-free animals. The treatment is carried out in a gas-tight chamber. The equipment is costly and the process needs careful handling because of the explosive nature and possible toxic effects of the gas. The ethylene oxide is usually diluted with CO_2 to a concentration of 10–20 per cent in order to reduce the risk of explosion. The time necessary for sterilization depends on temperature and pressure within the chamber, and may take several hours. The diet can be packed in plastic bags, since the gas can penetrate plastic film. Losses of vitamins of the B complex and of some amino acids have been reported in diets fumigated with ethylene oxide; toxic effects have also been observed in animals receiving diets from which ethylene oxide has not been completely removed.

Filtration. Soluble dietary supplements, that will not stand any of the treatments described above, may be sterilized by passing through a bacterial filter. The chemically defined water-soluble diets that have been devised for immunological studies, where it is necessary to use diets of low antigen content, can be sterilized by filtration.

STERILITY TESTS

The germ-free animal has already been defined as one which is free from any other detectable form of life. In order to ascertain that an animal is germ-free, exhaustive microbiological tests must be performed to demonstrate as far as possible the absence of contaminants. Tests should be made at intervals during the course of an experiment. Usually swabs are taken from the animals and their excreta, food, and water. Proof that the animals are germ-free depends on failure to culture any organisms from the swabs, hence the wider the range of cultural procedures to which the swabs are submitted, the more firmly will the claim to a germ-free status be established.

A full account of the methods used for detection of contaminants is beyond the scope of this chapter; the subject has been discussed at length by Wagner (1959). A procedure for routine sterility checks has been described by Coates *et al.* (1963). Briefly, some of the swabs are inoculated into several different liquid-culture media which are known to support the growth of aerobic and anaerobic bacteria. Others are used to spread agar plates. Incubation is carried out aerobically and anaerobically, at several different temperatures, and observation is continued for two weeks so that the presence of slow-growing organisms does not escape detection. Microscopic examination of smears stained by Gram's method is also made. A few organisms can usually be observed in smears under the microscope because some dead bacteria present in the sterilized diet pass through the gut intact. However, the appearance of large numbers in a smear indicates the presence of viable organisms, even thought the cultural tests may have been negative.

It is known that organisms exist which cannot be cultivated on artificial media, so that even if extensive microbiological tests prove negative, absolute freedom from contamination cannot be claimed. For example, it is so far impossible to demonstrate the absence of all viruses. For this reason any claim that animals are germ-free should always be accompanied by a statement of the tests used to detect possible contamination. In practice, the term 'germ-free' usually implies freedom from bacteria, yeasts, filamentous fungi, protozoa, endo-, and ectoparasites.

CHARACTERISTICS OF GERM-FREE ANIMALS

The ability of germ-free animals to live and reproduce indicates that they can successfully carry out the normal biological processes. In fact, in experiments with ageing mice the life span of a germ-free colony was longer than that of conventional controls. However, when an animal is born and reared in the absence of microbes its defence mechanisms are not challenged, with the result that certain of its characteristics differ from those of a conventional animal. The differences are particularly noticeable in the regions that are normally in close contact with micro-organisms, such as the skin, the lungs, and the alimentary tract. As might be expected, the lymphatic tissue and the number of lymphocytes is reduced; the production of gamma-globulin, with which antibodies are associated, is less. However, if a germ-free animal becomes infected its defence mechanisms respond, although rather more slowly than in conventional animals. Reductions in heart output, blood flow to the liver, and metabolic rate have also been reported, all of which might be the result of a reduced need for defensive activity.

The enlarged caecum which develops in germ-free rodents has already been discussed. Other changes in the alimentary tract have been observed in most species that have been reared germ-free. The wall of the intestine is noticeably thinner, the villi are smaller, and the whole surface area is less. It is not yet known how far such changes affect the processes of digestion and absorption, but there are indications that the absorptive capacity of the intestinal mucosa is slightly greater in germ-free animals. The contents of the alimentary tract are more fluid and the faeces softer than those of conventional animals. Measurements of other tissues, organs, and bones of germ-free animals have not shown any gross or consistent departures from the normal range.

Broadly speaking, the conventional animal can be regarded as exhibiting a mild state of inflammation of the organs that are normally in contact with microbes. It is exposed to the action of substances produced by organisms in the intestine and lacks others that are removed or destroyed by the flora. All these conditions are absent in the germ-free animal, and some of the consequent effects have been recognized, as described above. However, the germ-free animal has been available for study for only a relatively short time; much still remains to be learned about its morphological and physiological characteristics and how far they depart from the hitherto accepted normal.

REFERENCES

COATES, M. E., FULLER, R., HARRISON, G. F., LEV, M. and SUFFOLK, S. F., *Br. J. Nutr.* **17**, 141 (1963).

GUSTAFSSON, B. E., *Ann. N.Y. Acad. Sci.* **78**, 17 (1959).

LEV, M., *J. appl. Bact.* **25**, 30 (1962).

LEV, M., *J. appl. Bact.* **27**, 41 (1964).

PLEASANTS, J. R., *Ann. N.Y. Acad. Sci.* **78**, 116 (1959).

REYNIERS, J. A. and SACKSTEDER, M. R., *J. Nat. Cancer Inst.* **24**, 1045 (1960).

WAGNER, M., *Ann. N.Y. Acad. Sci.* **78**, 89 (1959).

Fuller accounts of this subject will be found in the following publications:

REYNIERS, J. A. (Ed.) *Lobund Reports*, No. 1 (1946), No. 2 (1949), No. 3 (1960). University of Notre Dame Press, Notre Dame, Indiana.

Recent Progress in Microbiology. Symposium V: Germ-free Animals. VIIth International Congress for Microbiology, Stockholm (1958). Almqvist and Wiksell, Stockholm.

Germ-free Vertebrates: Present Status, *Ann. N.Y. Acad. Sci.*, **78**, art. 1 (1959).

LUCKEY, T. D., *Germ-free Life and Gnotobiology*, Academic Press, New York and London (1963).

CHAPTER 27

The Use of Calculations in the Establishment of Breeding Programmes

The following questions have been used in recent years in the Associateship Examination of the Institute of Animal Technicians to show if candidates have: (*a*) a basic knowledge of the relevant data, and (*b*) a practical understanding of the particular specie. Here some examples are worked out showing the reasoning behind the answers.

When attempting to answer such questions, candidates must make assumptions and give averages. In doing so, it is in their own interest to use figures which are easy to handle, e.g. an average litter size of 10 is easier to manipulate than 9·85. This is acceptable to the examiners, providing the figures chosen fall within reasonable limits. Simplicity is the keynote of success with this type of question. Candidates should use the hypothetical 'average normal' rather than a real, but unusual or abnormal, figure.

Candidates should remember that examiners can award marks only for what is written on the answer paper, and cannot assume the candidate has any knowledge not included in the answer.

I. You are in charge of a breeding colony; how would you provide the following animals throughout the year?

(*a*) 400 mice per week

(*b*) 20 male guinea-pigs per week

(*c*) 10 ferrets per week

(*a*) **400 mice per week**

BREEDING SYSTEM CHOSEN—monogamous pairs, with the assumption that only 50 per cent post-partum mating will occur

LENGTH OF OESTROUS CYCLE—4 to 5 days

GESTATION PERIOD—3 weeks

AGE AT WEANING—3 weeks

Assume the period of time to the first weanings is approximately 7 weeks. Thereafter, if half the females mate at the post-partum oestrus they will produce a litter once every 3 weeks, while the remaining females will produce a litter once every 7 weeks. Then the *average* litter interval will be

$$\frac{3+7}{2} = 5 \text{ weeks}$$

Assume the average litter size is 8 for good, non-inbred mice.

Then, $\dfrac{\text{Average litter size}}{\text{Litter interval in weeks}} = \text{Number of young per female per week}$

in this case: $\dfrac{8}{5} = 1{\cdot}6$ young per female per week

$$\text{Number of breeding females required} = \frac{\text{Number of young}}{\text{Young per female per week}}$$

$$= \frac{400}{1{\cdot}6}$$

$$= 250 \text{ breeding females}$$

BUT to have a sufficient surplus for replacement males and females and a small surplus for seasonal and other variations, additional breeding pairs are required.

A breeding pair would be used for 6 months. Thus, about $\frac{1}{25}$ of the stock needs to be replaced each week, or in this case: $\dfrac{250}{25} = 10$ pairs. The 20 additional mice (10 males and 10 females) would be supplied by 12 to 13 pairs, and a further 12 to 13 pairs would allow for variations. That is, 25 additional pairs in all, or about 10 per cent of the total number of breeders needed to fulfil the order itself.

Therefore, *275 breeding females would be needed altogether*. If 55 pairs were set down each week for 5 weeks, then, starting 7 weeks later, a continuous supply of 400 mice could be weaned each week.

(*b*) 20 male guinea-pigs per week

LENGTH OF OESTROUS CYCLE—14 to 16 days

GESTATION PERIOD—66 days

AGE AT WEANING—10 to 14 days

NUMBER OF LITTERS PER YEAR—4 (taking advantage of post-partum mating)

AVERAGE LITTER SIZE—3·5

The number of young per female per year is $4 \times 3{\cdot}5 = 14$

Then the number of young per female per week is 0·25 (approx.)

20 are required, so, assuming an average sex ratio of 50: 50, the number of breeding females is $\dfrac{20 \times 2}{0{\cdot}25} = 160$

160 females will produce 40 young per week, of which 20 should be males.

A harem system of 1 male to 10 females would be used. About 75 per cent post-partum mating will be achieved with harems of this size. If an allowance is made for some variance in sex ratio, say about 10 per cent additional breeding females, then *18 pens of 1 male to 10 females would fulfil the order* and maintain the colony, weaners becoming available about 13 weeks from commencement.

In this instance no allowance need be made for replacement breeding stock, as all the females born are surplus, and only one male per month would be required for breeding. The economic breeding life is 18–24 months. Thus, one pen would be replaced each month.

If there is no outlet for the surplus females other than as replacement breeders, then they should be destroyed as early as is practicable, thus saving not only food and bedding but also space for housing them.

(*c*) **10 ferrets per week**

BREEDING SEASON—February to August

NUMBER OF LITTERS—normally 2 litters per season

AVERAGE LITTER SIZE—7

The assumption must be made that ferrets of variable age will be issued in the course of the year, and that this is acceptable to the experimenter.

Number of young per female per year is $7 \times 2 = 14$

Number of ferrets required per year is $10 \times 52 = 520$

Number of breeding females required is $\dfrac{10 \times 52}{14} = 37$

An answer of *40 to 45 breeding females would fulfil such an order* and provide replacement stock for the maintenance of the colony.

II. Devise a breeding programme to supply 40 female, 17-day-old, non-inbred mice every Wednesday for one year. (Show clearly the reasoning behind your method.)

This question requires particular knowledge, i.e. how to provide dated matings or young of a stated age on a specified date. The answer consists of two parts: (i) the calculation; (ii) the breeding method and the reasons for choosing it.

(i) **Calculation**

(*a*) How many litters are required to provide 40 female mice per week.

(*b*) The day on which these litters must be born.

(*c*) The number of breeding females required to be mated on that day.

(*a*) **How many litters are required:**

40 female mice are required

Average litter size is 8.

Assume sex ratio of 50 : 50.

Then, number of litters required is $\frac{40 \times 2}{8} = 10$ litters per week.

(*b*) **and** (*c*) **10 litters must be 17 days old each Wednesday**

These litters must be born 17 days before a Wednesday, i.e. on a Sunday. As the gestation period is 21 days, then 10 females must be mated each Sunday.

(ii) **Possible Breeding Methods**

It must be clearly shown to the examiner that the method chosen will fulfil the order economically, both in labour and in the number of animals to be maintained.

(*a*) **10 litters born every day of the year** all the year would give 28,000 mice in surplus per annum. This method is uneconomical and cannot be recommended.

(*b*) **Taking vaginal smears** and mating selected females that would be in heat on Sundays. This would involve taking 40 to 60 smears each week-end; a task which would be unpopular with technicians.

(*c*) **Assuming 20 per cent of mature, non-pregnant females will be in heat on any one day,** if sufficient males were added for one night only, theoretically, one-fifth of the females would be mated, so 50 to 60 females mated in this way may fulfil the order. However, females which are kept segregated from males cannot be relied upon to show a regular, 4 to 5-day cycle. In practice, either more than 60 females must be paired for one night in order to obtain 10 mated females or a modification of the Whitten Effect must be employed as follows:

Mate 20 females, in trios of 1 male to 2 females, 3 days prior to the required mating day, and 60 per cent of the females will actually be in heat that day. In this case set up 10 trios on Friday, and 12 of the females should be in heat on the Sunday. Dates of birth are recorded and litters born outside the required date are discarded. This method would work, but would be more expensive and wasteful than the following method.

(*d*) **Use of the Whitten Effect** (Ross, 1961). If castrated or caged males are placed in the colony cage with the females on Fridays and removed on Sundays and replaced with normal, uncaged males for one night only 60 per cent of the females will be on heat and will mate on that night. Here no female is wasted, as no female will be mated on any day other than the required one. This is the best method to use, because of its simplicity, economy in mice and space, and limited work-commitment at weekends.

From the calculation in (i) above, 10 females had to be mated every Sunday, so on Fridays place 1 male in a small wire container inside a colony cage containing 20 female mice. Also, set up 10 small breeding cages each containing 1 proven male. On Sundays remove the 20 conditioned females and place 2 with each male in the breeding cages. On Mondays remove the 10 males.

About 12 females should have mated. The other females can be seen to be non-pregnant in about 10 to 14 days time.

Therefore:

Week 1 20 females set up
Week 2 20 females set up
Week 3 12 females set up + 8 non-pregnant females from week 1
Week 4 12 females set up + 8 non-pregnant females from week 2
Week 5 12 females set up + 8 non-pregnant females from week 3
Week 6 12 females set up + 8 non-pregnant females from week 4

At the 7th week the weaned females set up at week 1 will be available for mating together with the non-pregnant females from week 5. There is no value in attempting to take advantage of post-partum matings, so females from which litters have been taken are returned to the pool of females. Less than 100 females and 10 to 12 males would easily fulfil the requirement.

Note. Obviously, a candidate should choose one method and state the reasons for his choice. He should not attempt to describe all possible methods.

III. A pharmacology laboratory requires the following animals each week:

(*a*) 850 male rats at 60–80 gm

(*b*) 400 guinea-pigs at 260–380 gm

Devise a continuous breeding programme to supply this number of animals.

This question is similar to example I, except that the numbers are much greater and the candidate is required to know the age of animals at the given weight and to make reference to storing the issue stock until it reaches the required weight.

(*a*) **850 male rats at 60–80 gm per week**

BREEDING SYSTEM CHOSEN—harem system, with 10 females to 1 male and boxing-out the pregnants

LENGTH OF OESTROUS CYCLE—4 to 5 days

GESTATION PERIOD—3 weeks

AGE AT WEANING—3 weeks

TIME TO PRODUCE A LITTER—7 weeks

Post-partum mating is not used, therefore each rat will produce a litter once every 7 weeks.

Number of young per female per week is $\frac{9}{7} = 1{\cdot}3$

Assuming a sex ratio of 50: 50, the number of young required each week is 850 × 2. Then, the number of breeding females required is

$$\frac{850 \times 2}{1{\cdot}3} = 1{,}307$$

The economic breeding life of rats is 6 months, so one-twenty-fifth of the stock must be replaced each week. That is, 53 females and 5 or 6 males are required. These females are readily available, as only males are issued. Therefore, 1,310 breeding females would be sufficient, but to allow for some variation, say 5 per cent, *1,400 females and 140 males would be required.*

Each week set up 20 colonies of 10 females and 1 male, and a continuous programme will have been established. Deliveries can start about 9 weeks after setting up the first group, as these rats have to be issued at 60–80 gm when they will be about 5 weeks old.

Sufficient space would be required to house up to 3 × 850 weaned, young males for issue, and, in addition, about 400 growing-on replacement breeding stock between 3 and 10 weeks of age, to be mated at the rate of 50 females per week on reaching 10 weeks of age.

The candidate should mention the economics of keeping females not required for issue or replacement stock. Such females could be culled at a few days of age, and the male pups could be pooled to re-form litters of about 10; also up to half the mothers could be returned to the mating cages. (Lane-Petter *et al.*, 1968.) This method would be more economical than keeping unwanted females.

(*b*) **400 guinea-pigs at 260–380 gms per week**

Breeding system chosen—harems of 10 females to 1 male
As in Example I, there are 0·25 young from each female each week.
Therefore,

$$\frac{\text{Number of young required}}{\text{Number of young per female per week}} = \text{Number of breeding females needed}$$

i.e.
$$\frac{400}{0{\cdot}25} = 1{,}600 \text{ breeding females}$$

Breeders are replaced every 20 months, therefore:

$$\frac{\text{Number of breeding females}}{\text{Breeding life (in months)}} = \begin{array}{c}\text{Number of breeding females needed}\\ \text{per month}\end{array}$$

i.e.
$$\frac{1{,}600}{20} = 80 \text{ per month, OR } 20 \text{ per week}$$

Thus, 2 new pens of 10 females and 1 male must be set up each week. Therefore, 22 additional guinea-pigs must be bred each week. These replacement breeders would be supplied by

$$\frac{22}{0{\cdot}25} = 88 \text{ females}$$

Thus, 1,688 breeding females are required altogether, but with an allowance of 5 per cent for production variabilities, *1,800 females would fulfil the order* and establish a continuous programme. These would occupy 180 pens.

The guinea-pigs would be about 8 weeks old at issue and, consequently, would need space for storing from weaning, at 2 weeks of age, until they were 8 weeks of age. That is, some 7 batches of 400 pigs, making a rearing stock of about 3,000 pigs (including replacement breeders) at any one time.

REFERENCES

LANE-PETTER, W., LANE-PETTER, M., and BOWTELL, C. W. (1968) 'Intensive breeding of rats. I. Crossfostering', *Lab. Anim.* **2**, 35.

ROSS, MARGARET (1961) 'Practical applications of the Whitten Effect on mice', *J. anim. Tech. Assoc.* **12**, No. 1, p. 21.

Index

Main passages indicated with bold figures, tables with italic figures, and illustrations with underlined figures.